Advancements in Power Management Systems for Hybrid Electric Vessels

Advancements in Power Management Systems for Hybrid Electric Vessels

Guest Editors

Peilin Xie

Sen Tan

Rosemary Norman

MDPI

Basel • Beijing • Wuhan • Barcelona • Belgrade • Novi Sad • Cluj • Manchester

Guest Editors
Peilin Xie
Department of Energy Technology
Aalborg University
Aalborg
Denmark

Sen Tan
Department of Energy Technology
Aalborg University
Aalborg
Denmark

Rosemary Norman
School of Engineering
Newcastle University
Newcastle
United Kindom

Editorial Office
MDPI AG
Grosspeteranlage 5
4052 Basel, Switzerland

This is a reprint of the Special Issue, published open access by the journal *Journal of Marine Science and Engineering* (ISSN 2077-1312), freely accessible at: https://www.mdpi.com/journal/jmse/special_issues/06Q72757GH.

For citation purposes, cite each article independently as indicated on the article page online and as indicated below:

Lastname, A.A.; Lastname, B.B. Article Title. *Journal Name* **Year**, *Volume Number*, Page Range.

ISBN 978-3-7258-4149-3 (Hbk)
ISBN 978-3-7258-4150-9 (PDF)
https://doi.org/10.3390/books978-3-7258-4150-9

Contents

Journal of
Marine Science and Engineering

MDPI

Editorial

Advancements in Power Management Systems for Hybrid Electric Vessels

Sen Tan [1,*], Peilin Xie [1] and Rose Norman [2]

1 Energy Department, Aalborg University, 9220 Aalborg, Denmark; pxi@energy.aau.dk
2 School of Engineering, Newcastle University, Newcastle upon Tyne NE1 7RU, UK; rose.norman@newcastle.ac.uk
* Correspondence: sta@energy.aau.dk

With the growing urgency of climate change, environmental regulations governing the maritime industry have become increasingly stringent, imposing significant restrictions on ship emissions. In response, the industry is shifting towards hybrid and fully electric vessels, reducing reliance on conventional diesel-based propulsion [1]. These advanced vessels integrate diverse energy sources, including fuel cells, photovoltaic systems (PV), batteries, and supercapacitors [2]. However, the integration of these heterogeneous energy sources, coupled with variations in electrical topologies [3], ship capacities [4], and operational conditions [5], introduces substantial complexities to the development and management of shipboard power systems (SPSs).

Unlike terrestrial power networks, SPSs must operate with high reliability under diverse and often unpredictable conditions. As localized microgrids that lack the support of a robust external power grid, ensuring stable power quality remains a critical challenge [6]. This necessitates a re-evaluation of key research questions that distinguish SPSs from conventional terrestrial energy systems:

1. What strategies can be employed to ensure the long-term reliability and stability of SPSs?
2. How can power quality be effectively maintained in SPSs across varying operational scenarios and routing conditions?
3. What methodologies can optimize the efficiency of SPSs to minimize emissions and operational costs?

This Special Issue aims to address these challenges by advancing research on critical component reliability, with a particular focus on batteries and fuel cells, power quality enhancements, and operational efficiency improvements. Emphasis is placed on state-of-the-art control methodologies, including Particle Swarm Optimization (PSO) and Extended Kalman Filter (EKF)-based control strategies, as well as optimization techniques and Energy Management Systems (EMSs) utilizing Model Predictive Control (MPC), multi-time scale approaches, and layered architectures. Additionally, this Special Issue explores novel diagnostic frameworks using Long Short-Term Memory (LSTM) networks and predictive analytics based on clustering techniques to enhance the monitoring and management of SPS.

The first theme addresses reliability issues in critical components of SPSs, specifically batteries and fuel cells, in terms of their prognostics and protection. Advanced diagnostic techniques play a key role in predicting component degradation, enabling proactive maintenance and extending system longevity.

A marine lithium-ion battery capacity prognostic method is presented in Contribution 1 based on LSTM and an improved Crested Porcupine Optimization (ICPO) algorithm under dynamic operating conditions. Key features are extracted from battery data to facilitate accurate capacity prognostics. Furthermore, Density-Based Spatial Clustering is introduced in Contribution 2 using voltage data to forecast variations in battery voltage, allowing for the early detection of potential faults. The results indicate that the DBSCAN clustering algorithm demonstrates superior effectiveness and accuracy in identifying irregular battery clusters. Contribution 3 proposes an attention-based prediction model utilizing Convolutional Neural Networks (CNNs) for long-term power allocation, optimizing the lifespan of fuel cells and lithium batteries by enhancing energy distribution strategies. Similarly, Contribution 4 demonstrates that a trained Back Propagation (BP) neural network can create an offline strategy library, providing intelligent energy distribution recommendations that effectively reduce lithium-ion battery degradation by 28%.

The second theme focuses on improving power quality in SPSs, particularly in terms of harmonic suppression and enhanced control performance, both of which are crucial for vessel servo systems. However, the inherent nonlinearity of motor systems presents significant challenges in achieving precise control.

To address this, Contribution 5 proposes an Extended Kalman Particle Filter (EKPF), which combines a particle filter to identify motor resistance and inductance, thereby improving high-precision control performance. Meanwhile, Contribution 6 introduces a fractional-order controller tuned by PSO and the Oustaloup approximation algorithm to suppress harmonics induced by motor operations, ensuring smoother performance.

The third theme addresses power efficiency challenges, primarily by optimizing fuel consumption through advanced Energy Management Systems (EMSs) under dynamic operating conditions.

For example, MPC is utilized in Contribution 7 to analyze working conditions and dynamically adjust EMS strategies, enabling the optimal management of generators and batteries while minimizing energy waste. To further enhance overall efficiency, the Equivalent Consumption Minimization Strategy (ECMS) is employed in Contribution 8 to balance power distribution among fuel cells, batteries, and ultracapacitors, ensuring optimal energy utilization. Moreover, Contribution 9 explores an advanced approach that considers the battery State of Charge (SOC) alongside power generation source characteristics across various operating modes. This approach results in significant improvements in system efficiency and operational reliability. Furthermore, a Bond graph is used in Contribution 10 to model the various energy sources of hybrid propulsion ships. Based on the proposed model, optimal operational scenarios and reduction ratios are then formulated for different maritime regions, thereby improving propulsion efficiency.

By integrating these advanced computational and control strategies, this Special Issue seeks to facilitate the transition towards more resilient, efficient, and sustainable shipboard power systems. Looking ahead, the future of SPSs is poised to integrate hydrogen-based technologies [7], providing a promising avenue for further reducing emissions and enhancing energy sustainability. However, this transition presents new challenges in hydrogen generation, onboard storage, transportation, and safe and efficient hydrogen consumption in marine environments. Furthermore, as hydrogen becomes a primary energy carrier, ensuring the reliability, efficiency, and stability of SPSs will require innovative control strategies, advanced diagnostic frameworks, and resilient system architectures. Addressing these emerging challenges will be essential to unlocking the full potential of hydrogen-powered SPSs and achieving a truly decarbonized maritime industry.

The evolution of SPSs towards full electrification represents a transformative milestone in sustainable maritime transport. Achieving this vision will require continuous innovation in energy storage, power management, and system resilience to facilitate widespread adoption and long-term viability. With advancements in battery and hydrogen technologies, future SPSs will not only improve operational efficiency but also significantly contribute to global decarbonization efforts, making maritime transport cleaner and more sustainable than ever before.

Funding: This work was supported by VILLUM FONDEN, Denmark, under the VILLUM Investigator Grant (no. 25920): Center for Research on Microgrids (CROM). We would also like to acknowledge the project FC-COGEN (no. 223853 under the EUDP grant) for sponsoring this research.

Conflicts of Interest: The authors declare no conflicts of interest.

List of Contributions

1. Song, Q.; Yang, X.; Tang, T.; Liu, Y.; Chen, Y.; Liu, L. Capacity Prognostics of Marine Lithium-Ion Batteries Based on ICPO-Bi-LSTM Under Dynamic Operating Conditions. *J. Mar. Sci. Eng.* **2024**, *12*, 2355.
2. Liu, Y.; Jin, H.; Yang, X.; Tang, T.; Song, Q.; Chen, Y.; Liu, L.; Jiang, S. Early Fault Diagnosis and Prediction of Marine Large-Capacity Batteries Based on Real Data. *J. Mar. Sci. Eng.* **2024**, *12*, 2253.
3. Zhou, X.; Yang, X.; Zhou, M.; Liu, L.; Niu, S.; Zhou, C.; Wang, Y. Multi-Temporal Energy Management Strategy for Fuel Cell Ships Considering Power Source Lifespan Decay Synergy. *J. Mar. Sci. Eng.* **2024**, *13*, 34.
4. Liu, L.; Yang, X.; Li, X.; Zhou, X.; Wang, Y.; Tang, T.; Song, Q.; Liu, Y. Prior Knowledge-Based Two-Layer Energy Management Strategy for Fuel Cell Ship Hybrid Power System. *J. Mar. Sci. Eng.* **2025**, *13*, 94.
5. Yuan, T.; Wang, T.; Bai, J.; Fan, J. Parameter Identification of Maritime Vessel Rudder PMSM Based on Extended Kalman Particle Filter Algorithm. *J. Mar. Sci. Eng.* **2024**, *12*, 1095.
6. Yuan, T.; Wang, T.; Fan, J.; Bai, J. Current Harmonic Suppression in Maritime Vessel Rudder PMSM Drive System Based on Composite Fractional-Order PID Repetitive Controller. *J. Mar. Sci. Eng.* **2024**, *12*, 1108.
7. Yan, Y.; Chen, Z.; Gao, D. Nonlinear Model Predictive Control Energy Management Strategy for Hybrid Power Ships Based on Working Condition Identification. *J. Mar. Sci. Eng.* **2025**, *13*, 269.
8. Truong, H.V.A.; Do, T.C.; Dang, T.D. Enhancing Efficiency in Hybrid Marine Vessels through a Multi-Layer Optimization Energy Management System. *J. Mar. Sci. Eng.* **2024**, *12*, 1295.
9. Choi, E.; Kim, H. Advanced Energy Management System for Generator–Battery Hybrid Power System in Ships: A Novel Approach with Optimal Control Algorithms. *J. Mar. Sci. Eng.* **2024**, *12*, 1755.
10. Moon, S.-W.; Ruy, W.-S.; Park, K.-P. A Study on Fishing Vessel Energy System Optimization Using Bond Graphs. *J. Mar. Sci. Eng.* **2024**, *12*, 903.

References

1. Xie, P.; Tan, S.; Bazmohammadi, N.; Guerrero, J.M.; Vasquez, J.C.; Alcala, J.M.; Carreño, J.E.M. A distributed real-time power management scheme for shipboard zonal multi-microgrid system. *Appl. Energy* **2022**, *317*, 119072. [CrossRef]
2. Perčić, M.; Vladimir, N.; Jovanović, I.; Koričan, M. Application of fuel cells with zero-carbon fuels in short-sea shipping. *Appl. Energy* **2022**, *309*, 118463. [CrossRef]
3. Wang, Y.; Mondal, S.; Satpathi, K.; Xu, Y.; Dasgupta, S.; Gupta, A.K. Multiagent distributed power management of DC shipboard power systems for optimal fuel efficiency. *IEEE Trans. Transp. Electrif.* **2021**, *7*, 3050–3061. [CrossRef]
4. Mahdi, H.; Hoff, B.; Ostrem, T. A review of power converters for ships electrification. *IEEE Trans. Power Electron.* **2022**, *38*, 4680–4697. [CrossRef]
5. Tu, Z.; Zhang, W.; Liu, W. Deep reinforcement learning-based optimal control of DC shipboard power systems for pulsed power load accommodation. *IEEE Trans. Smart Grid* **2022**, *14*, 29–40. [CrossRef]

6. Semwal, P.; Narayanan, V.; Singh, B.; Panigrahi, B.K. Performance evaluation of power quality in shipboard microgrid under different working conditions. *IEEE Trans. Ind. Appl.* **2024**, *60*, 2685–2696. [CrossRef]
7. Fan, F.; Aditya, V.; Xu, Y.; Cheong, B.; Gupta, A.K. Robustly coordinated operation of a ship microgird with hybrid propulsion systems and hydrogen fuel cells. *Appl. Energy* **2022**, *312*, 118738. [CrossRef]

Journal of
Marine Science and Engineering

MDPI

Article

Capacity Prognostics of Marine Lithium-Ion Batteries Based on ICPO-Bi-LSTM Under Dynamic Operating Conditions

Qijia Song, Xiangguo Yang *, Telu Tang, Yifan Liu, Yuelin Chen and Lin Liu

The School of Naval Architecture, Ocean and Energy Power Engineering, Wuhan University of Technology, Wuhan 430063, China; 299486@whut.edu.cn (Q.S.); 331552@whut.edu.cn (T.T.); 12023201239@whut.edu.cn (Y.L.); 299621@whut.edu.cn (Y.C.); 299437@whut.edu.cn (L.L.)
* Correspondence: yxglyr@whut.edu.cn

Abstract: An accurate prognosis of the marine lithium-ion battery capacity is significant in guiding electric ships' optimal operation and maintenance. Under real-world operating conditions, lithium-ion batteries are exposed to various external factors, making accurate capacity prognostication a complex challenge. The paper develops a marine lithium-ion battery capacity prognostic method based on ICPO-Bi-LSTM under dynamic operating conditions to address this. First, the battery is simulated according to the actual operating conditions of an all-electric ferry, and in each charge/discharge cycle, the sum, mean, and standard deviation of each parameter (current, voltage, energy, and power) during battery charging, as well as the voltage difference before and after the simulated operating conditions, are calculated to extract a series of features that capture the complex nonlinear degradation tendency of the battery, and then a correlation analysis is performed on the extracted features to select the optimal feature set. Next, to address the challenge of determining the neural network's hyperparameters, an improved crested porcupine optimization algorithm is proposed to identify the optimal hyperparameters for the model. Finally, to prevent the interference of test data during model training, which could lead to evaluation errors, the training dataset is used for parameter fitting, the validation dataset for hyperparameter adjustment, and the test dataset for the model performance evaluation. The experimental results demonstrate that the proposed method achieves high accuracy and robustness in capacity prognostics of lithium-ion batteries across various operating conditions and types.

Keywords: lithium-ion batteries; capacity prognostics; all-electric ships; feature extraction; bidirectional long short-term memory; improved crested porcupine optimizer

1. Introduction

In recent years, lithium-ion batteries, as a primary representative of clean energy, have found widespread applications in various fields, including electric ships (which encompass all-electric propulsion systems, conventional fuel-driven systems, and, as an intermediate solution, hybrid drive systems) [1], as well as electric vehicles, owing to their high energy density, absence of a memory effect, and overall reliability. Moreover, with the development of new technologies for mobile systems, reducing the mass of batteries has become more and more important [2], and lithium-ion batteries have been widely used in applications requiring a light weight and high performance due to their high energy density. For comparison purposes, Table 1 provides the key characteristics of several battery types used in electric ships and vehicles. However, over time, the aging mechanisms within the battery (e.g., loss of active substances) can result in performance degradation and may even lead to safety hazards. Therefore, there is an urgent need to establish an effective battery management system for the real-time monitoring of the lithium-ion battery lifespan [3]. Currently, battery capacity prediction methods can be broadly categorized into three types: experimental measurement-based methods, model-based methods, and data-driven methods.

Table 1. The key characteristics of several battery types used in electric ships and vehicles.

Battery	Capacity (Ah)	Energy Density (Wh/kg)	Mass (kg)	Dimensions (mm)	Charging Time (h)	Cycle Life (Cycles)
Lead Acid	50–200	30–50	20–60	200 × 300 × 200	6–8	300–500
Nickel–Metal Hydride	10–40	60–120	5–15	130 × 110 × 140	4–6	500–1000
Sodium Ion	10–50	90–160	3–12	120 × 100 × 150	2–5	500–1000
Flow Batteries	50–200	30–50	50–100	300 × 400 × 500	4–6	2000–5000

Experimental measurement-based methods utilize battery parameters (e.g., resistance, impedance, and capacity) derived from direct measurements to predict the capacity [4]. This approach, however, is costly, time-consuming, and may damage the battery, making it unsuitable for real-time applications. Model-based methods, including equivalent circuit models, empirical models, and electrochemical models, focus on the internal mechanisms of the battery. These methods require the development of mathematical models to describe the capacity degradation and use the model parameters to predict the battery's capacity [5]. However, accurately capturing battery characteristics and operating conditions is challenging, and the complex degradation behavior of lithium-ion batteries makes it difficult for model-based methods to fully characterize and quantify the degradation process.

With the accumulation of battery data and the rapid advancement of artificial intelligence, data-driven methods for battery capacity prediction have gained significant attention [6]. These methods primarily rely on capacity decay data and characterization data (e.g., current, voltage, and temperature) to train prediction models, which include techniques such as support vector machines [7], recurrent neural networks [8], least squares support vector machines [9], and long short-term memory networks [10]. These methods do not require an in-depth understanding of the battery's internal mechanisms and are therefore well-suited for predicting the capacity of lithium-ion batteries [11]. In addition, more and more researchers are now applying machine learning to a technique called digital twins to build a model with high accuracy for battery capacity prognostics [2]. However, the accuracy of data-driven methods depends significantly on the features extracted from the data and the way the model is trained [12]. To improve the prediction accuracy, it is essential to optimize three aspects: feature extraction, feature processing, and model training.

A. Feature Extraction

Simplifying battery charging and discharging data into a series of features can reduce the model's training burden and enhance its efficiency. However, under certain dynamic operating conditions, extracting effective features from the discharge process is challenging, whereas the charging process typically follows a more predictable pattern, making it easier to analyze. As such, it is more effective to extract features from charging data that reflect the degradation of battery capacity [13]. Moreover, the variety of charging methods leads to diverse ways of extracting features. To enhance the versatility of the prediction model, a standardized approach for feature extraction is required. Peng [14] and others developed a series of features that accurately represent the capacity degradation using a unified standard based on time, energy, and incremental capacity (IC) features. To predict the battery capacity degradation in electric vehicles (EVs), Deng et al. [15] extracted statistical features from the charging data, ensuring both methodological consistency and a comprehensive feature set. Guo et al. [16] combined rational analysis and principal component analysis (PCA) to derive features from charging data that are adaptable to various operating conditions, thus strengthening the versatility of their capacity prediction method. For high-precision capacity prediction across different lithium-ion battery datasets, Dai et al. [17] extracted six statistical features from charging data, determining the optimal feature combination by comparing various combinations of these features and thus reducing computational complexity. When effective features can be consistently extracted from charging data according to a unified standard, the method's versatility is proven, and the workload in feature extraction is minimized. However, extracting too few features may fail to capture the full degradation process, while too many features may result in redundancy, thereby increasing the computational burden and reducing model efficiency. Therefore, the extracted features must accurately reflect capacity degradation across different operating conditions while being computationally efficient.

B. Feature Processing

Selecting feature sequences that are highly correlated with the battery capacity can significantly improve the prediction's accuracy. In cases where two feature sequences are highly correlated with both the capacity and each other, redundancy can be reduced by eliminating one of the features, thus easing the computational burden. For example, in [18], Box–Cox transformation (BCT) was used to enhance the correlation between the extracted features and battery capacity. In [15], Pearson's correlation coefficient and gray correlation were employed to identify and remove redundant features, leading to the optimal set of features. In [14], principal component analysis (PCA) and empirical mode decomposition (EMD) were applied to the experimental curves of battery charging and discharging, as well as the incremental capacity curves, to extract features that strengthened the correlation between features and capacity. Furthermore, ref. [19] employed a two-step feature engineering approach—feature dimensionality reduction and seasonal fluctuation decoupling—to select the most relevant features for the capacity prediction while eliminating interfering components, thereby improving the model's prediction accuracy.

C. Model Training

The battery capacity prediction based on data-driven methods is influenced not only by the effectiveness of the extracted features but also by the choice of machine learning algorithms and the configuration of their hyperparameters. In [16], an adaptive RVM model based on PSO optimization was proposed, demonstrating high robustness and effectiveness for estimating the remaining capacity of lithium-ion batteries. Gong et al. [20] developed a battery capacity prediction model by combining empirical mode decomposition (EMD) and backpropagation with a long- and short-term cyclic memory network. In [21], a hybrid capacity estimation model was proposed by integrating the Arrhenius degradation equation and a lightweight Transformer architecture tailored for different operating conditions. Zhang et al. [22] employed a temporal convolutional network combined with Gaussian process regression to establish a novel capacity estimation method capable of automatically extracting capacity decay features from partial charging segments. Furthermore, improper hyperparameter settings can significantly degrade the performance of machine learning algorithms, thereby reducing the accuracy of capacity prediction. To address this, ref. [23] adopted an improved dung beetle optimization (IDBO) algorithm to optimize the hyperparameters of temporal convolutional networks (TCNs), obtaining optimal hyperparameter combinations quickly and accurately, which notably enhanced the accuracy of battery capacity predictions. It is important to note that if the test set is involved in hyperparameter tuning during the model training process, the model's performance on the test set may exceed its true capability, leading to evaluation errors. Thus, using the validation set for model tuning is recommended to preserve the independence of the test set.

Building upon these principles, this paper first extracts a series of features from battery data. The correlations between the extracted feature sequences, as well as between these features and the capacity sequences, are then analyzed. Features that exhibit a strong correlation with the capacity are retained, while redundant features are removed, resulting in an optimal feature set. Subsequently, the improved crested porcupine optimization (ICPO) algorithm is employed to optimize the hyperparameters of the bidirectional long short-term memory (Bi-LSTM) network, thus constructing the ICPO-Bi-LSTM model for accurate prognostics of the lithium-ion battery capacity. The dataset is divided into training, validation, and test sets in a specified ratio. The training and validation sets are used for model training, while the test set is reserved for the final performance evaluation. Finally, this paper investigates the impacts of working conditions, the dataset ratio, and the different models on the capacity prediction results by analyzing batteries discharged under complex and simple conditions, thus demonstrating the generality and robustness of the proposed ICPO-Bi-LSTM method.

The main contributions of this paper are as follows:

1. A unified statistical feature extraction method is proposed, i.e., calculating the mean, sum, and standard deviation values of current, voltage, energy, and power in the charging data for each charging and discharging cycle of a battery. These features apply to different batteries under complex and simple operating conditions, which solves the difficulty of needing to adjust the feature extraction method according to changes in battery conditions. The voltage difference between the battery before and after the simulated operating conditions in each cycle is extracted as another type of feature to fully reflect the capacity decay trend of the battery. The above-extracted features show a strong correlation with the battery capacity.
2. To overcome the challenge of determining the hyperparameters of the Bi-LSTM model, the improved crested porcupine optimization algorithm (ICPO) is proposed. This algorithm identifies the optimal hyperparameter combination and integrates the improved Chebyshev chaotic mapping initialization to ensure diversity within the initial population. This improves the algorithm's early-stage search speed and introduces a random difference variance strategy to avoid local optima, thereby enhancing the algorithm's overall efficiency.
3. The ICPO-Bi-LSTM model is developed using the optimal feature set to predict the capacity of lithium-ion batteries accurately. The dataset is divided into training, validation, and test sets, with the validation set being used for model training along with the training set. The test set is reserved exclusively for the final performance evaluation, preventing evaluation errors.

The remainder of the paper is organized as follows: Section 2 describes the experimental apparatus and dataset; Section 3 details the feature engineering process; Section 4 presents capacity prognostics based on the ICPO-Bi-LSTM method; Section 5 provides the results of battery capacity prognostics; and Section 6 concludes the paper.

2. Battery Data Analysis

2.1. Experimental Equipment

To comprehensively analyze the operational characteristics of lithium-ion batteries under real-world conditions, an experimental platform was developed to collect data from various battery types. The experimental equipment used in this study is the NEWARE CTE-4008D-5V30A tester, which is a battery test equipment manufactured by NEWARE, and its main function is to test the capacity, efficiency, cycle life, and other performances of the battery by simulating the battery charging and discharging process. It consists of a battery testing system, a host computer with software (BTS Client 8.0.0.516), and a battery under test. The physical schematic of the experimental apparatus is shown in Figure 1. After experiments were conducted with the NEWARE CTE-4008D-5V30A tester to obtain battery charge/discharge data, both model construction and battery capacity prognostics were carried out using the Python 3.11 (64-bit) platform.

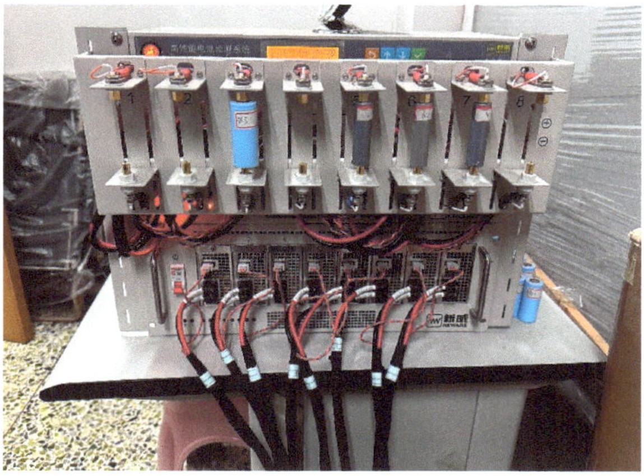

Figure 1. NEWARE CTE-4008D-5V30A tester.

2.2. Description of Experimental Data

This experiment used A123 APR18650M1A LFP/C (the manufacturer is A123 Systems LLC, Waltham, USA) batteries (B1) and OXUN IFR26650 LFP/C (the manufacturer is OXUN Energy, Changzhou, China) batteries (B2) to simulate real-world operating conditions. The research focused on the "Jun Lv Hao", a 300-passenger all-electric ferry operating in Wuhan. The ship's battery system comprises multiple clusters connected in parallel, offering a total capacity of 2240 kWh. The specific topology of the battery system is illustrated in Figure 2. The battery system of the "Jun Lv Hao" ship is divided into two sections (left and right), with each section containing six battery clusters. Once the six battery clusters are connected in parallel, they supply power to the pod and other loads via the ship's DMSB. The experiment was designed to replicate the actual operating conditions of the "Jun Lv Hao" and assess the capacity degradation of lithium-ion batteries under complex operational scenarios based on the rated capacity of the selected batteries. To verify the effectiveness of the battery capacity prediction method developed in this study, we also employed LISHEN LR18650LA NCM/C (the manufacturer is Tianjin Lishen Battery Joint-Stock Co., Ltd., Tianjin, China) batteries (B3) for a simplified discharge test under controlled conditions. Figure 3 displays the current variation curves observed during a typical "Jun Lv Hao" voyage and under simulated conditions. The specifications of the "Jun Lv Hao" are provided in Table 2.

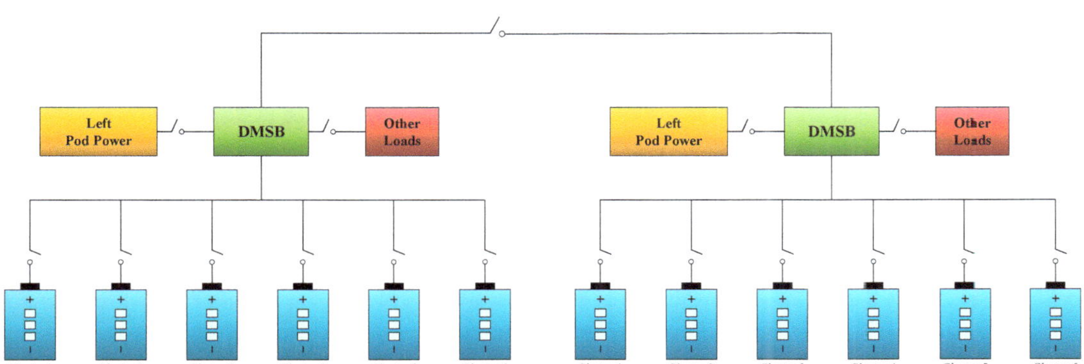

Figure 2. Network topology of the battery system.

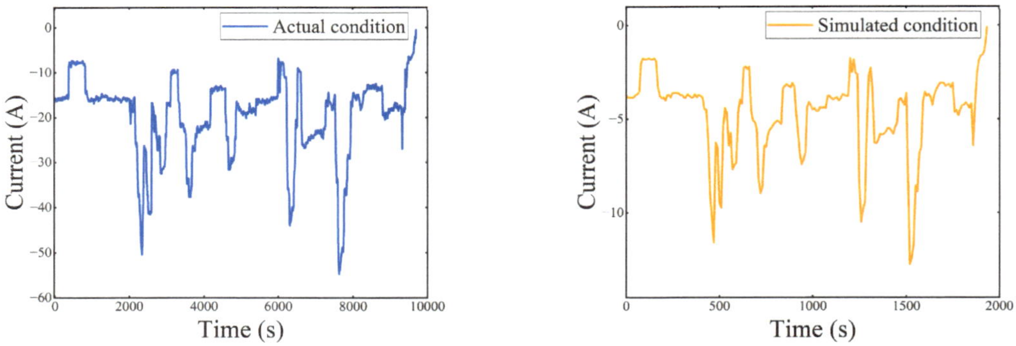

(**a**) (**b**)

Figure 3. The current variation curves. (**a**) Actual condition; (**b**) simulated condition.

Table 2. The specifications of the "Jun Lv Hao".

Parameter	Specification
Type	Passenger Ship
Length	55 m
Width	10 m
Draft Depth	1.6 m
Route	The Yangtze River in Wuhan City
Coordinates	114°17.414′ E″, 30°34.296′ N″
Operating Mode of the Propulsion System	All-Electric Propulsion System
Engaged Power of the Motor	1200 kW
Charging Time	2 h
Maximum Speed	10 knots/h
Cruising Range	118 km
Battery Capacity	2240 kWh
Battery Weight	25 tons
Number of Battery Clusters	12

The specific operation of the simulated working condition is as follows: Firstly, the capacity value released under the actual working condition is calculated by the ampere-time integration method, the capacity value is reduced by a certain number of times so that it does not exceed the rated capacity of the battery used in the experiment, and then the output current of the actual working condition is reduced by the same number of times. At the same time, the original sampling time of the working condition is 5 s, and this paper shortens the time to 1 s, which constitutes the simulated working condition. After that, the capacity released under the simulated condition is calculated again using the ampere–time integration method to ensure that the value is less than or equal to the rated capacity of the battery used in the experiment. Additionally, to ensure the safety and efficiency of the experiment, the charging currents and the duration of a single cycle were constrained. The specific charging and discharging protocols for the three batteries are outlined as follows:

B1,

① Charge the battery with a constant current of 7.7 A to a cut-off voltage of 3.26 V;

② Charge the battery with a constant current of 5.28 A to a cut-off voltage of 3.32 V;

③ Charge the battery with a constant current of 5.28 A to a cut-off voltage of 3.33 V;

④ Charge the battery with a constant current of 4.015 A to a cut-off voltage of 3.36 V,

⑤ Leave the battery to stand for 5 min;

⑥ Charge the battery with a constant current of 4 A at 3.6 V to a cut-off current of 0.4 A;

⑦ Leave the battery to stand for 5 min;
⑧ Apply the simulated working conditions shown in Figure 3b;
⑨ Discharge the battery with a constant current of 4.4 A to a cut-off voltage of 2 V;
⑩ Leave the battery to stand for 5 min;
⑪ Repeat the above steps (①–⑩) until 200 cycles are completed;
B2,
① Charge the battery with a constant current of 7.2 A at 3.65 V to a cut-off current of 0.72 A;
② Leave the battery to stand for 5 min;
③ Apply the simulated working conditions shown in Figure 3b;
④ Discharge the battery with a constant current of 14.4 A to a cut-off voltage of 2 V;
⑤ Leave the battery to stand for 5 min;
⑥ Repeat the above steps (①–⑤) until 276 cycles are completed;
B3,
① Charge the battery with a constant current of 8 A at 4.2 V to a cut-off current of 0.1 A;
② Leave the battery to stand for 10 min;
③ Discharge the battery with a constant current of 8 A to a cut-off voltage of 2.75 V;
④ Leave the battery to stand for 10 min;
⑤ Repeat the above steps (①–④) until 528 cycles are completed.

Figure 4 presents the capacity degradation curves for three types of batteries, as obtained from the experiments described earlier. Table 3 provides the details of the battery data experimentally obtained, which are relevant to the capacity prediction of lithium-ion batteries.

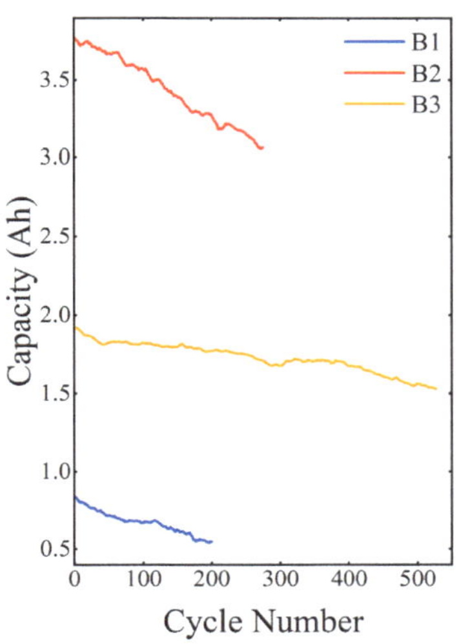

Figure 4. Capacity degradation curves for B1, B2, and B3.

Table 3. Items of battery data.

Items	Unit	Resolution
Current	A	0.0001 A
Voltage	V	0.0001 V
Energy	Wh	0.0001 Wh
Power	W	0.0001 W
Cycle	Time	1 Time

3. Feature Engineering

To build a data-driven model capable of accurately predicting the capacity degradation of lithium-ion batteries, a series of feature sequences are extracted from battery charge and discharge data in this section. It also uses a correlation analysis to identify and remove feature sequences that have a low correlation with the capacity sequence and eliminate redundant features that may increase the computational burden of the model. Finally, the optimal set of features that are highly correlated with the capacity sequence is obtained.

3.1. Feature Extraction

Charging data, including the current (I), voltage (U), energy (E), and power (P), can be obtained experimentally. In each cycle, the mean, sum, and standard deviation values (denoted by the subscripts ave, sum, and std, respectively) of the current, voltage, energy, and power are calculated, resulting in 12 features.

The above series of statistical features extracted from the charging data belong to the same type of features. To ensure that the extracted features can comprehensively reflect the complex and nonlinear degradation trend of the battery, the voltage difference of the battery before and after the simulated working condition (abbreviated as U_{dif}) is extracted as a feature in each charging and discharging cycle as well. The 13 features extracted are shown in Table 4.

Table 4. The features extracted.

Items	Charge Current	Charge Voltage	Charge Energy	Charge Power	Voltage Data Under Simulated Conditions
Features	I_{sum} I_{ave} I_{std}	U_{sum} U_{ave} U_{std}	E_{sum} E_{ave} E_{std}	P_{sum} P_{ave} P_{std}	U_{dif}

3.2. Correlation Analysis

In this paper, two metrics are employed, Spearman's correlation coefficient [24] and grey correlation [14], to assess the correlations between feature sequences and capacity sequences. The Spearman correlation coefficient reveals the strength of the monotonic relationship between the battery capacity sequence and the feature sequence, while the grey correlation more effectively distinguishes the degree of correlation between the capacity sequence and each feature sequence. Generally, the higher the correlation between the feature sequences and the capacity sequences, the greater the accuracy of the predicted capacity decay curves upon inputting the capacity sequences and the feature sequences into the data-driven model.

The Spearman correlation coefficient is calculated as follows:

$$r_s = 1 - \frac{6\sum_{i=1}^{n} d_i^2}{n(n^2-1)} \tag{1}$$

where r_s is the Spearman correlation coefficient, n is the length of the feature sequence, d_i is the rank difference between X_i and Y_i (where X_i and Y_i are the values in the ith position in the feature and capacity sequences, respectively), and the rank of a number refers to its position after all the numbers in the sequence have been ordered from smallest to largest, with ranks assigned in ascending order, i.e., 1, 2, …, n. Note that if there are ties in the data, the rank of each tied value is the arithmetic mean of the positions where they occur.

The steps for calculating the gray correlation are listed below.

(1) The capacity sequence and the feature sequence are normalized and the normalized expression is given as follows:

$$x' = \frac{x - x_{\min}}{x_{\max} - x_{\min}} \quad (2)$$

where x represents the original value, x' represents the normalized value, and $x_{\max}$ and $x_{\min}$ represent the maximum and minimum values in the same original sequence, respectively.

(2) Calculate the gray correlation coefficient between the normalized feature sequence and the capacity sequence using the following formula:

$$\xi_i(k) = \frac{\min_i \min_k |y(k) - x_i(k)| + \rho \min_i \min_k |y(k) - x_i(k)|}{|y(k) - x_i(k)| + \rho \max_i \max_k |y(k) - x_i(k)|} \quad (3)$$

where $\xi_i(k)$ represents the gray correlation coefficient of the ith feature sequence concerning the capacity sequence at the kth position, and $k = 1, 2, \ldots, n$ represents the total number of charging and discharging cycles of the battery, $y(k)$ represents the capacity sequence, $x_i(k)$ represents the feature sequence, and ρ is the distinguishing coefficient, which is typically set to 0.5.

(3) The gray correlation degree of the ith feature sequence concerning the capacity sequence can be obtained by calculating the average of the gray correlation coefficient using the following formula:

$$\gamma_i = \frac{1}{n} \sum_{k=1}^{n} \xi_i(k) \quad (4)$$

3.3. Features Selection

Using Equations (1)–(4), the feature sequences related to the capacity sequences, along with the Spearman correlation coefficients and gray correlations between features, can be calculated for the batteries under the two complex conditions. Additionally, the number of charging and discharging cycles is considered as a feature, as shown in Figure 5.

Feature sequences that exhibit a strong correlation with the capacity sequence and a low degree of autocorrelation among different feature sequences are selected to form the optimal feature set. In this study, feature sequences with an absolute Spearman correlation coefficient and gray correlation greater than 0.8 with the capacity sequence are considered highly correlated. If the absolute correlation coefficient between any pair of these highly correlated feature sequences exceeds 0.9, the sequences are considered highly autocorrelated, and one of the sequences should be removed to reduce redundancy [15]. The optimal feature sets for the two batteries, B1 and B2, for the capacity sequence can be derived from Figure 5, as shown in Table 5. In the optimal feature set for both batteries, common features are retained, while differing features are discarded to obtain a universally applicable feature set, consisting of I_{sum}, I_{ave}, E_{sum}, E_{ave}, E_{std}, P_{sum}, P_{ave}, and U_{dif}. This set is then used as input to the data-driven model for both batteries. Figure 6 illustrates the optimal feature set extracted from the charging data of both batteries, normalized according to Equation (2).

Table 5. The optimal feature sets of B1 and B2.

B1	I_{sum}	I_{ave}	U_{std}	E_{sum}	E_{ave}	E_{std}	P_{sum}	P_{ave}	U_{dif}
B2	I_{sum}	I_{ave}	U_{sum}	E_{sum}	E_{ave}	E_{std}	P_{sum}	P_{ave}	U_{dif}

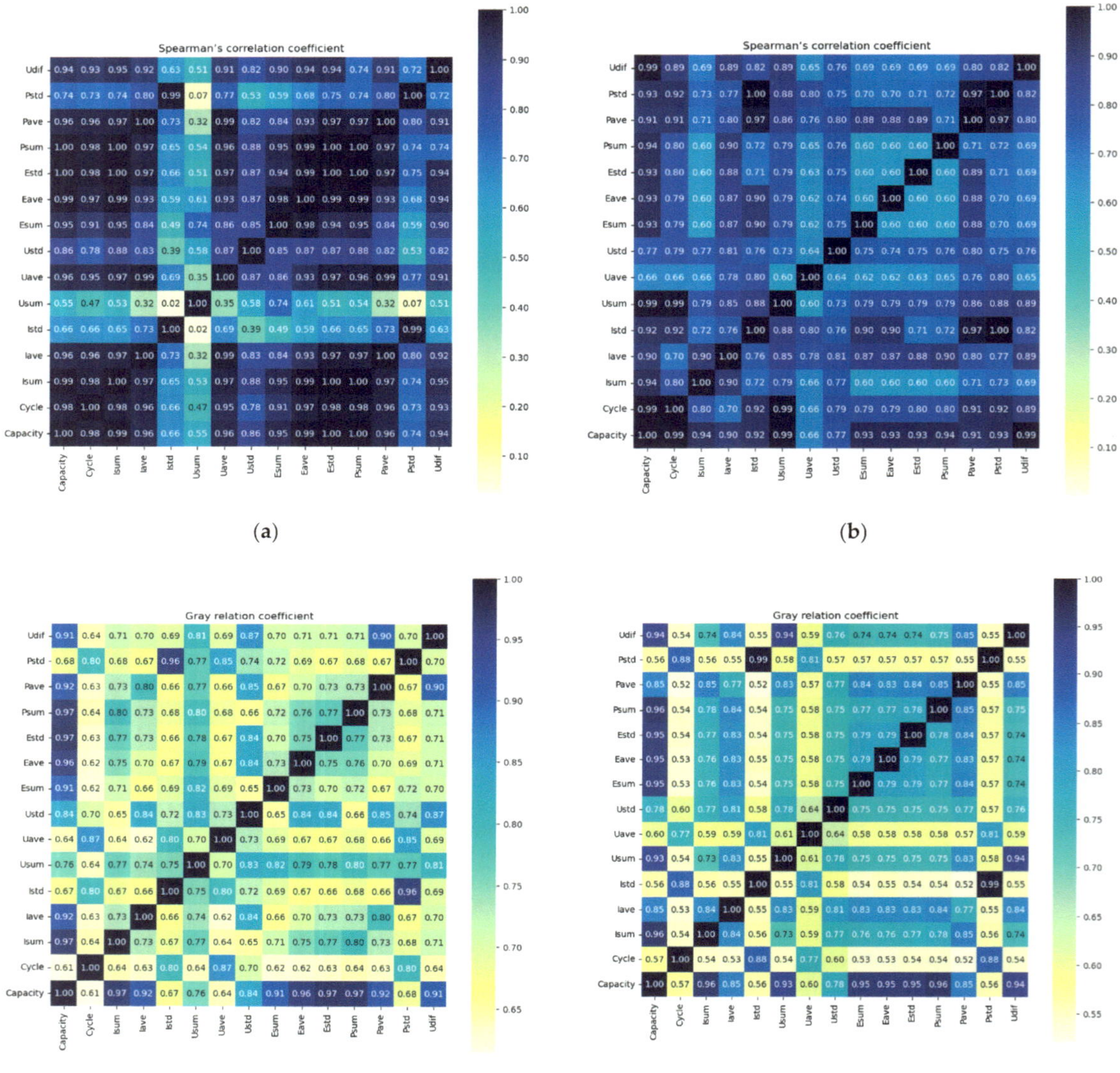

Figure 5. Correlation analysis results. (**a**) Spearman's correlation coefficients for B1; (**b**) Spearman's correlation coefficients for B2; (**c**) gray relation coefficients for B1; and (**d**) gray relation coefficients for B2.

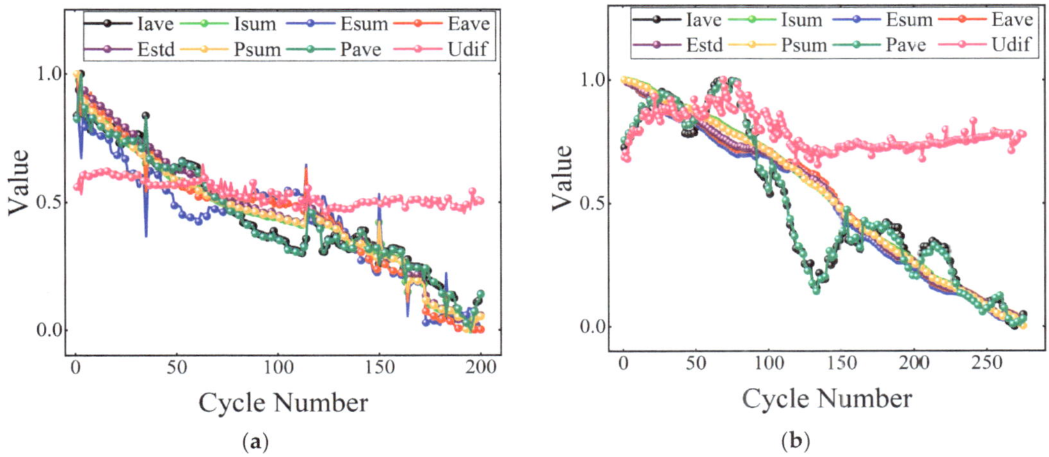

Figure 6. Normalized optimal feature set. (**a**) Features of B1; (**b**) features of B2.

4. Battery Capacity Degradation Prediction

4.1. Bidirectional Long Short-Term Memory Network (Bi-LSTM)

As an enhanced version of the traditional recurrent neural network (RNN), LSTM largely addresses the issues of gradient vanishing and gradient explosion that commonly occur in standard RNNs when handling time series data by introducing input gates, forget gates, output gates, and cell states [24]. Given that the behavior of the lithium-ion battery capacity over time is influenced by complex, dynamic patterns, LSTM is capable of modeling these temporal correlations more effectively than other algorithms, such as convolutional neural networks (CNNs) or simple recurrent neural networks (RNNs), which may struggle with long-term dependencies. The specific architecture of the LSTM model is illustrated in Figure 7.

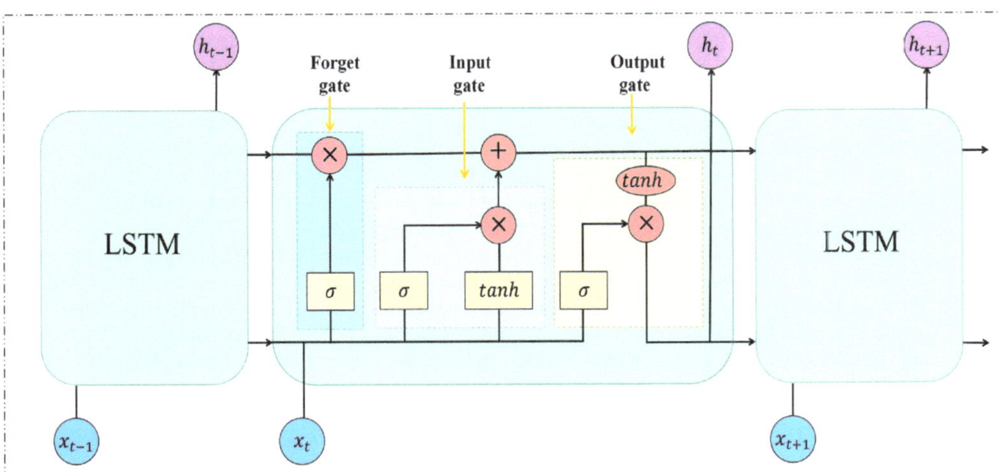

Figure 7. The structure of LSTM.

The core steps of LSTM are as follows:

(1) Decide which information is discarded by the cell state in the forget gate f_t

$$f_t = \sigma(W_f \cdot [\mathrm{h}_{t-1}, x_t] + b_f) \tag{5}$$

where σ is the sigmoid function, W_f and b_f are the weight matrix and bias vector of the oblivion gate, h_{t-1} is the output of the cell state at the moment $t-1$, and x_t is the input of the cell state at the moment t.

(2) Decide what information is stored in the internal state.

First, update the input gate i_t

$$i_t = \sigma(W_i \cdot [h_{t-1}, x_t] + b_i) \tag{6}$$

where W_i and b_i are the weight matrix and bias vector of the input gate, respectively.

Next, a candidate vector $\widetilde{C}_t$ is obtained.

$$\widetilde{C}_t = \tanh(W_c \cdot [h_{t-1}, x_t] + b_c) \tag{7}$$

where W_c and b_c are the weight matrix and bias vector of $\widetilde{C}_t$, respectively.

(3) Update the cellular state.

$$C_t = f_t * C_{t-1} + i_t * \widetilde{C}_t \tag{8}$$

where C_{t-1} is the cell state at the moment $t-1$.

(4) Obtain the output of LSTM.

$$o_t = \sigma(W_o \cdot [h_{t-1}, x_t] + b_o) \tag{9}$$

$$h_t = o_t * \tanh(C_t) \tag{10}$$

where o_t is the sigmoid layer of the output gate, h_t is the output of the LSTM at time t, and W_o and b_o are the weight matrix and bias vector of the output layer, respectively.

Bi-LSTM is an improved LSTM, which enhances the model's context-capturing capability by computing the input sequence in two directions (frontward and backward) separately [25]. In the capacity prediction of lithium-ion batteries, the performance of the battery does not only depend on the current charging and discharging state, temperature, voltage, etc., but is also affected by the historical state, as well as the future trend. Therefore, the bi-directional structure can better capture these long-term and short-term dynamic changes and improve the prediction accuracy. Moreover, Bi-LSTM can extract more features from complex data by considering forward and backward time series information. Especially for the battery capacity prediction when the aging trend of the battery needs to be modeled, the advantage of bi-directional information flow is even more obvious, which can capture the subtle changes and patterns in the time series more effectively than LSTM and thus improve the accuracy of prediction models. The specific structure of Bi-LSTM is shown in Figure 8. The input layer receives the input data and passes it to the forward layer and the backward layer. The forward layer processes the forward time series of the input data and the backward layer processes the reverse time series of the input data. These two layers learn the different contextual information of the sequence separately and merge their outputs. Eventually, the output layer generates the final prediction based on the forward and backward outputs.

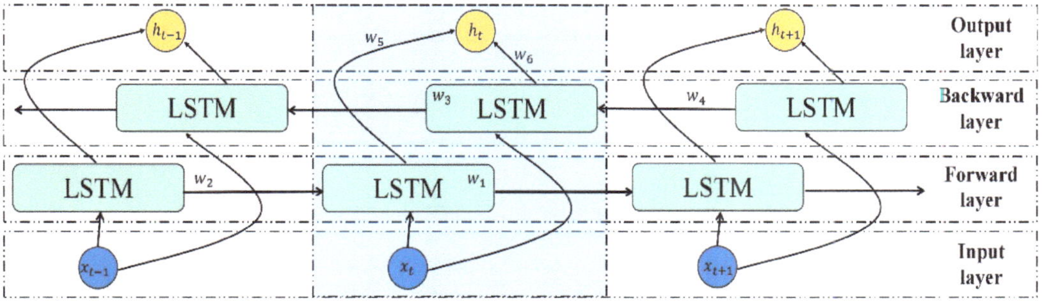

Figure 8. The structure of Bi-LSTM.

The formula for the forward layer is as follows:

$$L_t^f = \sigma(w_1 x_t + w_2 L_{t-1}^f + b_{L^f}) \tag{11}$$

where L_t^f is the output value of the forward layer, ω_1 and ω_2 are the weight matrices of the forward layer, and b_{L^f} is the bias vector of the forward layer.

The equation for the backward layer is as follows:

$$L_t^b = \sigma(w_3 x_t + w_4 L_{t+1}^b + b_{L^b}) \tag{12}$$

where L_t^b is the output value of the backward layer, ω_3 and ω_4 are the weight matrices of the backward layer, and b_{L^b} is the bias vector of the backward layer.

The final output value h_t is calculated as follows:

$$h_t = w_5 L_t^f + w_6 L_t^b \tag{13}$$

where ω_5 and ω_6 are two weight matrices.

The Bi-LSTM model employs a bi-directional structure that captures hidden information in time series data more efficiently, thereby enhancing the accuracy of lithium-ion battery capacity predictions. The Bi-LSTM model employed in this study consists of three layers, each containing two LSTMs, which extract deeper features from the data and thereby improve the prediction accuracy compared to the single-layer Bi-LSTM.

4.2. Crested Porcupine Optimizer (CPO)

The CPO algorithm mimics the four defense strategies of the crested porcupine, which are executed sequentially as the distance between the predator and the crested porcupine decreases, including sight, sound, odor, and physical attack, and accelerates the convergence of the algorithm by introducing a cyclic population reduction technique. The main steps of CPO are listed below.

(1) Population initialization

$$\vec{X}_i = \vec{L} + \vec{r} \times (\vec{U} - \vec{L}) | i = 1, 2 \ldots, N \tag{14}$$

where N denotes the number of populations, $\vec{X}_i$ denotes the ith candidate solution in the search space, $\vec{L}$ and $\vec{U}$ are the lower and upper bounds of the search, respectively, and $\vec{r}$ is a vector randomly initialized between 0 and 1.

(2) Cyclic population reduction technique

The cyclic population reduction technique means that some CPs are allowed to leave the population during the optimization process to accelerate the convergence speed, and then some CPs are added to the population to improve the population diversity and avoid falling into local minima, which ultimately achieves the purpose of accelerating the convergence speed while maintaining the population diversity. The specific mathematical model is as follows:

$$N = N_{\min} + (N\prime - N_{\min}) \times (1 - (\frac{t\% \frac{T_{\max}}{T}}{\frac{T_{\max}}{T}})) \tag{15}$$

where T is the number of cycles, t is the number of current function evaluations, $T_{\max}$ is the maximum number of function evaluations, % is the remainder operator, and $N_{\min}$ is the minimum number of individuals in the newly generated population. In the optimization search process, the number of populations N first reaches the maximum value and then gradually decreases until it reaches $N_{\min}$, which represents a cycle. Then, the above cycle is repeated many times throughout the optimality-seeking process until T times.

(3) The first defense strategy

The first defense strategy is visual intimidation, and the mathematical model is as follows:

$$\vec{x}_i^{t+1} = \vec{x}_i^{t} + \tau_1 \times \left| 2 \times \tau_2 \times \vec{x}_{CP}^{t} - \vec{y}_i^{t} \right| \tag{16}$$

where $\vec{x}_i^{t}$ is the position of the ith CP at the tth function evaluation and $\vec{x}_{CP}^{t}$ is the position of the current best CP, τ_1, τ_2 are two random numbers. $\vec{y}_i^{t}$ is calculated as follows:

$$\vec{y}_i^{t} = \frac{\vec{x}_i^{t} + \vec{x}_k^{t}}{2} \tag{17}$$

where $\vec{x}_k^{t}$ is the randomly chosen position of another CP.

(4) The second defense strategy

The second defense strategy is sound intimidation, which is mathematically modeled as follows:

$$\vec{x}_i^{t+1} = (1 - \vec{U}_1) \times \vec{x}_i^{t} + \vec{U}_1 \times (\vec{y} + \tau_3 \times (\vec{x}_{r_1}^{t} - \vec{x}_{r_2}^{t})) \tag{18}$$

where $\vec{U}_1$ is a random vector between 0 and 1, τ_3 is a random number, and $\vec{x}_{r_1}^{t}$ and $\vec{x}_{r_2}^{t}$ are randomly chosen positions of the two CPs.

(5) The third defense strategy

The third defense strategy is the odor attack, which is mathematically modeled as follows:

$$\vec{x}_i^{t+1} = (1 - \vec{U}_1) \times \vec{x}_i^{t} + \vec{U}_1 \times (\vec{x}_{r_1}^{t} + S_i^t \times (\vec{x}_{r_2}^{t} - \vec{x}_{r_3}^{t}) - \tau_3 \times \vec{\delta} \times \gamma_t \times S_i^t) \tag{19}$$

where S_i^t is a fitness function, $\vec{\delta}$ is a random vector, γ_t is a time-dependent factor, and $\vec{x}_{r_1}^{t}$, $\vec{x}_{r_2}^{t}$, and $\vec{x}_{r_3}^{t}$ are randomly selected positions of the three CPs.

(6) The fourth defense strategy

The fourth defense strategy is the physical attack, and the mathematical model is as follows:

$$\vec{x}_i^{t+1} = \vec{x}_{CP}^{t} + (\alpha(1 - \tau_4) + \tau_4) \times (\delta \times \vec{x}_{CP}^{t} - \vec{x}_i^{t}) - \tau_5 \times \delta \times \gamma_t \times \vec{F}_i^{t} \tag{20}$$

where α is a control parameter, τ_4 and τ_5 are random numbers, δ is a random vector, and $\vec{F}_i^{t}$ is a fitness-based factor.

CPO was validated using three CEC benchmarks (CEC2014, CEC2017, and CEC2020), and its performance was compared against that of three categories of existing optimization algorithms [26] as follows: (i) the most highly cited optimizers, including the Gray Wolf Optimizer (GWO), Whale Optimization Algorithm (WOA), Differential Evolution (DE), and Salp Swarm Algorithm (SSA); (ii) recently published algorithms, including the Gradient-Based Optimizer (GBO), African Vultures Optimization Algorithm (AVOA), Runge–Kutta Method (RUN), Equilibrium Optimizer (EO), Artificial Gorilla Troops Optimizer (GTO), and Slime Mold Algorithm (SMA); and (iii) high-performance optimizers, such as SHADE, LSHADE, AL-SHADE, LSHADE-cnEpSin, and LSHADE-SPACMA. The statistical analysis revealed that CPO can be regarded as a high-performance optimizer due to its significantly superior performance compared to all competing optimizers across the majority of the test functions in the three validated CEC benchmarks. Quantitatively, CPO achieved an improvement over rival optimizers, with percentages of up to 83% for CEC2017, 70% for CEC2017, 90% for CEC2020, and 100% for six real-world engineering problems.

4.3. Improved Crested Porcupine Optimizer (ICPO)

No single algorithm can be applied to all application scenarios with efficiency. When applied to the prediction of battery capacity, the CPO algorithm can be specifically adapted to enhance its capability in addressing the complex, dynamic characteristics of lithium-ion battery behavior, thereby improving the prediction accuracy. In the original CPO algorithm, the initialized population distribution is relatively random and poorly positioned, which may result in a slower global search during the early stages of the algorithm's iterations or lead to convergence at a local optimum in the later stages. Additionally, during the algorithm's iterations, if the position of the current optimal individual differs from that of the global optimal individual, as the number of iterations increases, individuals in the population may mistakenly converge toward the locally optimal region, resulting in premature convergence and a decrease in the accuracy of the global search. These limitations of the original CPO algorithm contribute to a reduced capacity prediction accuracy for lithium-ion batteries. Consequently, this paper improves the original CPO algorithm in two key areas: enhanced population initialization and a refined variation strategy, thereby establishing the ICPO algorithm, which offers high lithium-ion battery capacity prediction accuracy.

4.3.1. Improved Chebyshev Chaotic Mapping Initialization

Chebyshev chaotic mapping [27] is a widely used chaotic mapping method for population initialization in optimization algorithms, which is computed as follows:

$$x_{k+1} = \cos(k \cos^{-1}(x_k)) \tag{21}$$

where k is the order, which takes the value of 4 in this paper, and x_0 is a random number between -1 and 1.

However, the traditional Chebyshev chaotic mapping may still be unable to make the initial population fully cover the search space, which reduces the optimization effect, and so this paper makes the following improvements to the Chebyshev chaotic mapping [28]:

$$x_{k+1} = 1 - 2(\cos(2\mathrm{arccos}x_k))^2 \tag{22}$$

The initialization of the CP population after the introduction of the improved Chebyshev chaotic mapping is calculated as follows:

$$\vec{X}_i = \vec{L} + \frac{x_{k+1}}{2}\vec{r} \times (\vec{U} - \vec{L})|i = 1,2\ldots,N \tag{23}$$

The improved Chebyshev chaotic mapping enhances the population's dispersion and diversity while ensuring the randomness of the initial population distribution, greatly improving the algorithm's performance.

4.3.2. Random Differential Mutation Strategy

In each iteration of the CPO algorithm, a Random Differential Mutation strategy [29] is used to perform mutation operations on the population to generate candidate individuals with greater potential, thus increasing the diversity of the population, which can help the algorithm to jump out of the local optimum and optimize the effect of the optimization search. The formula for stochastic differential variation is as follows:

$$\vec{x}_i^{t+1} = r_1(\vec{x}_{CP}^{t} - \vec{x}_i^{t}) + r_2(\vec{x}_k^{t} - \vec{x}_i^{t}) \tag{24}$$

where r_1 and r_2 are random numbers between 0 and 1.

4.4. Developed ICPO-Bi-LSTM

In this paper, the ICPO algorithm is employed to optimize the hyperparameters of the Bi-LSTM model, such as the number of neurons in each layer, the learning rate, the number of iterations, and the dropout rate. Compared to the original CPO algorithm, the ICPO algorithm enhances the global search speed during the pre-iteration phase and addresses the issue of premature convergence. The flowchart of the ICPO-Bi-LSTM method is illustrated in Figure 9, and the primary steps are outlined below.

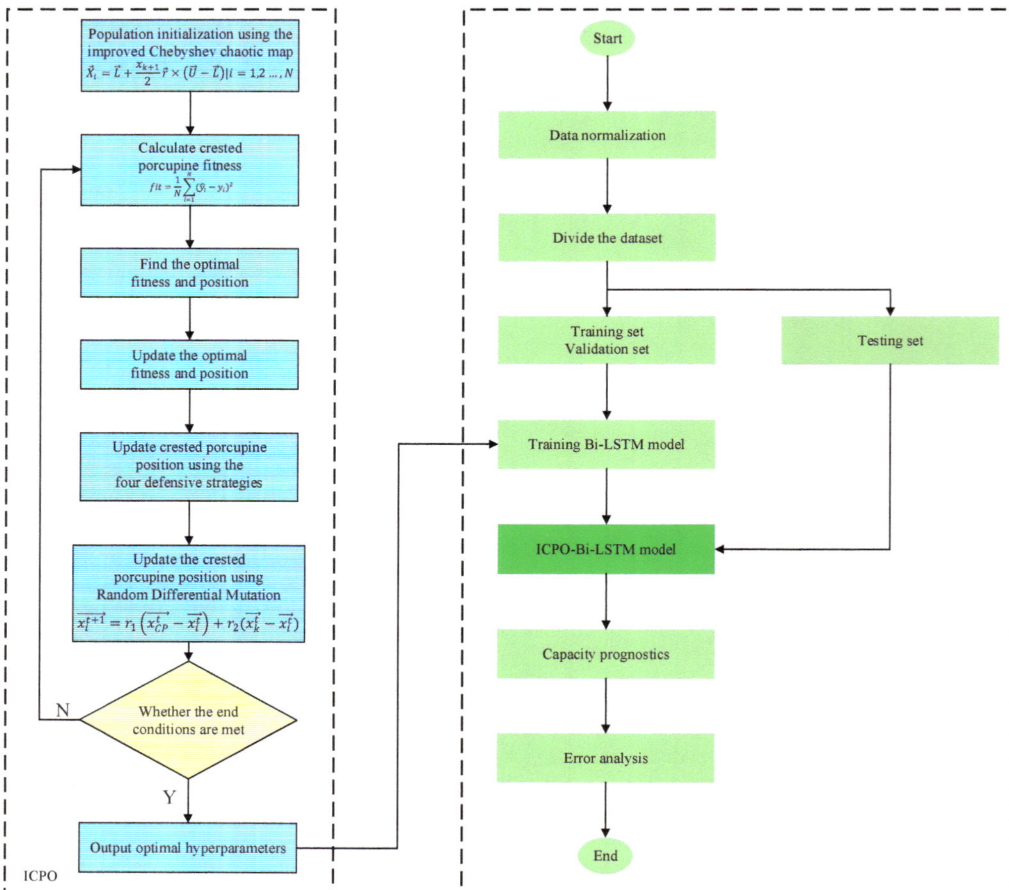

Figure 9. The flowchart of the developed ICPO-Bi-LSTM.

(1) Normalize both the capacity and the optimal set of battery characteristics according to Equation (2), and divide them into training, validation, and test sets based on a predefined ratio.
(2) Set the parameters of the ICPO algorithm and define the optimization ranges for the Bi-LSTM parameters. For instance, in the ICPO algorithm, the population size, maximum number of iterations, and optimization-seeking dimension are set to 100, 100, and 6, respectively; for Bi-LSTM, the search ranges for the number of neurons in each layer, learning rate, number of iterations, and dropout rate are defined as [1, 500], [0.0001, 0.001], [1, 200], and [0, 1], respectively.
(3) Introduce improved Chebyshev chaotic mapping to initialize the population, as described in Equation (23).
(4) Calculate the fitness value for each individual in the population and rank them based on their fitness values. The fitness function is determined using the following formula

$$fit = \frac{1}{N}\sum_{i=1}^{N} (\hat{y}_i - y_i)^2 \quad (25)$$

where N is the number of validation set samples, y_i is the actual value of the ith validation set sample, and $\hat{y}_i$ is the predicted value of the ith validation set sample.

(5) Obtain the fitness value and position corresponding to the individual with the lowest fitness value, as determined by sorting the fitness values of all individuals, and update these to the global best fitness and best position.
(6) Apply the four defensive strategies of the ICPO algorithm, along with the cyclic population reduction strategy, to update the positions of the individuals. Subsequently, update the positions further after each iteration by introducing random difference variation, as described in Equation (24).
(7) After each update, check the position of each individual and adjust it back within the boundaries using a random number if it exceeds the upper or lower limits.
(8) Calculate the fitness value for each individual after the update, and update the global best fitness and best individual position.
(9) Determine if the loop should terminate: if the maximum number of iterations is reached or the fitness value attains the minimum, output the optimal hyperparameter combination for the Bi-LSTM model. Otherwise, return to step (6).

4.5. Capacity Prognostics Based on ICPO-Bi-LSTM and Feature Extraction

In this section, we propose an integrated Li-ion battery capacity estimation framework based on the ICPO-Bi-LSTM method. As illustrated in Figure 10, the proposed framework for capacity prediction comprises three key components: data processing, model training, and capacity prediction.

Initially, a set of features is extracted from the data of the battery. Subsequently, the correlations between the features and the capacity, as well as the inter-feature correlations, are assessed using Spearman's correlation coefficient and a gray correlation analysis. Features exhibiting stronger correlations are selected, while redundant features are discarded to form an optimal feature set. This optimal feature set is then used as input for the ICPO-Bi-LSTM method. To avoid introducing bias in the performance evaluation, which could lead to erroneous predictions during online testing, the experimental data are divided into training, validation, and test sets. The training set is used for model parameter fitting, while the validation set is employed for iterative hyperparameter tuning to identify the optimal parameters for the Bi-LSTM model. Finally, the test set is used for the capacity prediction and performance evaluation.

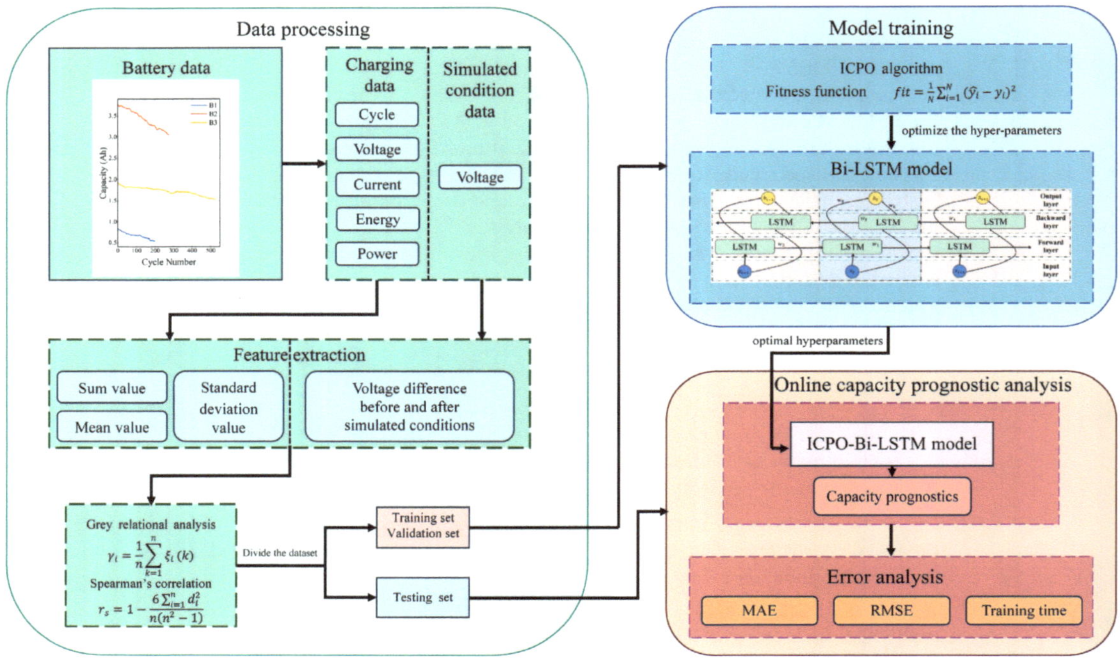

Figure 10. The framework of the capacity prognostic analysis based on ICPO-Bi-LSTM and feature extraction.

5. Results and Discussion

5.1. The Evaluation Criteria

In this paper, the mean absolute error (MAE) and root mean square error (RMSE) are used to evaluate the capacity prediction accuracy of the established ICPO-Bi-LSTM method, which are calculated as follows:

$$MAE = \frac{1}{N}\sum_{i=1}^{N}|\hat{y}_i - y_i| \tag{26}$$

$$RMSE = \sqrt{\frac{1}{N}\sum_{i=1}^{N}(\hat{y}_i - y_i)^2} \tag{27}$$

where N is the number of samples in the test set, $\hat{y}_i$ is the predicted value of the ith capacity predicted by the model, and y_i is the actual value of the ith capacity. If the values of MAE and $RMSE$ are smaller, the model's capacity prediction is better.

5.2. Capacity Prognosis Accuracy Using Different Batteries at Various Set Ratios

In this section, the B1 and B2 batteries are used to evaluate the accuracy of the ICPO-Bi-LSTM model's capacity prediction under dynamic operating conditions, while the B3 battery is employed to assess the model's generalizability under simple operating conditions. The training, validation, and test sets are split into ratios of 6:1:3 (case 1), 7:1:2 (case 2), and 8:1:1 (case 3) for all three batteries, respectively. Figure 11 illustrates the capacity prognosis results for the three batteries across different dataset ratios. It can be observed that the ICPO-Bi-LSTM method not only accurately predicts the capacity under varying training, validation, and test set ratios but also captures the capacity regeneration phenomenon during the aging process of the batteries with high precision. Furthermore, the predicted capacity closely matches the actual capacity for both simple and dynamic operating conditions, demonstrating that the method is robust and versatile in battery capacity prognostics under diverse discharge conditions.

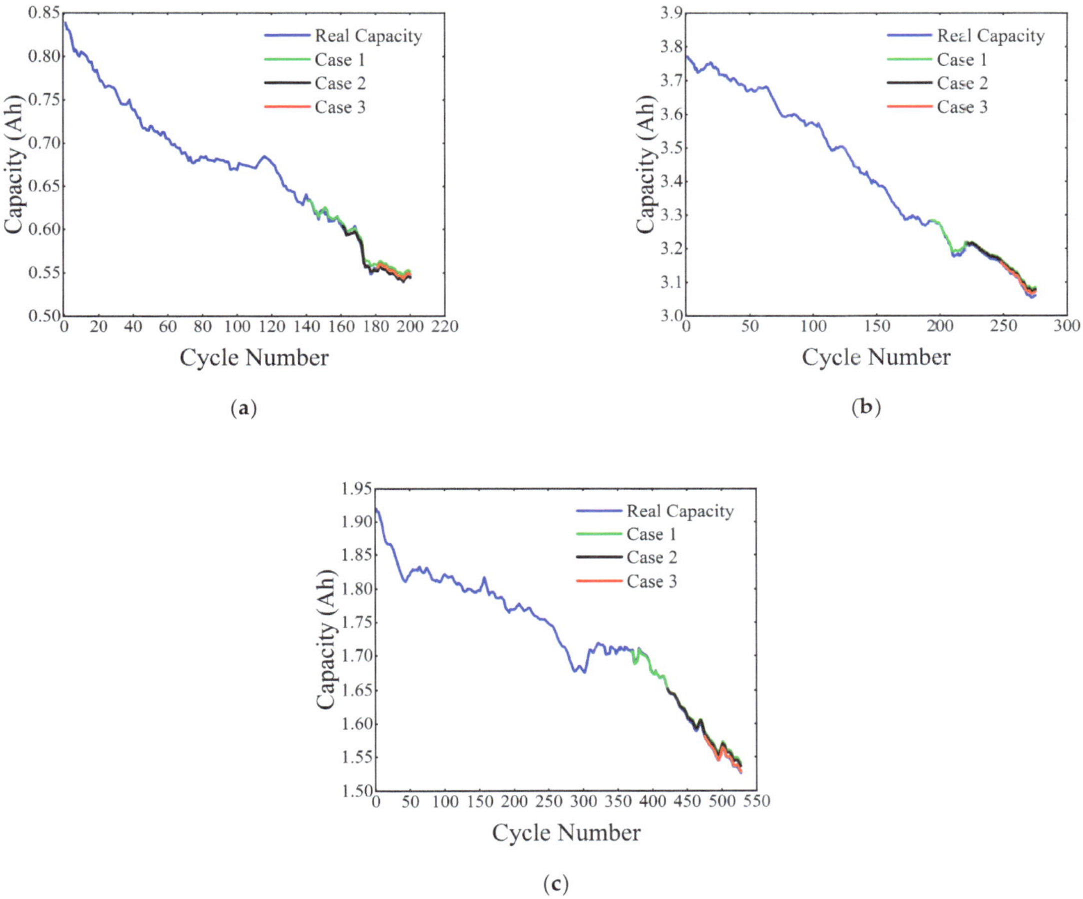

Figure 11. Capacity prognostics for different batteries at various set ratios: (**a**) B1; (**b**) B2; and (**c**) B3.

Table 6 displays the capacity prognosis errors for the three batteries across various dataset scales. It is noted that all error metrics remain within 1%, with the maximum values of the root mean square error (RMSE) and mean absolute error (MAE) recorded at 0.9622% and 0.9145%, respectively. Additionally, the estimation errors for all three batteries decrease as the proportion of data in the training and validation sets increases. This improvement is attributed to the larger amount of historical data on battery capacity degradation available in the training and validation sets, which enhances the accuracy of capacity prognostics using the ICPO-Bi-LSTM method.

Table 6. Capacity prognosis errors for different batteries at various set ratios.

Battery	Ratios	MAE (%)	RMSE (%)
B1	6:1:3 (case 1)	0.6827	0.8236
	7:1:2 (case 2)	0.4346	0.6027
	8:1:1 (case 3)	0.1841	0.2473
B2	6:1:3 (case 1)	0.8862	0.9541
	7:1:2 (case 2)	0.7627	0.8463
	8:1:1 (case 3)	0.5538	0.5973
B3	6:1:3 (case 1)	0.9145	0.9622
	7:1:2 (case 2)	0.6434	0.6912
	8:1:1 (case 3)	0.1314	0.1386

5.3. Results of Different Prediction Methods

To further validate the superiority of the proposed ICPO-Bi-LSTM method and the superiority of the ICPO algorithm over other algorithms, five models—LSTM, Bi-LSTM, CPO-Bi-LSTM, PSO-Bi-LSTM, and the developed method—are employed to predict the battery capacity and compare their performance. The dataset is divided into 80% for training, 10% for validation, and 10% for testing for all four methods. For CPO, PSO, and ICPO, the population size is set to 100, the number of iterations to 100, and the optimization dimension to six. The hyperparameters of the Bi-LSTM model without algorithmic optimization are configured with 200 neurons per layer, a learning rate of 0.0001, 150 iterations, and a dropout rate of 0.5. For the LSTM model without algorithmic optimization, the parameters are set as follows: 50 neurons in the hidden layer, a learning rate of 0.001, 50 iterations, and a dropout rate of 0.5.

Figure 12 presents the capacity prognosis results for the three types of batteries using the five methods described above. The capacity curves predicted by all five methods closely match the actual capacity curves, demonstrating that the feature set extracted in this study accurately reflects the capacity degradation process of lithium-ion batteries under varying operating conditions. Furthermore, the fitting accuracy of the capacity curves predicted by the five methods, ranked in descending order as ICPO-Bi-LSTM, CPO-Bi-LSTM, PSO-Bi-LSTM, Bi-LSTM, and LSTM, further highlights the superiority of the ICPO-Bi-LSTM model.

The MAEs and RMSEs of the five previously discussed methods for predicting the three cell capacities are presented as evaluation metrics in Table 7. In addition, to further validate the advantages of the low computational effort and high efficiency of the present ICPO-Bi-LSTM framework in model training, the time required to train the models using different algorithms is also listed in Table 7 as an evaluation metric. It can be seen that the final MAEs and RMSEs of all models are less than 1%, which further verifies that the optimal set of features extracted in this paper can effectively reflect the complex degradation trend of lithium-ion batteries. Moreover, the ICPO-Bi-LSTM method not only accurately predicts the battery capacity under simple and dynamic operating conditions but also minimizes the error generated by the prediction compared with other methods. On the other hand, the LSTM model requires manual setting of hyperparameters and cannot be dynamically adjusted according to the battery data, which limits its prediction capability and leads to lower accuracy. In contrast, the Bi-LSTM model adopts bi-directional training, which can effectively capture the hidden patterns in the battery time series data and thus improve the prediction performance, which further illustrates the advantage of Bi-LSTM over LSTM in battery capacity prediction. However, it still has the limitation of not being able to determine the optimal hyperparameters based on the data; the PSO algorithm can search for the optimal hyperparameter combinations of the Bi-LSTM model based on the data, and so the optimized Bi-LSTM model using the PSO greatly improves the prediction accuracy compared with the original Bi-LSTM model. On this basis, the original CPO algorithm can not only dynamically adjust the hyperparameters of the Bi-LSTM model according to the prediction error, which solves the problem that the optimal hyperparameters are difficult to determine, but also has a better optimization effect compared with the PSO algorithm, thus further improving the prediction accuracy. However, the original CPO algorithm has difficulty in determining the optimal hyperparameter combinations of the model due to the uneven distribution of the population and the possibility of falling into the local optimum, which leads to premature convergence of the algorithm, resulting in the limited prediction accuracy of the model. In contrast, the ICPO-Bi-LSTM method developed in this paper employs an improved Chebyshev chaotic mapping initialization to ensure a more uniform initial population distribution. This improvement helps to speed up the search rate in the initial stages of the CPO algorithm, as well as to perform a more thorough search in the later stages. In addition, the introduction of the Random Differential Mutation strategy after each iteration of the CPO algorithm not only shortens the optimization search time but also helps to prevent premature convergence, which is more conducive to determining the optimal hyperparameter combinations for the Bi-LSTM model, and these improvements greatly enhance the accuracy of the model in predicting the battery capacity. Finally, for the three batteries, training the Bi-LSTM model using the original CPO algorithm not only has a higher model prediction accuracy but also takes less time to determine the optimal hyperparameters of the model compared to training the Bi-LSTM model using the original PSO algorithm, which demonstrates the superiority of the CPO algorithm. The ICPO algorithm used in this paper not only optimizes the search results but also effectively shortens the search time due to the introduction of the improved Chebyshev chaotic mapping initialization and the Random Differential Mutation strategy, and thus outperforms the PSO algorithm and the original CPO algorithm in terms of both the prediction accuracy and model training time. Since the separate LSTM and Bi-LSTM models are not trained with an algorithm, there is no training time.

Table 7. Capacity prognosis errors and training times for different batteries using different methods.

Battery	Method	MAE (%)	RMSE (%)	Training Time (min)
B1	LSTM	0.7922	0.8793	
	Bi-LSTM	0.6827	0.7423	
	PSO-Bi-LSTM	0.5246	0.6433	23
	CPO-Bi-LSTM	0.3977	0.4813	18
	ICPO-Bi-LSTM	0.2433	0.2758	11
B2	LSTM	0.9273	0.9827	
	Bi-LSTM	0.8427	0.8817	
	PSO-Bi-LSTM	0.7062	0.7423	27
	CPO-Bi-LSTM	0.6213	0.6527	21
	ICPO-Bi-LSTM	0.5348	0.5777	14
B3	LSTM	0.5612	0.5837	
	Bi-LSTM	0.3633	0.3849	
	PSO-Bi-LSTM	0.2527	0.2734	35
	CPO-Bi-LSTM	0.2203	0.2276	25
	ICPO-Bi-LSTM	0.1346	0.1408	18

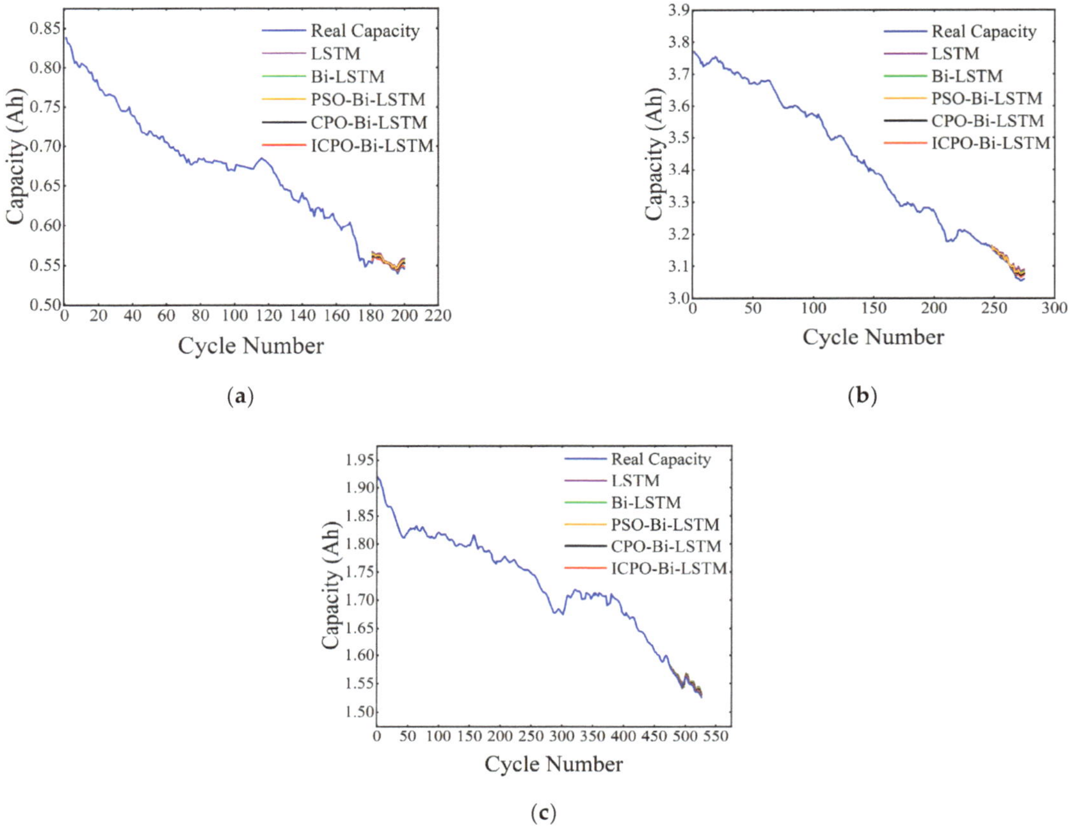

Figure 12. Capacity prognostics for different batteries using different methods. (**a**) B1; (**b**) B2; and (**c**) B3.

6. Conclusions

Capacity prognostics using data-driven methods can be inaccurate when the extracted features fail to sufficiently capture the degradation trend of lithium-ion batteries or when the model's hyperparameters are improperly specified. This paper proposes a capacity prognostic method for marine lithium-ion batteries, which extracts features from battery charging and discharging data collected under dynamic operating conditions and utilizes the ICPO-Bi-LSTM model for capacity prognostics. First, a series of features are extracted from the charging and discharging data to ensure the adequate capture of battery capacity degradation. Then, the gray correlation degree and Spearman correlation coefficient are calculated to select features that are strongly correlated with capacity, while eliminating redundant features to obtain an optimal feature set. Additionally, the issues of an uneven population distribution and slow search efficiency during the early stages of the original CPO algorithm are addressed. The tendency of the original CPO algorithm to prematurely converge and fall into local optima in later stages, which hinders the identification of the optimal hyperparameter combination, is also mitigated. To this end, an improved CPO algorithm is proposed, combining enhanced Chebyshev chaotic mapping and a Random Differential Mutation strategy, which improve the population initialization and iterative search strategies of the original CPO algorithm, respectively.

The accuracy of our method is validated by predicting the capacities of two battery models (B1 and B2), which are discharged under dynamic operating conditions but charged using different methods. Additionally, we evaluate the method's generalization ability using a third battery model (B3), which is discharged under simpler conditions. The experimental results confirm the method's capacity to predict battery capacity accurately across various training, validation, and test set ratios, as well as under different charging and discharging conditions, demonstrating both high accuracy and robustness. Specifically, the mean absolute error (MAE) and root mean square error (RMSE) of the predicted capacities for the different batteries across various dataset ratios are consistently below 1%. Furthermore, our method achieves the smallest MAE and RMSE values when compared to other methods (e.g., CPO-Bi-LSTM, PSO-Bi-LSTM, Bi-LSTM, and LSTM), and the ICPC algorithm used in our approach demonstrates the shortest model training time compared to the original CPO algorithm and the PSO algorithm, resulting in the most accurate and efficient capacity prediction. In all model comparison experiments, the maximum MAEs and RMSEs of the predicted capacities using this method remain consistently below 0.6%.

Future work will include extensive experiments using various marine battery models under diverse dynamic operating conditions to further validate the effectiveness of our proposed method. We will progress from controlled laboratory environments to real-world marine applications, assessing the method in increasingly complex and dynamic conditions. One major challenge is the variability in battery charging and discharging conditions, which could impact data integrity. To address this, we plan to enhance both data acquisition techniques and the robustness of the model. Furthermore, we will investigate the effect of ambient temperature on battery performance, particularly during the charging process. Additionally, we aim to integrate the proposed method into an online prediction system for continuous monitoring, with a focus on ensuring scalability and real-time performance.

Author Contributions: Conceptualization, Q.S.; methodology, Q.S.; software, T.T.; validation, T.T., Y.L. and Y.C.; formal analysis, Q.S; investigation, Q.S.; resources, X.Y.; data curation, Q.S.; writing—original draft preparation, Q.S.; writing—review and editing, X.Y.; visualization, L.L.; supervision, L.L.; project administration, X.Y.; funding acquisition, X.Y. All authors have read and agreed to the published version of the manuscript.

Funding: This work is supported by the China National Key Research and Development Project (Grant No. 2023YFB4301704) and the China National Key Laboratory of Electromagnetic Energy Technology Open Fund (Grant No. 61422172220403).

Institutional Review Board Statement: Not applicable.

Informed Consent Statement: Not applicable.

Data Availability Statement: Data will be made available upon request.

Conflicts of Interest: The authors declare no conflicts of interest.

References

1. Milićević, S.; Blagojević, I.; Milojević, S.; Bukvić, M.; Stojanović, B. Numerical Analysis of Optimal Hybridization in Parallel. Hybrid Electric Powertrains for Tracked Vehicles. *Energies* **2024**, *17*, 3531. [CrossRef]
2. Marinković, D.; Dezső, G.; Milojević, S. Application of machine learning during maintenance and exploitation of electric vehicles. *Adv. Eng. Lett.* **2024**, *3*, 132–140. [CrossRef]
3. Pang, Z.; Si, X.; Hu, C.; Du, D.; Pei, H. A Bayesian Inference for Remaining Useful Life Estimation by Fusing Accelerated Degradation Data and Condition Monitoring Data. *Reliab. Eng. Syst. Saf.* **2021**, *208*, 107341. [CrossRef]
4. Zhang, C.; Luo, L.; Yang, Z.; Zhao, S.; He, Y.; Wang, X.; Wang, H. Battery SOH Estimation Method Based on Gradual Decreasing Current, Double Correlation Analysis and GRU. *Green Energy Intell. Transp.* **2023**, *2*, 100108. [CrossRef]
5. Li, J.; Li, T.; Fang, D.; Wang, Y.; Guo, S.; Wang, Z.; Yu, Q. Internal Fault Diagnosis Method for Lithium Batteries Based on a Failure Physical Model. *Eng. Fail. Anal.* **2023**, *154*, 107714. [CrossRef]
6. Li, Y.; Liu, K.; Foley, A.M.; Zülke, A.; Berecibar, M.; Nanini-Maury, E.; Van Mierlo, J.; Hoster, H.E. Data-Driven Health Estimation and Lifetime Prediction of Lithium-Ion Batteries: A Review. *Renew. Sustain. Energy Rev.* **2019**, *113*, 109254. [CrossRef]
7. Zuo, H.; Liang, J.; Zhang, B.; Wei, K.; Zhu, H.; Tan, J. Intelligent Estimation on State of Health of Lithium-Ion Power Batteries Based on Failure Feature Extraction. *Energy* **2023**, *282*, 128794. [CrossRef]

8. Li, W.; Sengupta, N.; Dechent, P.; Howey, D.; Annaswamy, A.; Sauer, D.U. Online Capacity Estimation of Lithium-Ion Batteries with Deep Long Short-Term Memory Networks. *J. Power Sources* **2021**, *482*, 228863. [CrossRef]
9. Ma, Y.; Wu, L.; Guan, Y.; Peng, Z. The Capacity Estimation and Cycle Life Prediction of Lithium-Ion Batteries Using a New Broad Extreme Learning Machine Approach. *J. Power Sources* **2020**, *476*, 228581. [CrossRef]
10. Liu, K.; Shang, Y.; Ouyang, Q.; Widanage, W.D. A Data-Driven Approach with Uncertainty Quantification for Predicting Future Capacities and Remaining Useful Life of Lithium-Ion Battery. *IEEE Trans. Ind. Electron.* **2021**, *68*, 3170–3180. [CrossRef]
11. Deng, Z.; Hu, X.; Li, P.; Lin, X.; Bian, X. Data-Driven Battery State of Health Estimation Based on Random Partial Charging Data. *IEEE Trans. Power Electron.* **2022**, *37*, 5021–5031. [CrossRef]
12. Lu, Z.; Fei, Z.; Wang, B.; Yang, F. A Feature Fusion-Based Convolutional Neural Network for Battery State-of-Health Estimation with Mining of Partial Voltage Curve. *Energy* **2024**, *288*, 129690. [CrossRef]
13. Gou, B.; Xu, Y.; Feng, X. An Ensemble Learning-Based Data-Driven Method for Online State-of-Health Estimation of Lithium-Ion Batteries. *IEEE Trans. Transp. Electrific.* **2021**, *7*, 422–436. [CrossRef]
14. Peng, S.; Sun, Y.; Liu, D.; Yu, Q.; Kan, J.; Pecht, M. State of Health Estimation of Lithium-Ion Batteries Based on Multi-Health Features Extraction and Improved Long Short-Term Memory Neural Network. *Energy* **2023**, *282*, 128956. [CrossRef]
15. Deng, Z.; Xu, L.; Liu, H.; Hu, X.; Duan, Z.; Xu, Y. Prognostics of Battery Capacity Based on Charging Data and Data-Driven Methods for on-Road Vehicles. *Appl. Energy* **2023**, *339*, 120954. [CrossRef]
16. Guo, P.; Cheng, Z.; Yang, L. A Data-Driven Remaining Capacity Estimation Approach for Lithium-Ion Batteries Based on Charging Health Feature Extraction. *J. Power Sources* **2019**, *412*, 442–450. [CrossRef]
17. Dai, H.; Wang, J.; Huang, Y.; Lai, Y.; Zhu, L. Lightweight State-of-Health Estimation of Lithium-Ion Batteries Based on Statistical Feature Optimization. *Renew. Energy* **2024**, *222*, 119907. [CrossRef]
18. Zhang, Y.; Xiong, R.; He, H.; Pecht, M.G. Lithium-Ion Battery Remaining Useful Life Prediction with Box–Cox Transformation and Monte Carlo Simulation. *IEEE Trans. Ind. Electron.* **2019**, *66*, 1585–1597. [CrossRef]
19. Zhang, D.; Wang, Z.; Liu, P.; She, C.; Wang, Q.; Zhou, L.; Qin, Z. A Multi-Step Fast Charging-Based Battery Capacity Estimation Framework of Real-World Electric Vehicles. *Energy* **2024**, *294*, 130773. [CrossRef]
20. Cheng, G.; Wang, X.; He, Y. Remaining Useful Life and State of Health Prediction for Lithium Batteries Based on Empirical Mode Decomposition and a Long and Short Memory Neural Network. *Energy* **2021**, *232*, 121022. [CrossRef]
21. Chen, Z.; Wang, Z.; Wu, W.; Xia, T.; Pan, E. A Hybrid Battery Degradation Model Combining Arrhenius Equation and Neural Network for Capacity Prediction under Time-Varying Operating Conditions. *Reliab. Eng. Syst. Saf.* **2024**, *252*, 110471. [CrossRef]
22. Zhang, R.; Ji, C.; Zhou, X.; Liu, T.; Jin, G.; Pan, Z.; Liu, Y. Capacity Estimation of Lithium-Ion Batteries with Uncertainty Quantification Based on Temporal Convolutional Network and Gaussian Process Regression. *Energy* **2024**, *297*, 131154. [CrossRef]
23. Fu, J.; Wu, C.; Wang, J.; Haque, M.M.; Geng, L.; Meng, J. Lithium-Ion Battery SOH Prediction Based on VMD-PE and Improved DBO Optimized Temporal Convolutional Network Model. *J. Energy Storage* **2024**, *87*, 111392. [CrossRef]
24. Bouchareb, H.; Saqli, K.; M'sirdi, N.K.; Oudghiri Bentaie, M. Lithium-Ion Battery Health Management and State of Charge (SOC) Estimation Using Adaptive Modelling Techniques. *Energies* **2024**, *17*, 5746. [CrossRef]
25. Li, D.; Yang, C.; Li, Y. A Multi-Subsystem Collaborative Bi-LSTM-Based Adaptive Soft Sensor for Global Prediction of Ammonia Nitrogen Concentration in Wastewater Treatment Processes. *Water Res.* **2024**, *254*, 121347. [CrossRef]
26. Abdel-Basset, M.; Mohamed, R.; Abouhawwash, M. Crested Porcupine Optimizer: A New Nature-Inspired Metaheuristic. *Knowl.-Based Syst.* **2024**, *284*, 111257. [CrossRef]
27. Tavazoei, M.S.; Haeri, M. Comparison of Different One-Dimensional Maps as Chaotic Search Pattern in Chaos Optimization Algorithms. *Appl. Math. Comput.* **2007**, *187*, 1076–1085. [CrossRef]
28. Gu, J.; Hong, W.; Liang, T. Improvable Chebyshev Chaotic Sequence and Performance Analysis. *Journal of Military Communications Technology* **2006**, *27*, 43–46.
29. Li, Y.; Sun, K.; Yao, Q.; Wang, L. A Dual-Optimization Wind Speed Forecasting Model Based on Deep Learning and Improved Dung Beetle Optimization Algorithm. *Energy* **2024**, *286*, 129604. [CrossRef]

Journal of
Marine Science and Engineering

MDPI

Article

Early Fault Diagnosis and Prediction of Marine Large-Capacity Batteries Based on Real Data

Yifan Liu, Huabiao Jin, Xiangguo Yang *, Telu Tang, Qijia Song, Yuelin Chen, Lin Liu and Shoude Jiang

School of Naval Architecture, Ocean and Energy Power Engineering, Wuhan University of Technology, Wuhan 430070, China; l2023201239@whut.edu.cn (Y.L.); yx_jhb@163.com (H.J.); 331552@whut.edu.cn (T.T.); songqijia@whut.edu.cn (Q.S.); 299621@whut.edu.cn (Y.C.); 299437@whut.edu.cn (L.L.); 345388@whut.edu.cn (S.J.)
* Correspondence: yxglyr@whut.edu.cn

Abstract: The inconsistency of battery voltages in all-electric ships is a significant issue for electric vehicle battery systems, leading to numerous safety concerns during vessel operation. Therefore, timely fault diagnosis and accurate fault prediction are crucial for the safe operation of ships. This study examines the fault alarm system of marine battery management systems in conjunction with the unique operating conditions of ships, focusing on the system's latency. To facilitate prompt fault detection, a fault diagnosis method based on the Density-Based Spatial Clustering of Applications with Noise (DBSCAN) algorithm is proposed, utilizing the voltage data of battery clusters. Results indicate that the DBSCAN clustering algorithm demonstrates superior effectiveness and accuracy in identifying irregular battery clusters. Furthermore, the fault prediction method based on the iTransformer model is introduced to forecast variations in battery cluster voltages. Experimental findings suggest that this model can effectively predict consistency faults and over-/under-voltage conditions based on battery cluster voltage values and corresponding fault thresholds.

Keywords: marine lithium-ion battery; fault diagnosis; DBSCAN algorithm; iTransformer model; real-world driving data

1. Introduction

Entering this century, as the global energy crisis worsens and environmental problems intensify, addressing energy depletion and environmental degradation has become a top priority worldwide. The gradual enforcement of the International Maritime Organization's (IMO) sulfur cap regulation has placed new technological and environmental demands on the global shipping industry, making energy conservation and emission reduction an inevitable trend for ships. Compared to traditional fuel-powered vessels, all-electric ships, with their zero emissions, high transmission efficiency, lower operating costs, and higher technological value, are increasingly gaining prominence in the industry and showing broad potential for application [1,2]. The battery system is a core component of all-electric ships, and its performance largely determines the ship's power, safety, and endurance during navigation [3,4]. Unlike electric vehicles, marine battery systems face far more challenging operating conditions due to the complex and variable marine environment [5,6]. Frequent temperature changes, humidity, vibrations, and prolonged high-load operations significantly increase the risk of system failure. In extreme conditions, such failures could even lead to thermal runaway, posing serious threats to the safe operation of the vessel. Recent studies have shown that voltage anomalies in batteries are a key factor triggering system malfunctions [7–10]. These anomalies can typically be classified into four categories: overvoltage, undervoltage, rapid voltage fluctuations, and poor voltage consistency. Among these, inconsistency is one of the most common issues in battery management systems [11,12]. It results in uneven capacity utilization, where weaker batteries reach their charge or discharge limits first, reducing the efficiency of other batteries and lowering overall system capacity utilization. Furthermore, inconsistency accelerates system aging, leading to the premature failure of underperforming batteries, shortening the overall lifespan of the battery pack and increasing maintenance costs. After identifying the inconsistencies, it is necessary to isolate certain battery cells or perform balancing charge and discharge operations to improve the overall performance and safety of the battery pack. Therefore, promptly detecting and addressing voltage anomalies, particularly inconsistency issues, is crucial to ensuring the safe and efficient operation of all-electric ships. This not only helps meet industry standards for energy savings and emissions reduction but also plays a vital role in guaranteeing the safety of ships navigating complex marine environments.

The main lithium battery fault diagnosis algorithms today are typically categorized into knowledge-based, model-based, and data-driven approaches [13]. Knowledge-based lithium battery fault diagnosis algorithms draw on historical data to extract fundamental insights about battery behavior, which are then used for diagnosis by comparison with a knowledge base [14]. For instance, expert systems use fuzzy logic to establish a knowledge base that improves through self-optimization during diagnostics [15].

Model-based lithium battery fault diagnosis algorithms develop models to simulate lithium-ion battery behavior, then compare the predicted and actual values to generate residuals for fault detection. Kumara et al. [16] developed a connection fault diagnosis algorithm using a Luenberger observer, creating a first-order equivalent circuit model to simulate fault scenarios and using the observer to generate residual signals. LIN et al. [17] applied hybrid system theory to design automata for capturing both the continuous and discrete states of lithium battery packs, using dual extended Kalman filters to estimate parameters and diagnose sensor and relay faults.

Data-driven methods analyze extensive offline and online operational data to establish input-output mappings and extract diagnostic features without needing detailed battery models, though data quality is critical. Numerous global new energy data centers now host extensive datasets on electric vehicle performance, supporting data-driven fault diagnosis validation [18]. Xia et al. [19] used correlations between adjacent cell voltages to diagnose faults, while Kang et al. [20] employed a joint fault diagnostic method, combining staggered voltage measurement topology and modified correlation analysis to detect internal and external short-circuit faults. Sun et al. [21] used wavelet transforms to denoise voltage data and calculated Shannon entropy to diagnose faults, validating the method's effectiveness through battery vibration tests. To address challenges with traditional Shannon entropy, such as computational complexity and high hardware demands, Wang et al. [22] applied an improved Shannon entropy method, using a sliding window for iterative entropy calculations, and introduced a Z-score-based strategy for predictive maintenance. Liu et al. [23] used entropy weighting to assign objective weights to battery voltage, identifying anomalies through battery scoring. To address early fault-detection issues, where fault characteristics may be subtle, Hong et al. [24] proposed an enhanced multi-scale entropy method to accurately predict fault timing and locations, helping prevent thermal runaway.

With advancements in artificial intelligence and machine learning, effective battery fault diagnosis and prediction now rely on inputting feature factors and labels into models. For example, Qiu et al. [25] applied a nonlinear autoregressive exogenous (NARX) neural network for voltage prediction and fault diagnosis, while Fang et al. proposed a noise-applied DBSCAN clustering algorithm for fault diagnosis, paired with least-squares support vector regression (LS-SVR) for cell voltage change prediction [26]. However, much of this research focuses on electric vehicles, with few studies specifically addressing the unique requirements of all-electric vessels.

Existing battery voltage prediction methods have been validated primarily on electric vehicles, with little testing on all-electric ships. This paper addresses this gap by introducing a new approach for predicting voltage and detecting potential abnormal voltage fluctuations based on actual data from electric ships. The paper aims to make three notable contributions and improvements to current technology:

1. Adapting Fault Detection for All-Electric Ships: We adapt fault detection methods traditionally used for electric vehicles to all-electric ships, using real operational data to examine battery inconsistency.
2. Analyzing Ship Fault Alarm Mechanisms: This study investigates the delay issues in alarm communications, proposing a voltage anomaly diagnosis method based on battery clusters, specifically tailored to the operational context of ships.
3. Introducing an iTransformer-Based Fault Prediction Method: We propose a fault prediction method using the iTransformer algorithm to forecast trends in battery cluster behavior, revealing potential hazards associated with inconsistency faults.

The remainder of this paper is organized as follows: Section 2 gives a brief introduction of the real-world driving data details. Section 3 analyzes the alarm mechanism of the ship's battery management system and uses real data to illustrate the delay in alarms, from which a DBSCAN-based identification method is proposed. Section 4 utilizes iTransformer for voltage prediction. Section 5 summarizes the main conclusions of the paper.

2. Data Description and Preprocessing

In actual ship operations, battery voltage can experience significant and random fluctuations due to unpredictable environmental and operational factors, creating a more complex operating environment compared to electric vehicles. The dataset used in this study originates from a monitoring platform provided by a domestic research institute. This platform tracks the real-time status of over ten electric ships, delivering timely feedback to the vessels and their operators. Additionally, the platform provides IT support and data services for electric ship companies and government agencies, enabling the collection of extensive operational data and expanding the application potential of machine learning techniques, such as the Transformer network architecture proposed in this paper.

The data examined in this study specifically pertains to the Junlu, an all-electric ship developed collaboratively by the China State Shipbuilding Corporation's 712th and 702nd Research Institutes. Capable of carrying up to 300 passengers, it is the first large passenger ship in China to meet the *Inspection Guidelines for All-Battery Electric Ships* set by the China Classification Society. The ship's battery system consists of multiple clusters connected in parallel, providing a total capacity of 2240 kWh; the specific battery system topology is shown in Figure 1. The battery system in the "Junlvhao" ship is divided into two parts (left and right), each containing 6 battery clusters. After the 6 battery clusters are connected in parallel, they supply power to the pod and other loads through the ship's DMSB. Additional specifications are presented in Table 1. Data collection spanned from October 2022 to October 2023, including essential information such as timestamps, state of charge (SoC), battery cluster voltage lists, battery consistency status, fault alarms, alarm severity, total current, voltage, and power, as shown in Figure 2.

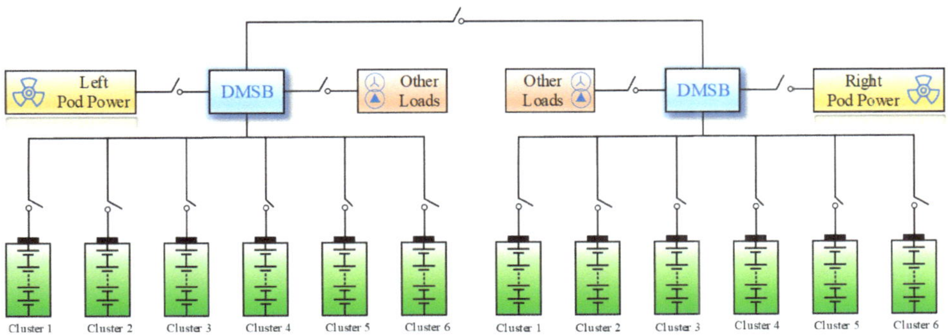

Figure 1. Network topology of the battery system.

Table 1. The specifications of the electric ship studied.

Parameter	Specification
Type	Passenger Ship
Length	55 m
Width	10 m
Draft Depth	1.6 m
Coordinates	114–17.414 E, 30–34.296 N
Destination	Wuhan Port
Battery Capacity	2240 kWh
Maximum Speed	10 knots/h
Number of Battery Clusters	12
Nominal Voltage	3.2 V
Rated Capacity	280 Ah

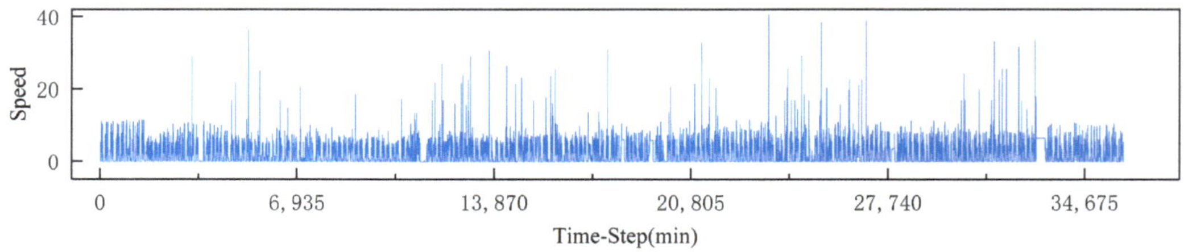

(**a**)

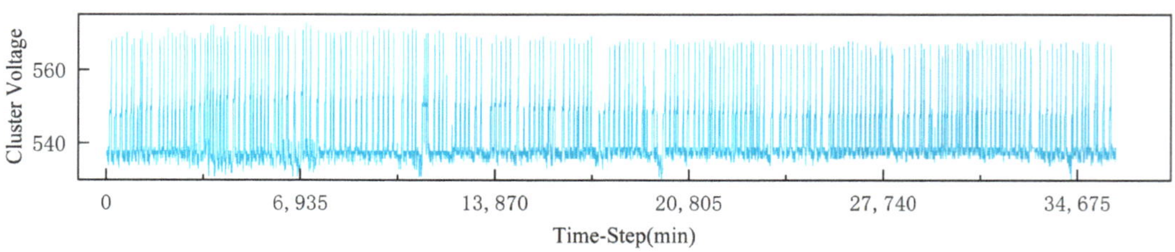

(**b**)

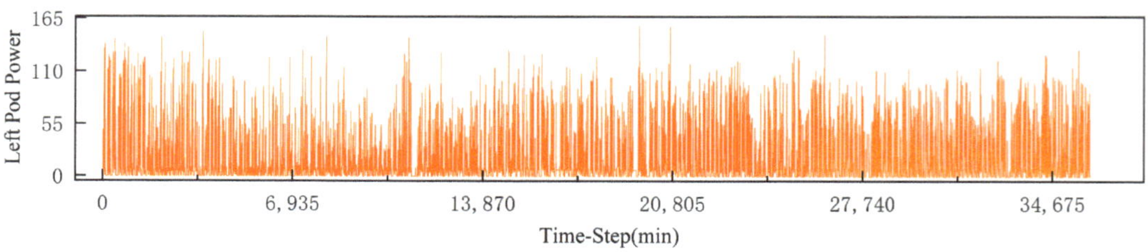

(**c**)

Figure 2. Some of the data used in this article. (**a**) is the electric boat speed, (**b**) is the battery cluster voltage sampled by the electric boat BMS, and (**c**) is the right pod power.

In the diagnostic phase, the DBSCAN clustering method is employed for fault detection and localization, while the iTransformer model is used for the prediction algorithm. The operational principles, specific parameters, and further technical details of these methods are discussed in subsequent sections.

For fault diagnosis and localization, the DBSCAN clustering method is utilized, while the prediction algorithm is based on the iTransformer model. The system principles, detailed parameters, and further technical specifics of these methods are discussed in the following sections.

3. Inconsistency Fault Analysis and Diagnosis

3.1. Inconsistency Fault Analysis

3.1.1. Fault Alarm Mechanism in Marine BMS

In practical applications, the BMS alarm system primarily responds to faults caused by individual battery cells, as the terminal voltage of individual cells is easily measurable. To further diagnose the fault, it is essential to establish the voltage differential threshold used for fault detection. In the battery management system, each data sample records the relevant information of individual cells, including their voltage data. By comparing the calculated feature factors with pre-calibrated thresholds, any value exceeding the threshold is flagged as a fault, and the corresponding alarm thresholds and levels are outlined in Table 2. These fault thresholds are set by the manufacturer based on specific conditions and are defined according to national standards. In our study, the design manual of the battery system from the Junlvhao vessel was used as the source of these thresholds.

Table 2. Common faults and thresholds.

Fault Type	Level	Threshold
Cell overvoltage (V)	1	3.5
	2	3.6
	3	3.65
Cell Undervoltage (V)	1	3.1
	2	3.0
	3	2.8
Cell Voltage Deviation (mV)	1	350
	2	400
	3	500
Cluster Overvoltage (V)	1	3.55 * N
	2	3.6 * N
	3	3.65 * N
Cluster Undervoltage (V)	1	3.1 * N
	2	3.0 * N
	3	2.8 * N

The battery management system (BMS) adopts a distributed design, as shown in Figure 3. The distributed BMS consists of the Battery Assembly Unit (BAU), Battery Control Unit (BCU), and Battery Management Unit (BMU), which work together to monitor the battery pack's status and ensure the safe operation of the battery. The BAU is the system-level control unit responsible for the monitoring, management, and control of the entire system. It directly monitors the status of each battery module within the battery system and can also monitor the battery pack's information. The BCU is a module-level control unit, with each battery module having a corresponding BCU. The BCU is responsible for monitoring the status of individual cells within the battery module and reporting status information to the higher-level control unit. The BMU is the cell-level control unit, responsible for monitoring the status of individual cells and communicating this information to the BCU, which in turn passes it to higher-level units. The entire system communicates through a CAN network. Therefore, when a fault occurs at the cell level, the fault reporting process follows a hierarchical flow: from the BMU to the BCU to the BAU.

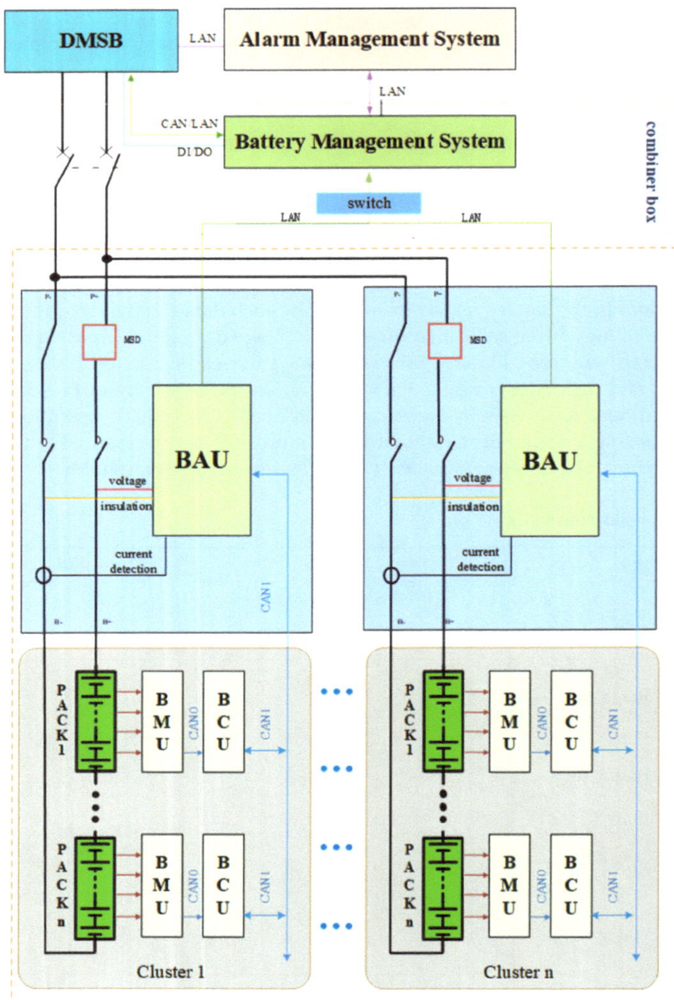

Figure 3. Alarm flow chart of BMU, BCU, and BAU in Marine BMS.

When using the CAN network for BMS communication on ships, significant delays often occur due to the ship's complex system architecture and challenging operating environment. First, ships have many interconnected subsystems, with considerable physical distances between them. Coupled with a large number of nodes, this setup often results in data congestion and delays. Second, environmental factors such as temperature, humidity, and salt fog significantly impact the battery system, increasing the monitoring demand and necessitating redundant design features, which in turn prolong the data processing time. Additionally, interference from high-power electromagnetic equipment on board may affect the stability of communication, and the shielding measures implemented to ensure signal integrity can further contribute to delays.

In summary, due to the design of the ship's BMS, the individual cell fault alarm mechanism may have inherent delays. To address this issue, this study proposes a fault detection method based on the voltage of the battery pack, which directly uses data from the BAU monitoring the battery pack voltage. This method ensures higher real-time performance and can improve the overall fault detection process.

3.1.2. Fault Fragment Analysis

Battery inconsistency faults typically persist over an extended period, manifesting as an increasing voltage difference between cells and battery clusters. This voltage disparity may lead to the progression of faults from an initial level 1 to levels 2 or even 3. For instance, in Figure 4, an alarm event that occurred at 10:00 PM on 1 January 2023, clearly illustrates the temporal relationship between voltage difference changes and alarm states. Initially, the figure shows that the trend of voltage difference changes began to emerge before the alarm was triggered, indicating that voltage anomalies often serve as early warning signals for potential faults. In the alarm segment at 10:38 PM, it was observed that the voltage difference increased rapidly to its peak; however, the alarm did not trigger immediately but only after the voltage difference exceeded a specific threshold. This phenomenon indicates that the alarm mechanism relies not solely on instantaneous changes in voltage difference but rather on the cumulative effect of voltage differences over time. Additionally, after the voltage returns to normal levels, the alarm state persists for a period to ensure that all anomalies have been resolved. This process highlights the persistence of voltage anomalies in battery clusters and their critical role in fault detection. Therefore, it is evident that the occurrence of faults is often accompanied by abnormal voltage changes in battery clusters, which frequently precede the triggering of alarm states. These early voltage changes provide crucial insights for the timely detection and diagnosis of battery faults. Consequently, continuous monitoring of voltage differences in battery clusters is essential within battery management systems, as it enhances the accuracy and responsiveness of fault diagnostics.

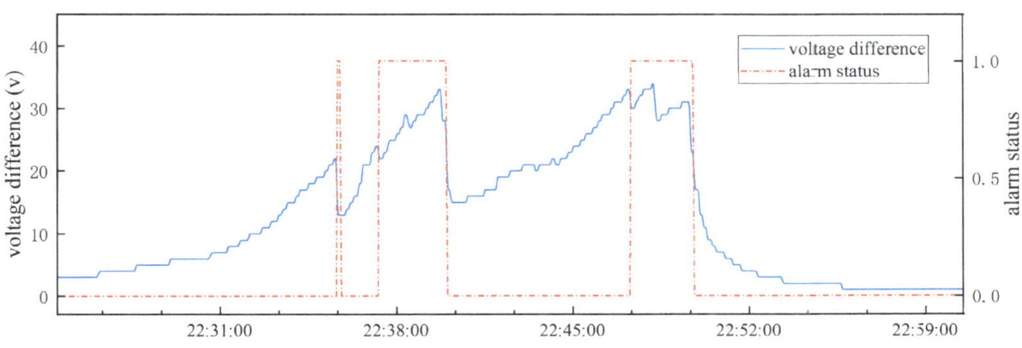

Figure 4. Time-stamped ship alarm status and differential signals on 1 January 2023.

3.2. Inconsistency Fault Diagnosis

Based on Table 2, overvoltage and undervoltage faults in individual battery cells can be detected by setting specific voltage thresholds using raw data. However, as analyzed above, this method does not provide immediate fault detection, leading to delays in alarms and assessments. To address this issue, this study proposes directly detecting anomalies at the battery cluster level, significantly reducing detection latency. A DBSCAN-based clustering method is introduced to automatically identify and classify faults across multiple battery cluster voltage curves. Unlike K-means, this method uses outlier detection to identify abnormal clusters.

K-means is a widely used unsupervised learning algorithm primarily employed for clustering tasks, with the goal of partitioning a dataset into K clusters such that data points within each cluster are as similar as possible, while data points in different clusters are as distinct as possible. The core idea of K-means is to iteratively update the centroid of each cluster, optimizing the cluster assignments over time. However, K-means requires a predefined number of clusters, K, which is often difficult to determine in practice. For example, when diagnosing voltage anomalies in battery cells using K-means, if K is set to 2, certain special cases may lead to underdiagnosis, requiring an increase in K to achieve more accurate clustering results. In cases where the voltage anomaly distribution is variable, it becomes challenging to accurately identify all anomalous cells with a fixed number of cluster centers. Therefore, the K-means-based anomaly detection method has certain limitations in such scenarios. The DBSCAN algorithm does not require setting the number of clusters, making it more effective in identifying the occurrence of multiple fault situations.

The DBSCAN (Density-Based Spatial Clustering of Applications with Noise) clustering algorithm, developed by Ester M., Kriegel H.P. et al. in 1996, is a density-based method that classifies data points with similar characteristics based on distances between them, automatically dividing data into distinct clusters. Widely applied in fields such as battery voltage balancing and thermal runaway diagnostics, DBSCAN is highly effective for identifying patterns and anomalies. A cluster is formed when data points satisfy a minimum point threshold (minpts) within a specified neighborhood (eps).

The algorithm's operation, shown in Figure 5, classifies data points as core points, boundary points, or noise points. Initially, minpts is set to 5, and orange-marked core points are identified. The algorithm then expands clusters by transferring from one core sample to neighboring points to locate additional core points, completing the clustering task. In the figure, light red points have only three neighbors and thus do not qualify as core points; however, they remain within reach of the core point density, classifying them as boundary points. Blue points, which cannot be reached from any core point, are considered noise.

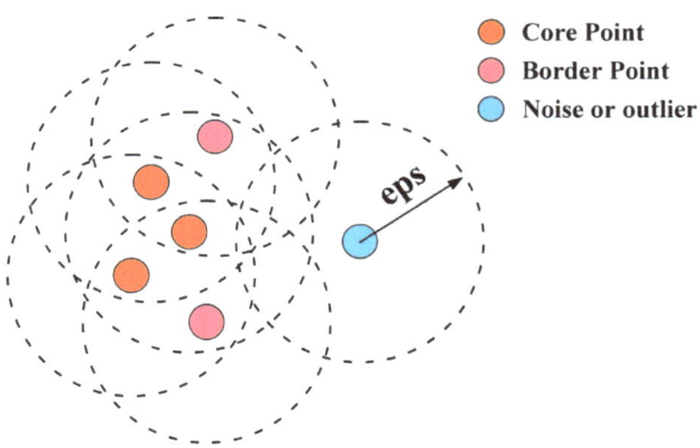

Figure 5. Schematic diagram of DBSCAN clustering method.

Before implementing DBSCAN, it is essential to intelligently set the neighborhood radius (eps) and minimum neighborhood sample count (minpts), as these parameters directly impact clustering outcomes. The detailed steps for implementing the DBSCAN algorithm are outlined as follows:

Step 1: Select a Core Point, p: A data point is randomly selected as a core point, p, to begin forming a cluster.

Step 2: Form Cluster C: Using core point p as the starting point, a cluster, C, is created by adding p and all data points within a specified radius, r (i.e., the r-neighborhood), to the cluster.

Step 3: Expand Cluster C: For each unprocessed point in the neighborhood, the cluster is expanded. If a neighboring point, q, also qualifies as a core point, all points within q's neighborhood that are not yet assigned to another cluster are added to C, continuing the cluster expansion.

Step 4: Repeat Expansion: This process is repeated until all points that can be included have been assigned to a cluster, ensuring that every sample point is processed.

Step 5: Output Clusters and Noise Point: Finally, the algorithm outputs the set of clusters, C, and any points that could not be assigned to a cluster are labeled as noise and placed in the noise set O.

Based on data from January 2023, this study applied the DBSCAN clustering method to analyze voltage anomalies in battery clusters within the marine BMS. Figure 6 illustrates the correspondence between noise points identified by the method and actual fault points. In the figure, blue solid points represent detected noise or anomaly points, while orange hollow points denote actual labeled fault points. The figure shows that, in most cases, potential faults within the battery clusters were detected early, as these anomalies were marked as noise points, often preceding alerts from the ship's BMS. Additionally, during the period from 3 January to 18 January, several potential fault points were detected, with timing differences from the BMS fault alarms. Around 23 January, noise points identified by the algorithm appeared noticeably earlier than actual fault alerts, demonstrating the clustering method's significant predictive capability.

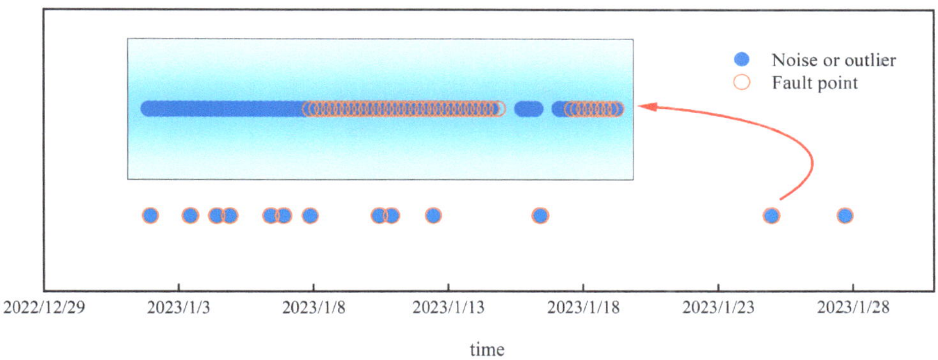

Figure 6. Comparison of fault points and true labels based on DBSCAN.

To conclude, the DBSCAN algorithm has proven effective in fault detection, as it can identify potential issues in the battery system before the BMS issues an alert. This early-warning capability provides valuable response time for maintenance personnel, enabling timely interventions and reducing potential losses associated with delayed fault reporting.

To evaluate DBSCAN's effectiveness in identifying abnormal voltage in battery clusters, this study introduces the F1 Score, a common classification metric. The F1 Score provides a straightforward assessment of the DBSCAN algorithm's performance in detecting voltage anomalies. Since fault anomalies are typically rare in battery systems, and sample classes are imbalanced, relying solely on precision or recall can be misleading. The F1 Score calculation follows these steps:

- True Positive (TP): The number of correctly clustered samples for a specific class.
- False Positive (FP): The number of samples incorrectly clustered into a specific class.
- False Negative (FN): The number of samples that belong to a specific class but were not correctly identified.

Using these definitions, the formulas for precision, recall, and F1 Score are as follows:

$$\text{Precision} = \frac{\text{TP}}{\text{TP} + \text{FP}} \tag{1}$$

$$\text{Recall} = \frac{\text{TP}}{\text{TP} + \text{FN}} \tag{2}$$

$$\mathrm{F1} = 2 \times \frac{\mathrm{Precision} \times \mathrm{Recall}}{\mathrm{Precision} + \mathrm{Recall}} \tag{3}$$

As shown in Figure 7, the DBSCAN clustering method's results for detecting voltage anomalies in battery clusters demonstrate a consistently high detection precision, with precision values remaining at 1.0 across all months, indicating excellent accuracy in identifying voltage anomalies. The recall rate, however, shows some fluctuations, particularly in September (0.92) and November (0.92946), where recall rates are relatively lower. This variability may relate to operational factors, such as climate, sea conditions, and maintenance schedules, which could increase the complexity of anomaly patterns and contribute to missed detections.

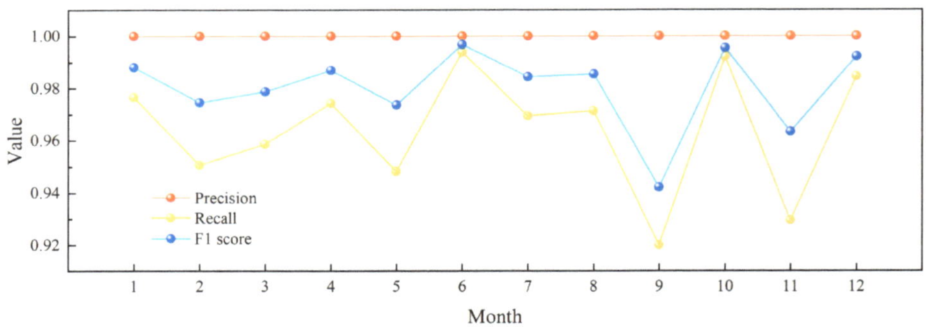

Figure 7. DBSCAN clustering result evaluation indicator value in each month.

The F1 score, which integrates both precision and recall, generally ranges between 0.9423 and 0.99692, indicating stable and strong overall model performance. Notably, the F1 score approaches 1.0 in June, October, and December, reflecting optimal detection performance during these months. In summary, the model performs well throughout most of the year, effectively detecting voltage anomalies, though a few missed detections may occur in certain months.

4. Inconsistency in Fault Prediction Based on iTransformer

The primary indicator for detecting inconsistency faults in large-capacity battery clusters on ships is the voltage residual of the battery pack. In the initial stages of a fault, before individual cell inconsistency alarm signals are triggered, the voltage range gradually expands over time. Typically, in maritime applications, reporting of individual cell faults is subject to delays. However, since the Battery Control Unit (BCU) directly monitors the battery cluster voltage without passing through the CAN0 network, using cluster voltage as a consistency indicator avoids such delays, providing a more immediate reflection of the overall health of the battery pack. Thus, in addition to accurately diagnosing faults in ship batteries, effectively predicting faults at the battery pack level would significantly enhance operational safety. To address this, this study proposes an iTransformer-based predictive method for early warning of voltage anomalies and forecasting fault progression trends.

The proposed iTransformer fault prediction method can be supported by the cloud-based big data platform for fully electric ships. This platform enables real-time information exchange between the ship and the platform, meaning the ship can upload relevant data to the platform, and the platform can return the predicted signals to the ship. Cloud-based cyber-physical systems and platform technologies provide an excellent environment for creating self-improving models that can effectively and efficiently enhance the safety of the ship during operation. Figure 8 demonstrates how we utilize the cloud data platform for fault prediction.

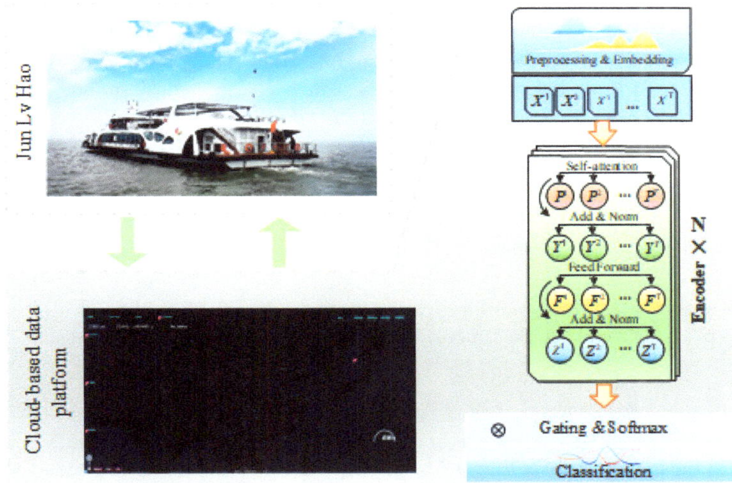

Figure 8. Cloud-based data platform for iTransformer fault prediction.

Our iTransformer plays a critical role in predicting voltage anomalies in large-capacity ship battery clusters. Given the influence of environmental factors, predicting the voltage of ship battery clusters requires capturing the complex dependencies in a multivariate time series. Traditional Transformer architectures often struggle with long time-series data due to performance declines and increased computational complexity. By utilizing an inverted dimensional structure, iTransformer more accurately captures the interdependencies among battery cells, providing greater stability in handling multivariate data. This capability aids in the earlier identification of inconsistency faults within battery clusters and enables effective forecasting of future voltage trends, ultimately enhancing the operational safety of ship battery systems.

4.1. Transformer Architecture

The Transformer model is a widely popular deep learning architecture, first introduced in the groundbreaking paper * "Attention is All You Need". * Unlike traditional sequence models such as LSTMs, the Transformer captures dependencies between input and output elements using a self-attention mechanism, allowing it to process the entire sequence in parallel rather than sequentially across time steps. This capability enables the Transformer to capture long-range dependencies more efficiently, significantly enhancing model performance. The schematic diagram of the relevant transformer principles is shown in Figure 9. The following sections outline the key components of the Transformer model.

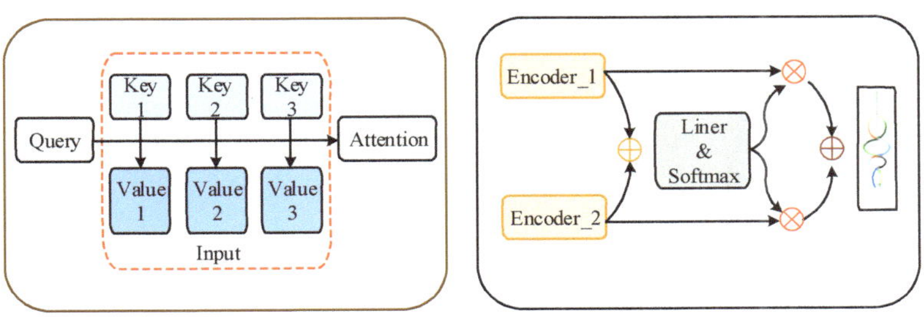

Self-attention

Gate transformer networks

Figure 9. Transformer-related structural design diagram.

4.1.1. Positional Encoding

In the Transformer model, the embedding process typically combines token and positional embeddings to capture sequence order. In this study, positional and operational embeddings are used to represent the position of battery charge and discharge curves within the sequence, as shown by the following formula:

$$\begin{aligned} \mathrm{PE}_{pos,2i} &= sin\left(\frac{pos}{10{,}000^{\frac{2i}{d_{\mathrm{model}}}}}\right) \\ \mathrm{PE}_{pos,2i+1} &= cos\left(\frac{pos}{10{,}000^{\frac{2i}{d_{\mathrm{model}}}}}\right) \end{aligned} \tag{4}$$

The PE refers to the Positional Embedding matrix, where pos indicates a specific position, i represents a particular dimension, and d_model denotes the dimension of the model.

4.1.2. Self-Attention Mechanism

The self-attention mechanism is central to the Transformer model, enabling each position (token) in the input sequence to relate to all other positions, capturing dependencies throughout the sequence. The self-attention mechanism follows these main steps:

1. Calculate Query, Key, and Value Vectors: For each position in the input sequence, a linear transformation generates a query vector (Q), a key vector (K), and a value vector (V), using the following formulas:

$$\begin{aligned} Q_i &= W_Q x_i \\ K_i &= W_K x_i \\ V_i &= W_V x_i \end{aligned} \tag{5}$$

where W_Q, W_K, and W_V are learned weight matrices that map the input into subspaces for query, key, and value. The **Q** represents the token seeking information, the **K** contains the information, and the **V** holds the data used for the output.

2. Compute Attention Scores: The attention scores are obtained by calculating the dot product between the query and key vectors, representing the relevance of each position to others in the sequence.

$$Attention(Q_i, K_j) = \frac{Q_i K_j^T}{\sqrt{d_k}} \tag{6}$$

The **Q** is compared to **K** using a similarity measure (usually dot product) to calculate how much attention one token should pay to another.

3. Apply Softmax Function: To normalize the attention scores into probabilities, they are passed through a softmax function, ensuring that the weights sum to 1:

$$\alpha_{ij} = \mathrm{softmax}\left(\frac{Q_i K_j^T}{\sqrt{d_k}}\right) \tag{7}$$

The attention scores are normalized with softmax, turning them into probabilities so that they sum to 1, determining the attention weight each token should have.

4. Weighted Sum to Obtain Output: Finally, the output for each query position is a weighted sum of all value vectors V, with weights determined by the attention scores:

$$Output_i = \sum_j \alpha_{ij} V_j \tag{8}$$

4.1.3. Multi-Head Attention

The multi-head attention mechanism extends self-attention by allowing the model to learn different types of dependencies in parallel across multiple subspaces. The formula for multi-head attention is as follows:

$$\text{MultiHead}(Q, K, V = \text{Concat}(\text{head}_1, \text{head}_2, \ldots, \text{head}_h)W_O \tag{9}$$

Each head is calculated similarly to single-head self-attention:

$$\text{head}_i = \text{Attention}(Q_i, K_i, V_i) \tag{10}$$

where W_O is the weight matrix for the final linear transformation, and concatenation combines the outputs from each attention head. Multiple attention mechanisms (heads) run in parallel, each focusing on different parts of the input. The results are combined to capture different relationships between tokens.

4.1.4. Feed-Forward Network

The feed-forward network further processes and transforms feature representations. The formula is as follows:

$$\text{FFN}(x) = \max(0, xW_1 + b_1)W_2 + b_2 \tag{11}$$

where W_1 and W_2 are weight matrices, b_1 and b_2 are biases, and ReLU is the activation function. It transforms and refines data through two layers with a ReLU activation to learn complex patterns.

4.1.5. Layer Normalization

To accelerate model training and improve stability, the Transformer applies layer normalization between the input and output of each sub-layer. Layer normalization helps mitigate issues like gradient vanishing or explosion during training and promotes faster convergence. The layer normalization formula is:

$$\text{LayerNorm}(x) = \frac{x - \mu}{\sigma + \epsilon} \cdot \gamma + \beta \tag{12}$$

where μ and σ are the mean and standard deviation of the input, ϵ is a small constant for numerical stability, and γ and β are learnable parameters. It normalizes the output to stabilize training, improving convergence and reducing learning rate dependency.

4.1.6. Residual Connections

Residual connections directly pass the input of each layer to the next layer to avoid information loss and ensure better gradient backpropagation. Specifically, input x is added to the output of the layer:

$$\text{Output} = \text{LayerNorm}(x + \text{SubLayer}(x)) \tag{13}$$

4.1.7. Final Output

The decoder's final output is passed through a linear layer, followed by a softmax function to produce predicted values.

$$\text{Target}(y|x) = \text{softmax}(HW + b) \tag{14}$$

The Transformer combines self-attention, multi-head attention, feed-forward networks, positional encoding, layer normalization, and residual connections to build a highly parallelized model architecture. The self-attention mechanism allows the model to focus on multiple positions within the input sequence, capturing long-range dependencies, while the multi-head mechanism enhances flexibility and expressiveness. The feed-forward network provides nonlinear transformation, positional encoding manages sequence order, and layer normalization and residual connections ensure stability and efficient training. Together, these components make the Transformer a powerful and widely adopted model.

4.2. iTransformer Architecture

The iTransformer model is an adaptation of the classic Transformer architecture, designed specifically for time series prediction using an inverted dimension approach. Unlike the traditional structure, the improved model inverts the time series dimensions, enabling the attention mechanism and feed-forward network to operate across different dimensions, capturing correlations and trends in multivariate time series more effectively. The schematic diagram of the relevant iTransformer principles is shown in Figure 10. Key iTransformer functions compared to the Transformer are as follows:

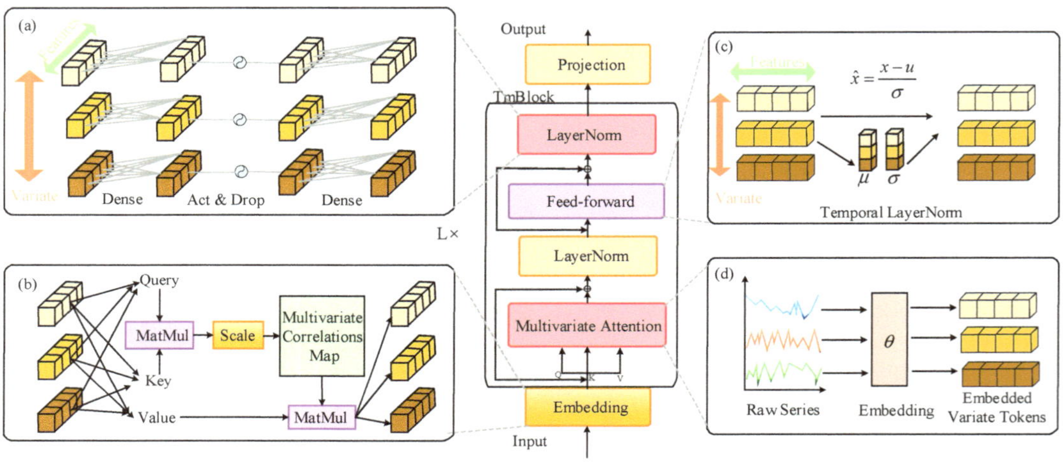

Figure 10. iTransformer-related structural design diagram.

4.2.1. Inverted Dimension Design and Embedding Process

In traditional architectures, the input time series $X \in R^{T \times N}$, which consists of **T** time steps and **N** variables, is embedded so that multiple variables at each time step form a single time-step token. This approach focuses primarily on dependencies across time steps, potentially overlooking correlations among variables. In contrast, the iTransformer uses an inverted dimension design, embedding each variable's entire time series as a separate variable token, thus allowing the attention mechanism to capture inter-variable relationships. The embedding process is as follows:

$$h_0^n = \text{Embedding}(X_{:,n}),\ n = 1, 2, \ldots, N \tag{15}$$

where $X_{:,n}$ represents the entire time series for the n-th variable, and h_0^n is the representation after embedding each variable.

4.2.2. Inverted Application of Self-Attention Mechanism

In traditional models, the self-attention mechanism operates along the time dimension, capturing temporal dependencies by calculating the relationships between each time step. In the iTransformer, however, self-attention is applied across the variable dimension. By calculating correlations among different variables, the model can capture inter-variable dependencies. The formula is as follows:

$$Q_n, K_n, V_n = h_n W_Q, h_n W_K, h_n W_V \tag{16}$$

$$\text{Attention}(Q_n, K_n, V_n) = \text{softmax}\left(\frac{Q_n K_n^T}{\sqrt{d_k}}\right) V_n \tag{17}$$

4.2.3. Application of Feed-Forward Network (FFN) in the Time Dimension

The FFN is applied to each variable's time-series representation, allowing the model to better capture temporal changes for each variable. The formula is:

$$H_{l+1}^n = FFN(H_l^n) \tag{18}$$

4.2.4. Omitting Positional Encoding

Since the iTransformer applies attention across the variable dimension rather than the time dimension, positional encoding is omitted. Temporal sequence information is implicitly captured through the FFN and attention mechanism.

4.2.5. Multivariable Correlation Handling

A key innovation of this model is its ability to specifically address variable correlations in multivariate time series. By applying attention across the variable dimension, the model generates a multivariable correlation map, which helps capture the influence among different variables. The attention score matrix **A** is calculated as follows:

$$A_{ij} = \text{softmax}\left(\frac{Q_i K_j^\top}{\sqrt{d_k}}\right) \tag{19}$$

This matrix describes the correlation between variables **i** and j, which is critical for accurate multivariate time-series prediction.

4.3. Feature Selection

This section addresses the selection of features most relevant to predicting battery cluster voltage. These features represent various operating conditions and state variables, such as temperature, load, current, and voltage history. Although the self-attention mechanism adjusts weights based on input features, selecting a higher-correlation feature set improves model convergence and performance. The Pearson Correlation Coefficient (PCC) and Spearman Coefficient are used for feature selection. PCC measures linear correlation with battery cluster voltage, while the Spearman Coefficient captures nonlinear relationships. Together, these metrics comprehensively identify features with strong relevance to the target variable.

4.3.1. Pearson Correlation Coefficient

PCC is a standard measure of linear correlation between two variables. It is used to select features with strong linear relationships to battery cluster voltage. The PCC formula is:

$$r_{x,y} = \frac{\sum(x_i - \overline{x})(y_i - \overline{y})}{\sqrt{\sum(x_i - \overline{x})^2}\sqrt{\sum(y_i - \overline{y})^2}} \tag{20}$$

where $r_{x,y}$ is the correlation coefficient between feature **x** and voltage y. PCC values range from −1 to 1, with values closer to 1 or −1 indicating a stronger linear correlation.

4.3.2. Spearman Coefficient

The Spearman Coefficient, a rank-based correlation measure, is suitable for selecting features with nonlinear relationships. It assesses correlation by comparing variable ranks, calculated as follows:

$$\rho_{x,y} = 1 - \frac{6\sum d_i^2}{n(n^2-1)} \tag{21}$$

where $\rho_{x,y}$ is the Spearman coefficient, d is the rank difference between feature x and voltage y, and n is the sample size. Like PCC, Spearman values range from −1 to 1, with values closer to 1 or −1 indicating a stronger correlation.

The combined use of PCC and Spearman can provide a comprehensive assessment of both linear and nonlinear relationships. PCC effectively evaluates linear relationships, while Spearman is suitable for identifying monotonic relationships in the data. Therefore, the combination of both can offer a thorough evaluation of the various associations between variables. Additionally, this approach enhances robustness, as PCC requires specific data distribution, particularly linear relationships and normality, while Spearman does not require normal distribution and is better suited for handling nonlinear or non-normally distributed data. Thus, the joint use of these two methods can offer more accurate correlation analysis across different types of data.

4.3.3. Feature Selection Based on Real Ship Data

In battery voltage prediction for electric vehicle BMSs, voltage characteristics or fluctuations during operation or charging may be influenced by various external factors, including meteorological conditions, vessel operational states, and inherent battery characteristics. To enhance prediction accuracy, this study comprehensively considers these three dimensions—meteorological factors, vessel operating conditions, and battery system characteristics—forming a more robust framework for predicting ship battery voltage.

Regarding meteorological factors, environmental conditions significantly impact battery voltage, especially in maritime applications. However, due to the variability and unpredictability of marine weather, variables such as humidity, precipitation, atmospheric pressure, temperature, visibility, and wind speed may not fully represent actual operational conditions, as they lack direct correlations with a vessel's dynamic state and do not sufficiently capture voltage fluctuation patterns. To address this limitation, this study incorporates a feature combination of "pod power + speed". This pairing effectively reflects the vessel's power demand and operating state in complex maritime environments, indirectly capturing real-time environmental effects on battery voltage. For example, pod power and speed indicate current power output demands and operational conditions, providing insights beyond environmental factors alone. Consequently, this feature combination enables more precise voltage predictions under varying conditions.

At the battery system level, several features closely related to battery voltage were selected, including voltage, probe temperature, state of charge (SOC), and current. These variables directly represent the real-time status and health of the battery system. While driver behavior may influence battery voltage fluctuations—especially as operational inputs can alter battery load (e.g., sudden acceleration or deceleration)—this impact is typically indirect and difficult to quantify. Furthermore, complex interactions exist between driver behavior and environmental factors; for instance, different driver actions in the same environment can uniquely impact the battery system, making precise modeling challenging. In light of this, this study excludes driver behavior as a factor and instead focuses on meteorological factors, vessel operating conditions, and battery characteristics, reducing model complexity and improving predictive accuracy. Ultimately, the selected features include SOC, temperature, current, total voltage, left pod power, right pod power, and speed.

As shown in Figure 11, An analysis of the Pearson correlation coefficient (PCC) and Spearman coefficient for features related to battery cluster voltage prediction reveals a strong positive correlation between current and battery cluster voltage, with a PCC of

0.905716 and a Spearman coefficient of 0.823703. Therefore, current should be prioritized as a key feature. The power of the left and right pods also demonstrates significant negative correlation, with Spearman coefficients of −0.78632 and −0.78287, respectively, indicating strong nonlinear effects and establishing them as important predictive indicators. The Spearman coefficient between the total voltage of the battery system and the battery cluster voltage is 0.697869, indicating a complex nonlinear relationship, which should be considered as an auxiliary feature. A negative linear relationship exists between speed and probe temperature with battery cluster voltage; however, the low Spearman coefficients suggest weak nonlinear effects, allowing these features to be considered as secondary. The state of charge (SOC) shows low correlation, n, with battery cluster voltage, indicating that it may not play a significant role in voltage prediction.

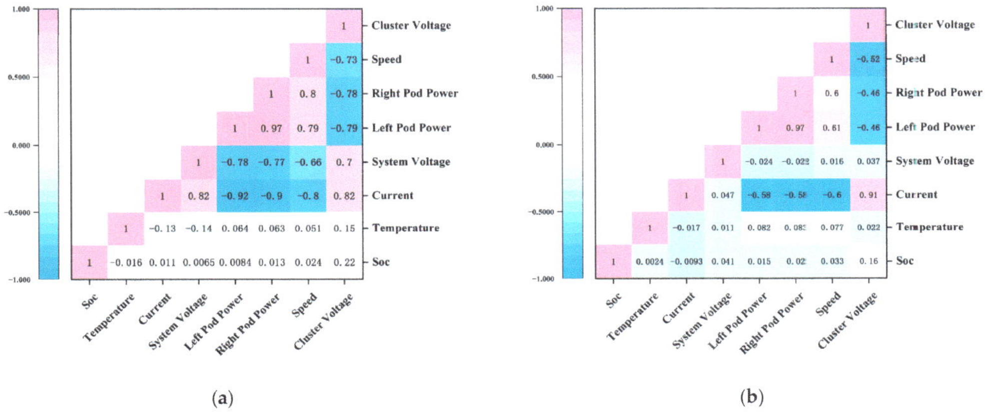

Figure 11. PCC (**a**) and Spearman (**b**) correlation coefficient calorific value map.

Although the analysis using PCC and Spearman coefficients revealed a strong correlation between battery cluster voltage and the left and right pod power, potential multicollinearity between these features may lead to information redundancy. Multicollinearity can affect model stability and interpretability; feature redundancy was further evaluated by calculating the Variance Inflation Factor (VIF). VIF is a tool that assesses linear relationships among features, calculated as follows:

$$VIF_i = \frac{1}{1 - R_i^2} \tag{22}$$

where R_i^2 represents the coefficient of determination, showing how well a feature can be predicted by the other features. A VIF > 10 typically indicates a strong linear relationship with other features, suggesting redundancy. Our calculations yielded VIF values of 15.438989 for both left and right pod power, significantly exceeding the threshold of 10, which suggests overlapping information. As a result, only one feature, left pod power, was retained to reduce redundancy. The final selected features were left pod power, probe temperature, current, system voltage, and speed.

To comprehensively evaluate model performance, three standard regression metrics were employed:

1. Root Mean Square Error (RMSE): RMSE measures the standard deviation of prediction errors, reflecting the magnitude of prediction error. The formula is:

$$\text{RMSE} = \sqrt{\frac{1}{n}\sum_{i=1}^{n}(\hat{y}_i - y_i)^2} \tag{23}$$

2. Mean Absolute Error (MAE): MAE measures the average absolute difference between predictions and actual values, representing the actual magnitude of prediction errors. The formula is:

$$\mathrm{MAE} = \frac{1}{n}\sum_{i=1}^{n}\left|\hat{y}_i - y_i\right| \tag{24}$$

3. Mean Absolute Percentage Error (MAPE): MAPE evaluates prediction error as a percentage of actual values, showing the relative magnitude of prediction error. The formula is:

$$\mathrm{MAPE} = \frac{100\%}{n}\sum_{i=1}^{n}\left|\frac{\hat{y}_i - y_i}{y_i}\right| \tag{25}$$

4.4. Prediction Results and Discussion

In this study, the electric ship has 12 battery clusters. Training and building separate models for each cluster would be time-consuming, and invoking multiple models simultaneously during real-time voltage prediction could substantially reduce prediction efficiency. To address this issue, the data from all 12 clusters were sequentially combined into a new, unified battery cluster voltage dataset. This combined data were then used to train a generalized voltage prediction model. By incorporating the voltage information from each individual cluster, this approach enabled the creation of a well-calibrated model capable of predicting voltage across all clusters simultaneously. The prediction and fault diagnosis responses were evaluated using an alarm scenario that occurred around 9:50 AM on 3 January 2023, as shown in Figure 12.

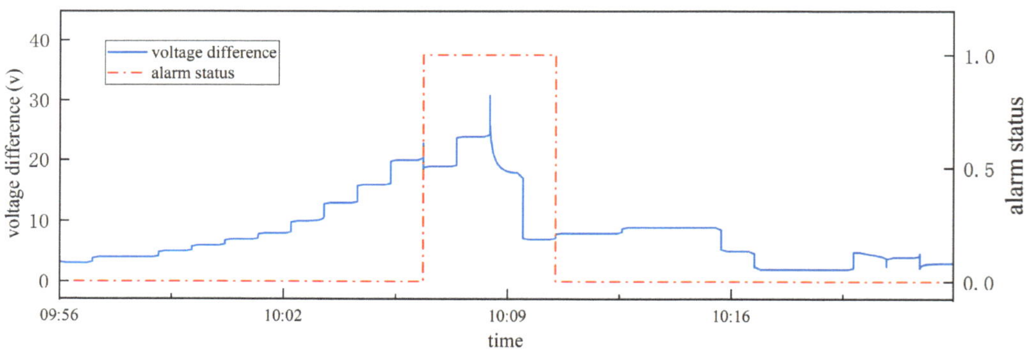

Figure 12. Voltage difference in the battery cluster and fault alarm status for the fault segment on 3 January 2023.

In Figure 12, the voltage difference fluctuated over time, remaining small between 09:56 and 10:05 before gradually increasing, peaking at a significant voltage difference around 10:10, and then slowly declining. The alarm status is indicated by a dashed line, with 0 representing no alarm and 1 indicating an active alarm. Whenever the voltage difference increased, the alarm status shifted from 0 to 1, signifying that the system detected an anomaly. This also illustrates the inherent delay in the alarm response.

This study combined Transformer and iTransformer models to predict short-term voltage range variations in segments with active alarms. Each input sequence consisted of voltage difference data across 30 consecutive time points, which were preprocessed as input to the model, with the goal of predicting the next 10 time points' voltage values. During training, the Mean Squared Error (MSE) was used as the loss function, and the model was optimized with the Adam optimizer. Learning rates were adjusted to ensure that the model converged quickly and stabilized at an optimal prediction performance. Transformer-based prediction results and relative error are shown in Figure 13 The analysis shows that the model effectively captures the overall trend of voltage changes, particularly in periods of gradual voltage increase. However, the model's response lagged during sharp declines in voltage difference, resulting in a spike in error, with prediction errors reaching up to 15 V during sudden events and alarms.

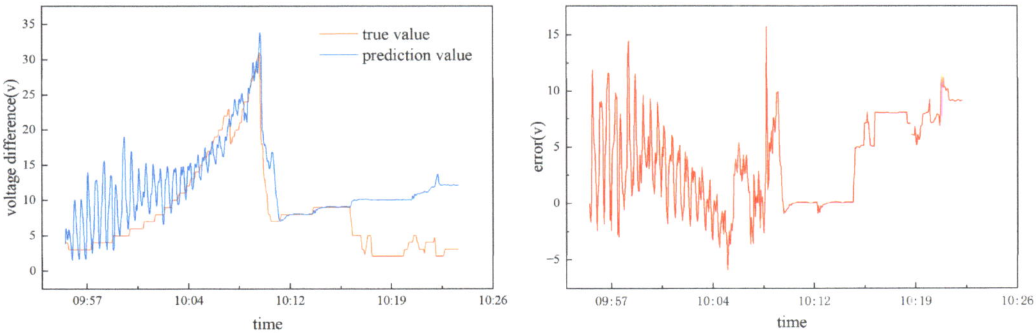

Figure 13. Voltage prediction results and errors based on the transformer model.

Applying the iTransformer model to the same alarm segment with identical sequences, the voltage data from the previous 20 time points were used as the training sample to predict the following 10 time points. The prediction results and errors for the sliding data window over the fault segment are shown in Figure 14. The iTransformer model captured fluctuations more accurately, with reduced amplitude and frequency of oscillations, indicating improved handling of noise and minor local variations. Overall, errors remained within a 5 V range, demonstrating strong predictive accuracy when voltage differences were relatively stable. Table 3 summarizes the model's prediction performance across all battery clusters, showing that iTransformer outperformed Transformer in RMSE, MAE, and MAPE, with predictions closer to actual values, confirming its superior predictive capability.

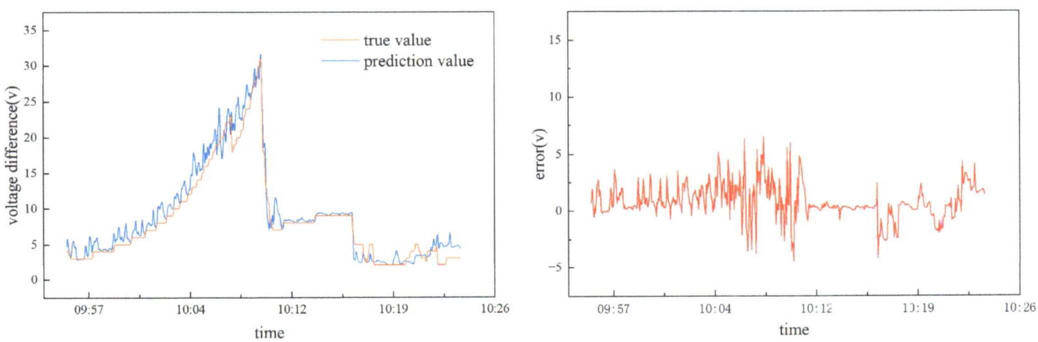

Figure 14. Voltage prediction results and errors based on the iTransformer model.

Table 3. Model's prediction performance.

Prediction Method	RMSE	MAE	MAPE (%)
Transformer	1.192	0.840	0.16
iTransiformer	0.390	0.343	0.03

To predict fault alarms, the predicted results were compared with fault detection thresholds, enabling fault identification and providing short-term predictions 90 s in advance. In fault prediction analysis, FP, FN, TP, and TN are calculated. From these, accuracy, recall, and F1 score are derived, which help us evaluate the model's performance in predicting faults. Based on fault alarms recorded during a year of operations in the ship data utilized in this study, and using iTransformer and threshold analysis, the following evaluation metrics were obtained in Table 4.

Table 4. Model fault prediction evaluation index.

Performance Metric	Value
Accuracy	93.25%
Precision	94.28%
Recall	95.47%
F1 Score	94.87%

5. Conclusions

This paper utilizes real driving data from the all-electric ship Junlv to diagnose and predict inconsistent faults under complex operating conditions. Initially, an analysis of the actual driving data revealed that the fault alarms of the marine BMS exhibit latency and that faults often coincide with anomalies in battery cluster voltage. A fault diagnosis method based on DBSCAN clustering of battery cluster voltage and a fault prediction method using the iTransformer model are proposed. Based on real driving data, the DBSCAN method effectively identifies the locations of fault units. Throughout the year of study, the accuracy of this clustering method was consistently 1, with a recall rate around 0.96 and an F1 score around 0.98, significantly outperforming the K-means clustering algorithm. This demonstrates the superiority of the DBSCAN model in the context of this research. Additionally, the iTransformer prediction method can diagnose faults up to 90 s in advance. Its RMSE for voltage prediction is 0.390, its MAE is 0.343, its MAPE is 0.03%, and the F1 score for fault diagnosis is 94.87%. Overall, this work demonstrates the potential of integrating real data with deep learning modeling to achieve accurate predictions of real-world physical problems characterized by hidden physics and lacking predefined initial or boundary conditions. Timely and accurate detection and prediction of battery fault risks under complex operating conditions are crucial for ensuring the safe operation of battery systems in real all-electric ship environments. Future research will focus on further optimizing detection and prediction models, as well as assessing multiple faults, including consistency faults caused by capacity, state of charge (SoC), and internal resistance.

Author Contributions: Conceptualization, X.Y. and Y.L.; methodology, H.J. and Y.L.; software, Y.L.; validation, Y.L., H.J. and T.T.; formal analysis, Y.L., H.J. and L.L.; investigation, Q.S.; resources, H.J. and Q.S.; data curation, Y.C.; writing—original draft preparation, X.Y. and S.J.; writing—review and editing, H.J. and Y.L.; visualization, Y.L. and H.J.; supervision, X.Y.; project administration, X.Y.; funding acquisition, X.Y. All authors have read and agreed to the published version of the manuscript.

Funding: This work is supported by the China National Key Research and Development Project (Grant No: 2023YFB4301704), the National Natural Science Foundation of China (Grant No: 52271329) and the China National Key Laboratory of Electromagnetic Energy Technology Open Fund (Grant No: 61422172220403).

Institutional Review Board Statement: Not applicable.

Informed Consent Statement: Not applicable.

Data Availability Statement: Data will be made available on request.

Conflicts of Interest: The authors declare no conflicts of interest.

References

1. Fang, S.; Wang, Y.; Gou, B.; Xu, Y. Toward Future Green Maritime Transportation: An Overview of Seaport Microgrids and All-Electric Ships. *IEEE Trans. Veh. Technol.* **2020**, *69*, 207–219. [CrossRef]
2. Sulligoi, G.; Vicenzutti, A.; Menis, R. All-Electric Ship Design: From Electrical Propulsion to Integrated Electrical and Electronic Power Systems. *IEEE Trans. Transp. Electrif.* **2016**, *2*, 507–521. [CrossRef]

3. Wang, J.; Lu, S.; Wang, Y.; Li, C.; Wang, K. Effect Analysis on Thermal Behavior Enhancement of Lithium–Ion Battery Pack with Different Cooling Structures. *J. Energy Storage* **2020**, *32*, 101800. [CrossRef]
4. Li, W.; Xiao, M.; Peng, X.; Garg, A.; Gao, L. A Surrogate Thermal Modeling and Parametric Optimization of Battery Pack with Air Cooling for EVs. *Appl. Therm. Eng.* **2019**, *147*, 90–100. [CrossRef]
5. Wang, Z.; Carriveau, R.; Ting, D.S.-K.; Xiong, W.; Wang, Z. A Review of Marine Renewable Energy Storage. *Int. J. Energy Res.* **2019**, *43*, 6108–6150. [CrossRef]
6. Lucà Trombetta, G.; Leonardi, S.G.; Aloisio, D.; Andaloro, L.; Sergi, F. Lithium-Ion Batteries on Board: A Review on Their Integration for Enabling the Energy Transition in Shipping Industry. *Energies* **2024**, *17*, 1019. [CrossRef]
7. Hu, X.; Zhang, K.; Liu, K.; Lin, X.; Dey, S.; Onori, S. Advanced Fault Diagnosis for Lithium-Ion Battery Systems: A Review of Fault Mechanisms, Fault Features, and Diagnosis Procedures. *IEEE Ind. Electron. Mag.* **2020**, *14*, 65–91. [CrossRef]
8. Sun, Z.; Han, Y.; Wang, Z.; Chen, Y.; Liu, P.; Qin, Z.; Zhang, Z.; Wu, Z.; Song, C. Detection of Voltage Fault in the Battery System of Electric Vehicles Using Statistical Analysis. *Appl. Energy* **2022**, *307*, 118172. [CrossRef]
9. He, H.; Zhao, X.; Li, J.; Wei, Z.; Huang, R.; Jia, C. Voltage Abnormality-Based Fault Diagnosis for Batteries in Electric Buses with a Self-Adapting Update Model. *J. Energy Storage* **2022**, *53*, 105074. [CrossRef]
10. Hong, J.; Wang, Z.; Yao, Y. Fault Prognosis of Battery System Based on Accurate Voltage Abnormity Prognosis Using Long Short-Term Memory Neural Networks. *Appl. Energy* **2019**, *251*, 113381. [CrossRef]
11. Hua, Y.; Zhou, S.; Cui, H.; Liu, X.; Zhang, C.; Xu, X.; Ling, H.; Yang, S. A Comprehensive Review on Inconsistency and Equalization Technology of Lithium-Ion Battery for Electric Vehicles. *Int. J. Energy Res.* **2020**, *44*, 11059–11087. [CrossRef]
12. Tian, J.; Fan, Y.; Pan, T.; Zhang, X.; Yin, J.; Zhang, Q. A Critical Review on Inconsistency Mechanism, Evaluation Methods and Improvement Measures for Lithium-Ion Battery Energy Storage Systems. *Renew. Sustain. Energy Rev.* **2024**, *189*, 113978. [CrossRef]
13. Jiang, L.; Deng, Z.; Tang, X.; Hu, L.; Lin, X.; Hu, X. Data-Driven Fault Diagnosis and Thermal Runaway Warning for Battery Packs Using Real-World Vehicle Data. *Energy* **2021**, *234*, 121266. [CrossRef]
14. Zhao, Z.; Liu, P.X.; Gao, J. Model-Based Fault Diagnosis Methods for Systems with Stochastic Process—A Survey. *Neurocomputing* **2022**, *513*, 137–152. [CrossRef]
15. Xiong, R.; Sun, W.; Yu, Q.; Sun, F. Research Progress, Challenges and Prospects of Fault Diagnosis on Battery System of Electric Vehicles. *Appl. Energy* **2020**, *279*, 115855. [CrossRef]
16. Kumara, P.A.; Cahyadi, A.I.; Wahyunggoro, O. Fault Detection Algorithm on Lithium-Polymer (Li-Po) Battery Based on Luenberger Observer. In Proceedings of the 2021 International Seminar on Machine Learning, Optimization, and Data Science (ISMODE), Jakarta, Indonesia, 29–30 January 2022; pp. 108–113.
17. Lin, T.; Chen, Z.; Zheng, C.; Huang, D.; Zhou, S. Fault Diagnosis of Lithium-Ion Battery Pack Based on Hybrid System and Dual Extended Kalman Filter Algorithm. *IEEE Trans. Transp. Electrific.* **2021**, *7*, 26–36. [CrossRef]
18. Zhang, X.; Chen, S.; Zhu, J.; Gao, Y. A Critical Review of Thermal Runaway Prediction and Early-Warning Methods for Lithium-Ion Batteries. *Energy Mater. Adv.* **2023**, *4*, 0008. [CrossRef]
19. Xia, B.; Shang, Y.; Nguyen, T.; Mi, C. A Correlation Based Fault Detection Method for Short Circuits in Battery Packs. *J. Power Sources* **2017**, *337*, 1–10. [CrossRef]
20. Kang, Y.; Duan, B.; Zhou, Z.; Shang, Y.; Zhang, C. A Multi-Fault Diagnostic Method Based on an Interleaved Voltage Measurement Topology for Series Connected Battery Packs. *J. Power Sources* **2019**, *417*, 132–144. [CrossRef]
21. Sun, Z.; Liu, P.; Wang, Z. Real-Time Fault Diagnosis Method of Battery System Based on Shannon Entropy. *Energy Procedia* **2017**, *105*, 2354–2359. [CrossRef]
22. Wang, Z.; Hong, J.; Liu, P.; Zhang, L. Voltage Fault Diagnosis and Prognosis of Battery Systems Based on Entropy and Z -Score for Electric Vehicles. *Appl. Energy* **2017**, *196*, 289–302. [CrossRef]
23. Liu, P.; Sun, Z.; Wang, Z.; Zhang, J. Entropy-Based Voltage Fault Diagnosis of Battery Systems for Electric Vehicles. *Energies* **2018**, *11*, 136. [CrossRef]
24. Hong, J.; Wang, Z.; Chen, W.; Wang, L. Multi-fault Synergistic Diagnosis of Battery Systems Based on the Modified Multi-scale Entropy. *Int. J. Energy Res.* **2019**, *43*, 8350–8369. [CrossRef]
25. Marcicki, J.; Onori, S.; Rizzoni, G. *Nonlinear Fault Detection and Isolation for a Lithium-Ion Battery Management System*; American Society of Mechanical Engineers Digital Collection: New York, NY, USA, 2011; pp. 607–614.
26. Fang, W.; Chen, H.; Zhou, F. Fault Diagnosis for Cell Voltage Inconsistency of a Battery Pack in Electric Vehicles Based on Real-World Driving Data. *Comput. Electr. Eng.* **2022**, *102*, 108095. [CrossRef]

Journal of
Marine Science and Engineering

MDPI

Article

Multi-Temporal Energy Management Strategy for Fuel Cell Ships Considering Power Source Lifespan Decay Synergy

Xingwei Zhou, Xiangguo Yang, Mengni Zhou *, Lin Liu, Song Niu, Chaobin Zhou and Yufan Wang

The School of Naval Architecture, Ocean and Energy Power Engineering, Wuhan University of Technology, Wuhan 430063, China; 345550@whut.edu.cn (X.Z.); yxglyr@whut.edu.cn (X.Y.); 299437@whut.edu.cn (L.L.); ns0522@whut.edu.cn (S.N.); chaobinzhou@whut.edu.cn (C.Z.); wangyuf@whut.edu.cn (Y.W.)
* Correspondence: zhoumengni@whut.edu.cn

Abstract: With increasingly stringent maritime environmental regulations, hybrid fuel cell ships have garnered significant attention due to their advantages in low emissions and high efficiency. However, challenges related to the coordinated control of multi-energy systems and fuel cell degradation remain significant barriers to their practical implementation. This paper proposes an innovative multi-timescale energy management strategy that focuses on optimizing the lifespan decay synergy of fuel cells and lithium batteries. The study designs an attention-based CNN-LSTM hybrid model for power prediction and constructs a two-stage optimization framework: The first stage employs Model Predictive Control (MPC) for long-term power planning to optimize equivalent hydrogen consumption, while the second stage focuses on real-time power allocation considering both power source degradation and system operational efficiency. The simulation results demonstrate that compared to single-layer MPC and the Equivalent Consumption Minimization Strategy (ECMS), the proposed method exhibits significant advantages in reducing single-voyage costs, minimizing differences in power source degradation rates, and alleviating power source stress. The overall performance of this strategy approaches the global optimal solution obtained through Dynamic Programming, comprehensively validating its superiority in simultaneously optimizing system economics and durability.

Keywords: fuel cell ships; multi-timescale energy management; lifespan decay synergy; equivalent hydrogen consumption; power prediction

1. Introduction

As a primary mode of maritime transportation, ships serve as a foundational support for human exploration of ocean resources. However, traditional propulsion systems relying on fossil fuels pose significant environmental pollution challenges [1]. The International Maritime Organization (IMO) has emphasized a future focus on the efficient utilization of clean energy in shipping. In response, shipping enterprises worldwide are actively advancing the integration of renewable energy technologies such as solar power, wind energy, and fuel cells to reduce carbon emissions and enhance energy efficiency [2,3]. Despite their advantages, such as high power density and efficiency [4], fuel cells face challenges including slow startup times and limited operational lifespan. The development of energy storage technologies has provided solutions for power dispatch in shipboard electrical systems [5]. Hybrid energy storage systems, combining energy-dense and power-dense storage components, significantly improve the performance and reliability of renewable energy-based ship power systems [6].

Advanced energy management strategies are essential for multi-energy hybrid power systems to achieve optimal power and torque distribution among various power sources while coordinating electric propulsion systems and ensuring the efficient synergistic operation of multiple power sources. According to the source and implementation method of actual energy allocation control, control strategies can be classified into three categories: rule-based control strategies oriented to engineering applications, optimal control strategies focusing on optimization modeling and solving, and learning-based control strategies driven by data [7]. Rule-based control strategies primarily rely on predetermined rules based on engineering experience, characterized by their simple structure and ease of implementation. The rule-based control strategy in reference [8], which utilizes batteries and shuts down engines at the minimum allowable power based on specific fuel consumption, demonstrated superior cost-effectiveness among all tested control strategies. In comparison, energy management strategies based on global optimization represent a more sophisticated approach, leveraging optimal control theory and artificial intelligence optimization techniques to design optimal energy allocation schemes for specific navigation conditions, thereby achieving a balance among multiple objectives. An optimization strategy was developed using dynamic programming (DP) in reference [9] that considered both battery degradation and electricity costs, significantly outperforming rule-based methods. While global optimization control strategies possess theoretical optimality, with DP strategies frequently serving as benchmarks for evaluating the global optimality of other algorithms, their practical application is somewhat limited due to the stochastic and unpredictable nature of vessel operating conditions. The ECMS, initially proposed by Paganelli [10], employs an equivalence factor to convert electrical energy consumption into fuel consumption, transforming global optimization problems into instantaneous optimization problems [11]. An efficient Energy Management System based on ECMS was proposed in reference [12], maintaining fuel cell system efficiency above 60% under most operating conditions while effectively suppressing fluctuations in fuel cell power output. Compared to ECMS and DP, MPC demonstrates unique advantages in hybrid power systems by predicting system behavior over a future time horizon while balancing computational efficiency and real-time capability. In marine hybrid power systems, predictive control achieves an energy balance among multiple power sources through online rolling optimization of multivariable problems. Reference [13] introduced an MPC-based coordinated control strategy that exhibited significant advantages in reducing fluctuations while maintaining autonomous operation. Furthermore, the energy management strategy designed based on nonlinear MPC models in [14] reduced fuel consumption and carbon emissions within the optimization period while maintaining robust disturbance rejection capabilities.

Furthermore, Multi-stack Fuel Cell Systems (MFCS) have attracted increasing attention due to their advantages in reduced hydrogen consumption, minimal degradation, and enhanced durability [15]. Power distribution strategies for MFCS primarily focus on system efficiency and hydrogen consumption optimization [16,17]. A hierarchical EMS was proposed for hybrid multi-stack fuel cell (FC) systems in the reference [18]. The first layer employs Sequential Quadratic Programming to determine power distribution ratios among multiple FC stacks while maintaining the battery State of Charge within acceptable bounds. The second layer optimizes an objective function based on overall system efficiency using genetic algorithms, achieving optimal power distribution among different FC stacks within the multi-stack system. Research findings demonstrate that this hierarchical EMS significantly reduces system hydrogen consumption. However, fuel cells are complex multi-physical systems characterized by relatively slow dynamic response characteristics and operational parameters that evolve dynamically with environmental variations and cell aging processes. This complexity can lead to performance inconsistencies among individual stacks within MFCS [19]. Consequently, strategies that solely prioritize system efficiency enhancement or fuel consumption optimization may not effectively sustain long-term system performance.

Ship power demands exhibit significant variations across different operational modes, including departure, cruising, port entry, and berthing, resulting in substantial load fluctuations and uncertainties [20]. These characteristics make long-term stable power demand prediction challenging, limiting forecasting capabilities to short-term load predictions. In MPC-based energy management strategies, load power prediction accuracy substantially influences power distribution effectiveness. ECMS requires dynamic calculations based on real-time data, introducing considerable computational complexity that increases controller burden and demands high hardware specifications. This computational load becomes particularly pronounced under high-frequency operational variations. Moreover, ECMS emphasizes instantaneous optimal distribution while inadequately addressing long-term energy system states (such as battery and supercapacitor State of Charge levels), potentially leading to the excessive utilization of energy storage devices during prolonged operation, thereby compromising system longevity and energy allocation flexibility. Furthermore, the EMS primarily focuses on optimizing either fuel economy or individual power source lifetime degradation [21,22]. However, such strategies fail to adequately account for the disparate degradation characteristics between fuel cells and power batteries. When one power source prematurely reaches its end-of-life due to excessive utilization, it disrupts the hybrid power system's equilibrium, forcing the remaining power source to bear additional loads, thereby accelerating its performance deterioration. This non-synergistic operation not only compromises the vessel's economic benefits but also leads to reduced durability of the entire power system, making it challenging to achieve an optimal balance between fuel economy and system durability.

Based on the aforementioned analysis, this paper proposes a multi-temporal energy management strategy for fuel cell ships that considers power source lifespan decay synergy. The primary innovations and contributions of this research include:

- A novel multi-temporal two-layer energy management strategy that achieves synergistic optimization across multiple time scales through an innovative hierarchical architecture. The upper-level controller manages long-term power distribution planning, while the lower-level controller executes precise real-time high-frequency power regulation, effectively balancing system economics and durability.
- An attention-enhanced CNN-LSTM power prediction model that leverages the complementary advantages of CNN in feature extraction and LSTM networks in temporal sequence modeling. The incorporation of attention mechanisms further enhances prediction accuracy, providing reliable decision support for energy management strategy implementation.
- Integration of differential fuel cell degradation characteristics and coordinated power source lifetime management into the optimization framework: The strategy considers varying degradation patterns among fuel cells and incorporates a lifetime-synchronized power distribution approach, achieving synchronized degradation control between fuel cells and power batteries.

The paper is structured as follows: Section 2 elaborates on the topological design of the hybrid ship power system. Section 3 presents a comprehensive examination of the theoretical framework and implementation methodology for the proposed multi-temporal energy management system. Section 4 validates the effectiveness of the proposed strategy through simulation experiments and provides comprehensive comparisons with existing methods. Section 5 summarizes the main research findings and innovative contributions.

2. Modeling of Fuel Cell Ship Propulsion System

2.1. Hybrid Power System Topology for Marine Applications

This study focuses on the "Hydrogen Vessel No. 1" fuel cell ship, which integrates fuel cells, lithium batteries, and supercapacitors as power sources. To enhance the economic viability and reliability of fuel cell vessels, a multi-stack fuel cell hybrid power system was designed based on the original power system architecture, with the primary power source parameters detailed in Table 1. The system employs a composite energy storage system (ESS) to replace the original single lithium battery configuration. This composite ESS comprises supercapacitors and lithium batteries, where lithium batteries offer high energy density but experience significant performance degradation over time and cycling. The degradation of lithium batteries primarily stems from complex electrochemical reactions during operation, encompassing irreversible processes such as electrode material structural changes, interfacial film growth, and active material loss [23,24]. These degradation mechanisms necessitate the careful consideration of performance deterioration characteristics in lithium battery applications. In contrast, supercapacitors, utilizing physical adsorption principles for energy storage, demonstrate exceptional cycling stability with negligible capacity degradation [25]. Their operational mechanism, avoiding chemical bond breaking and reformation, enables higher charge–discharge currents and extended operational lifetimes. The hybrid power system implements a fully active topology, as illustrated in Figure 1. The fuel cell subsystem consists of cell stacks and auxiliary equipment, including air compressors, hydrogen circulation pumps, and cooling water pumps. The dual fuel cell system employs a parallel topology, enabling independent control of each fuel cell stack and enhancing system stability. Each fuel cell connects to the DC bus through a unidirectional DC/DC converter, while lithium batteries and supercapacitors interface through bidirectional DC/DC converters. Within this architecture, the Energy Management System (EMS) coordinates the operation of fuel cells, lithium batteries, and supercapacitors to optimize electrical energy utilization efficiency and effectively reduce vessel operational costs.

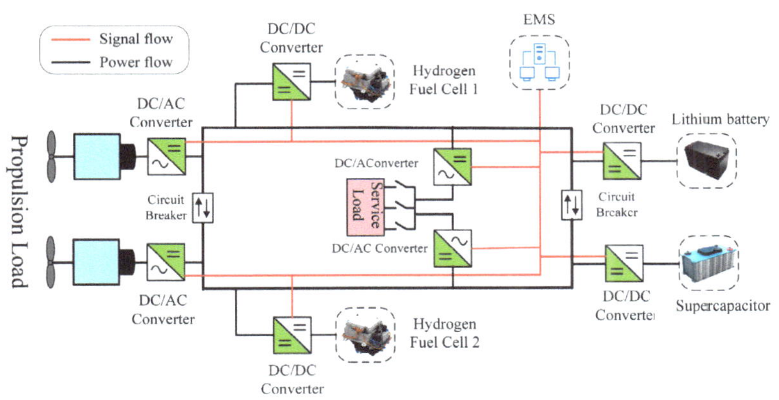

Figure 1. Power system topology of the hybrid ship.

Table 1. Power device parameters.

Power Supply Device	Parameter	Value
Fuel cell	Rated power (kW)	120
Lithium battery	Capacity (Ah)	120
	Rated voltage (V)	500
	Initial SOC (%)	60
Supercapacitor	Capacity (kWh)	5
	Rated voltage (V)	500
	Initial SOC (%)	60

2.2. Fuel Cell Model

This study employs the HYT-21-1346 Proton Exchange Membrane Fuel Cell (PEMFC). The fuel cell model is developed based on experimentally derived $P_{fc} - C_{fc}$ and $P_{fc} - \eta_{fc}$ characteristic curves. Let P_{fc} represent the fuel cell output power after DC/DC conversion, η_{fc} represent the system efficiency, and C_{fc} represent the actual hydrogen consumption rate. The relationship between these three parameters can be expressed as:

$$C_{fc} = \frac{P_{fc}}{\eta_{fc} LHV_{H_2}} \tag{1}$$

where LHV_{H_2} is the lower heating value of hydrogen, which equals 120 MJ/kg. With the prolonged operation of the system, fuel cell performance gradually deteriorates, leading to a decline in system durability. The fuel cell lifetime significantly dominates the total operational costs of the system. Generally, the factors contributing to fuel cell performance degradation can be categorized into four main aspects: start–stop cycles, high-load operation, low-load operation, and load fluctuations [26,27]. Therefore, fuel cell performance degradation patterns can be quantified by analyzing operational data during system runtime. The degradation rate of the fuel cell is defined as follows:

$$D_{fc} = \frac{d_1 T_{low} + d_2 T_{high} + d_3 N_{tran} + d_4 N_{ss}}{V_{init} \times 10\%} \tag{2}$$

where T_{low}, T_{high}, N_{tran}, and N_{ss} denote the operational durations at low and high power states, cumulative power variations, and start–stop cycles of the fuel cell, respectively; $d_1 \sim d_4$ represents the degradation rates under these four operating conditions (as shown in Table 2); and V_{init} is the initial single-cell voltage of the fuel cell, where the end-of-life criterion is defined as a 10% voltage degradation from the initial value.

Table 2. Degradation rate of fuel cell.

Operating Conditions	Degradation Rate
Low power operation	$d_1 = 10.17$ μV/h
High power operation	$d_2 = 11.74$ μV/h
Variable load	$d_3 = 0.0441$ μV/kW
On-Off	$d_4 = 23.91$ μV/cycle

Due to the dynamic power output characteristics of fuel cells under varying operating conditions, the performance degradation among individual stacks often exhibits non-uniform patterns [28]. To investigate the impact of fuel cell performance degradation, two operational scenarios were established for comparison: Fuel Cell 1 represents a new stack without degradation (healthy state S_{OH}= 1), while Fuel Cell 2 operates in a state of health S_{OH}= 0.7. Here, the state of health is characterized by the voltage degradation level at the rated current, where the maximum allowable voltage drop is defined as 10% relative to the initial output voltage. The system efficiency and hydrogen consumption rate curves under different SOH values were calculated using the degradation formula proposed in [29]. The fuel cell system efficiency comprises three components, defined as follows:

$$\eta_{fc} = \eta_{fuel} \cdot \eta_{conv} \cdot \eta_{elec} \tag{3}$$

$$\varepsilon_{conv} = 0.9 + 0.1 \cdot S_{OH} \tag{4}$$

where η_{fuel} is the fuel utilization rate, which can be approximated as 100% due to the hydrogen circulation mode implemented through hydrogen injectors and hydrogen circulation pumps, η_{fuel}= 1; η_{conv} represents the conversion efficiency, defined as the ratio between the electrical energy generated by the fuel cell and the chemical energy of the consumed hydrogen. This efficiency decreases with fuel cell degradation, and its degradation level can be quantified by the coefficient of variation given in Equation (4); η_{elec} denotes the electrical efficiency, which is determined by the DC/DC converter efficiency and parasitic power consumption of auxiliary equipment. Its degradation level can be expressed by the coefficient of variation as defined in Equation (5).

$$\varepsilon_{elec} = \frac{\eta_{DC/DC} - \frac{E_{auz}}{\varepsilon_{conv} \cdot E_{stack}}}{\eta_{DC/DC} - \frac{E_{auz}}{E_{stack}}} \tag{5}$$

where $\eta_{DC/DC}$ represents the conversion efficiency of the DC/DC converter, E_{auz} denotes the parasitic power consumption of auxiliary equipment, and E_{stack} is the stack output power. The degradation level of the fuel cell system efficiency can be characterized by the combined degradation of η_{conv} and η_{elec}:

$$\eta_{fc\_degraded} = \varepsilon_{elec} \cdot \varepsilon_{conv} \cdot \eta_{fc} \tag{6}$$

where $\eta_{fc\_degraded}$ represents the fuel cell system efficiency after performance degradation. The degraded fuel cell system efficiency is calculated based on Equations (3)–(6). The hydrogen consumption rate at various net power outputs of the fuel cells is derived from Equation (1). The system efficiency curves and hydrogen consumption rate curves of the fuel cells are illustrated in Figure 2.

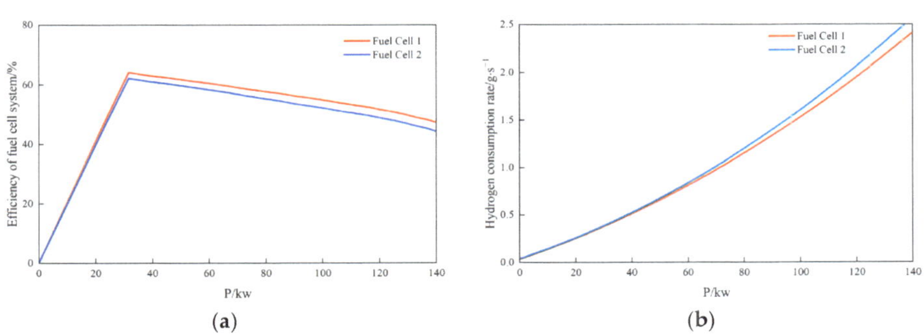

Figure 2. (**a**) Efficiency; (**b**) consumption rate.

2.3. Lithium Battery Model

The lithium battery possesses dual functionality in energy storage and charge–discharge capabilities. As an energy storage component, the battery capacity is determined by the total charge it can transfer. The battery capacity Q can be calculated using the following expression:

$$Q = I \cdot t \tag{7}$$

Equation (7) indicates that a battery with capacity Q can sustain a constant discharge current I for a duration t. The State of Charge (SOC) of the battery is defined as the ratio between the remaining capacity and the total capacity, expressed as:

$$SOC = 1 - \frac{\int I(t)dt}{Q} \tag{8}$$

In this study, the Rint equivalent circuit model is adopted for the traction battery, which consists of an ideal voltage source in series with internal resistance, as illustrated in Figure 3:

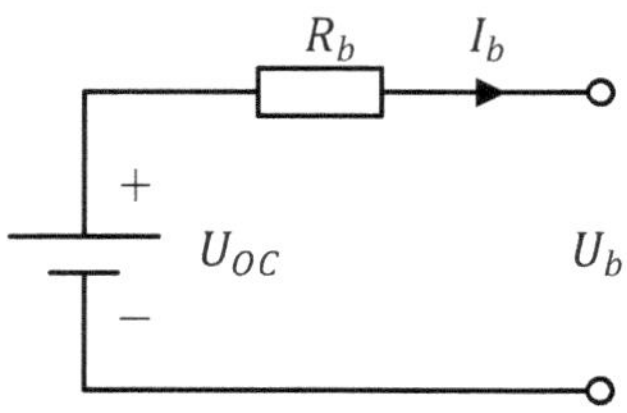

Figure 3. Rint equivalent circuit diagram of battery.

According to the Rint equivalent circuit model, the mathematical expression for the battery output power is given as:

$$P_{batt} = U_{oc}I_b - R_bI_b^2 \tag{9}$$

where U_{oc} denotes the open-circuit voltage, while I_b and R_b represent the battery current and internal resistance, respectively. Both the open-circuit voltage and internal resistance are functions of the battery SOC. According to Equation (9), the current I_b is derived as:

$$I_b = \frac{U_{oc} - \sqrt{U_{oc}^2 - 4R_bP_{batt}}}{2R_b} \tag{10}$$

The dynamic behavior of the lithium battery SOC can be described by the following expression, derived from the combination of Equations (8) and (10):

$$SOC(k+1) = SOC(k) - \frac{U_{oc} - \sqrt{U_{oc}^2 - 4R_bP_{batt}}}{2QR_b} \tag{11}$$

The degradation characteristics of lithium batteries are governed by their SOC and transient power profiles. High current operations, especially during charging processes, lead to accelerated lifetime reduction. The rate of performance degradation exhibits a positive correlation with the magnitude of SOC variations. The corresponding semi-empirical degradation model [30] is formulated as:

$$D_{batt} = \frac{1}{Q_{batt}} \int_0^t |F(SOC_{batt})G(i_{batt})I_{batt}(t)|dt \tag{12}$$

where Q_{batt} represents the battery capacity, $F(SOC_{batt})$ is a weighting function that penalizes battery degradation due to excessive SOC utilization, and $G(i_{batt})$ is a weighting function that penalizes battery degradation caused by excessive transient power. These functions are expressed as follows:

$$F(SOC_{batt}) = 1 + 3.25(1 - SOC_{batt})^2 \tag{13}$$

$$\begin{cases} G(I_{batt}) = 1 + 0.45\frac{I_{batt}}{I_{batt\_nom}}, i_{batt} \geq 0 \\ G(I_{batt}) = 1 + 0.55\frac{|I_{batt}|}{I_{batt\_nom}}, i_{batt} < 0 \end{cases} \tag{14}$$

where I_{batt_nom} represents the rated current of the battery, and the degradation function D_{batt} ranges between 0 (beginning of life) and 1 (end of life).

2.4. Supercapacitor Model

Among the diverse equivalent circuit models available for supercapacitors, this investigation utilizes a widely adopted simplified model for analytical modeling, as depicted in Figure 4:

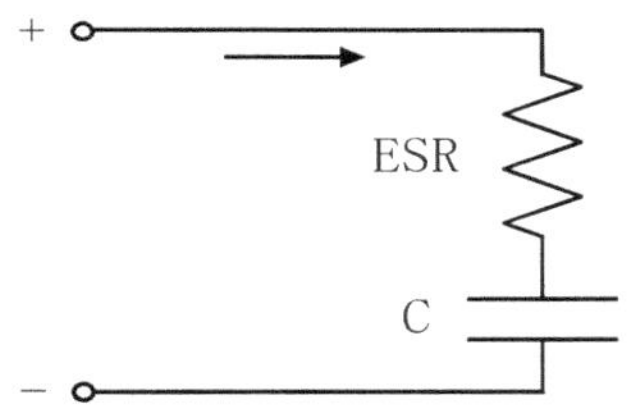

Figure 4. Equivalent circuit of supercapacitor.

Based on the equivalent circuit, the mathematical expressions for the output power and State of Charge (SOC) of the supercapacitor can be formulated as:

$$P_{sc} = U_{sc}I_{sc} - I_{sc}^2R_{sc} \tag{15}$$

$$SOC = \frac{U_{sc} - U_{scmin}}{U_{scmax} - U_{scmin}} \tag{16}$$

where P_{sc} represents the supercapacitor output power, U_{sc} denotes the terminal voltage, I_{sc} is the supercapacitor current, R_{sc} represents the supercapacitor resistance, and U_{scmax} and U_{scmin} denote the maximum and minimum operating voltages of the supercapacitor, respectively.

3. Proposed Energy Management Strategy

This section presents a two-stage energy management strategy based on Model Predictive Control (MPC) with multiple time scales. The strategy aims to optimize power distribution among fuel cells, lithium batteries, and supercapacitors based on power demand data during navigation (as illustrated in Figure 5), ensuring the economically efficient operation of the ship's hybrid power system while satisfying operational constraints. The hierarchy of the proposed algorithm is shown in Figure 6. The comprehensive methodology of the proposed energy management strategy proceeds as follows: Initially, future power load profiles are predicted based on historical power data, followed by power allocation for the predicted load data. The resulting reference power values are then dynamically computed in conjunction with real-time load data. The strategy primarily comprises a prediction phase and two optimization phases: In the first optimization phase, an MPC controller optimizes the fuel cell power and lithium battery charge–discharge power based on low-frequency power components over extended time scales, minimizing equivalent hydrogen consumption. The optimized battery and supercapacitor charge–discharge powers, along with the fuel cell power, serve as reference inputs for the lower-level real-time power optimization controller. In the second phase, the lower-level real-time power optimization controller performs secondary optimization of high-frequency power components based on the reference values from the upper-level MPC. The optimized results are then implemented as actual control outputs. Figure 7 illustrates the framework of the proposed energy management strategy.

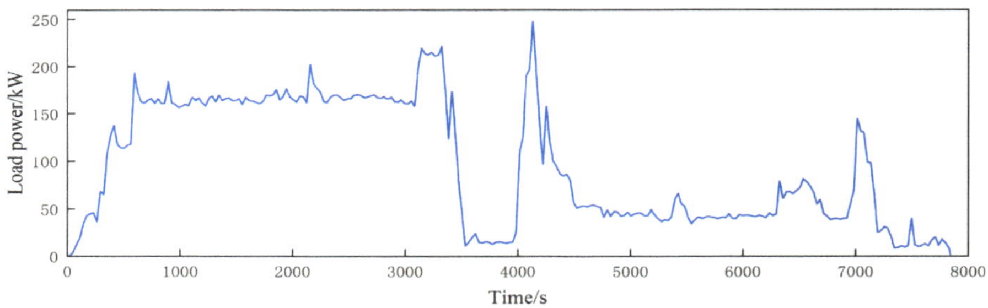

Figure 5. Fuel cell ship load power data.

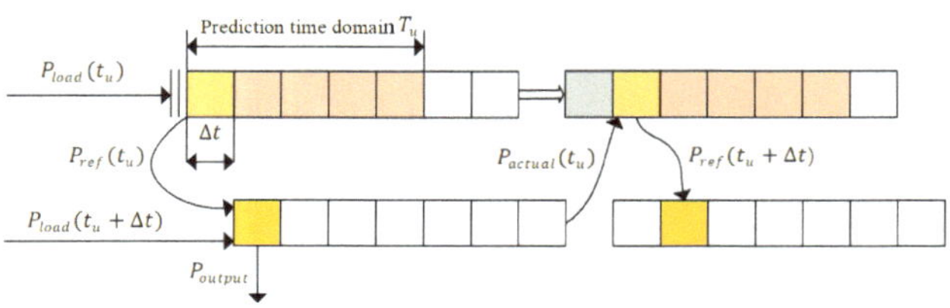

Figure 6. The algorithm hierarchy of the proposed algorithm.

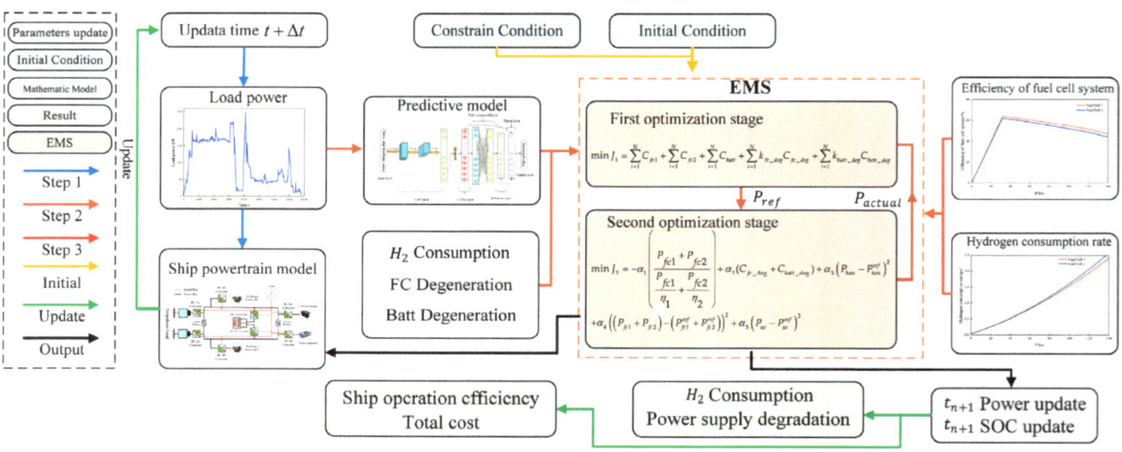

Figure 7. Energy management strategy framework.

3.1. CNN-LSTM Prediction Model with Attention Mechanism

This section presents an attention-enhanced CNN-LSTM model for power prediction. The model incorporates an attention-based CNN architecture that extends the standard CNN framework with parallel attention pathways for salient feature extraction. The attention pathway employs expanded input dimensions to broaden the receptive field, thereby comprehensively capturing temporal contextual information and effectively learning the significance of local sequence features. By amplifying the weights of crucial temporal features while suppressing the influence of non-essential features, the attention module effectively addresses the limitations of traditional models in discriminating temporal feature importance. The multi-scale input approach, where both the standard CNN module and attention mechanism module process input sequences of varying lengths, enables the more robust extraction of short-sequence features. In terms of feature extraction, the LSTM architecture extracts coarse-grained features from the fine-grained features obtained from the front end, providing refined processing of multi-dimensional features. This architectural design mitigates the memory loss and gradient vanishing problems typically associated with extended step lengths. The attention-enhanced CNN-LSTM model achieves comprehensive temporal data characterization [31] through the fusion of both coarse and fine-grained features.

The architecture of the attention-enhanced CNN-LSTM model comprises five primary layers: the input layer, CNN layer, LSTM layer, attention layer, and output layer, as illustrated in Figure 8. Historical load data serve as the input to the CNN layer, where convolution operations are performed to increase the depth and compress the parameter quantity while pooling operations reduce feature dimensionality. The fully connected layer transforms features into a one-dimensional structure, completing feature vector extraction. The LSTM and attention layers learn the internal variation patterns of the load from the extracted features, thereby enabling predictive functionality. The output layer generates the prediction results. The model has been trained using historical navigation data from the vessel and subsequently employed to predict the power profile shown in Figure 5, with a sampling interval of 2 s and a prediction horizon of 20 s. The prediction results are presented in Figure 9.

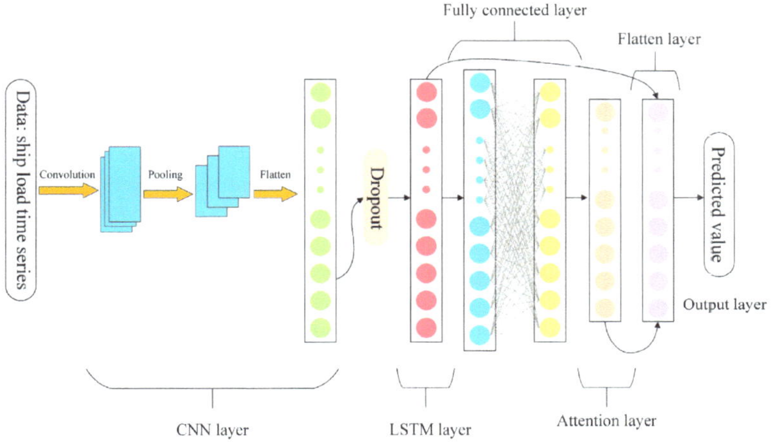

Figure 8. CNN-LSTM based on attention mechanism model framework.

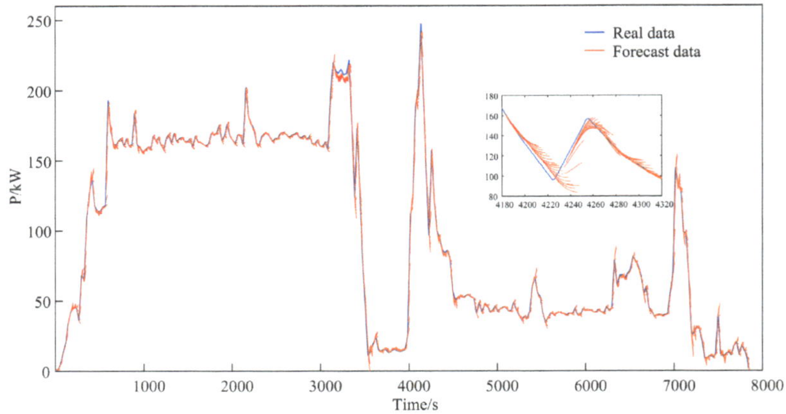

Figure 9. Load power prediction results.

As demonstrated in Figure 8, the predicted power demand curve closely aligns with the actual power demand profile. The Root Mean Square Error (RMSE) is employed as the evaluation metric for assessing the accuracy of the ship's load power prediction. RMSE represents the square root of the ratio between the squared deviations of predicted values from actual values and the number of observations, as expressed in Equation (17):

$$\begin{gathered} RMSE = \frac{\sum_{k=1}^{n} RMSE(k)}{n} \\ RMSE(k) = \sqrt{\frac{\sum_{i=1}^{t_p} (p(k+i) - p_0(k+i))^2}{t_p}} \end{gathered} \tag{17}$$

where n represents the number of sampling points in the operating condition, $RMSE(k)$ denotes the value at time k within prediction horizon t_p,$p(k+i)$ represents the predicted value at i seconds after time k, and $p_0(k+i)$ indicates the actual value at i seconds after time k. The prediction results for horizons of 10 s, 20 s, and 40 s were analyzed, with the outcomes presented in Table 3. From the perspective of RMSE, prediction horizons of 10 s, 20 s, and 40 s correspond to error values of 1.01 kW, 1.87 kW, and 3.98 kW, respectively. In terms of average prediction time, the computational requirements are approximately 20.12 ms, 57.34 ms, and 150.56 ms, respectively. While the 10 s prediction horizon exhibits the lowest RMSE and computational overhead, it proves insufficient to fully capture power variation trends. Conversely, although the 40 s horizon offers an extended prediction range, its significant error accumulation and 150 ms computational latency make it unsuitable for real-time control requirements. The 20 s prediction horizon, however, demonstrates an optimal balance across all metrics: its RMSE is only marginally higher than that of the 10 s horizon, its 57.34 ms computational latency remains acceptable for real-time systems, and it provides an adequate prediction range to support energy management system decisions. Therefore, this study adopts a 20 s prediction horizon, a selection that ensures both prediction accuracy and compliance with the real-time requirements of marine power systems.

Table 3. Different time domain prediction results.

Prediction Time Domain	10 s	20 s	40 s
REMS	1.01	1.87	3.98
Average time per prediction(ms)	20.12	57.34	150.56

3.2. Model Predictive Control Principles

Model Predictive Control (MPC), emerging as an innovative computational control algorithm in the late 1970s, is also known as receding horizon control [32]. The operational mechanism of MPC can be succinctly characterized as follows: At each sampling instant, the current control action is determined by solving a finite-horizon open-loop optimal control problem, utilizing the current process state as the initial condition for the optimization. Only the first element of the optimal control sequence is implemented. At the subsequent sampling instant, this process is repeated, with new measurements serving as the initial conditions for predicting future system dynamics, thereby refreshing and resolving the optimization problem. The fundamental architecture of MPC comprises four essential components: the prediction model, feedback correction, receding horizon optimization, and reference trajectory [33].

The operational principle of MPC is schematically illustrated in Figure 10. Let H_p denote the prediction horizon, H_q represent the control horizon, x_r designate the reference trajectory for state variable x, and x_p represent the predicted trajectory for state variable x. Generally, constraints must be imposed on the predicted trajectory, which are formulated as follows:

$$min J_k = \sum_{t=k}^{k+p} L(x(t), u(t)) \tag{18}$$

$$\begin{cases} x_{min}(t) \leq x(t) \leq x_{max}(t) \\ u_{min}(t) \leq u(t) \leq u_{max}(t) \\ k \leq t \leq k+p \end{cases} \tag{19}$$

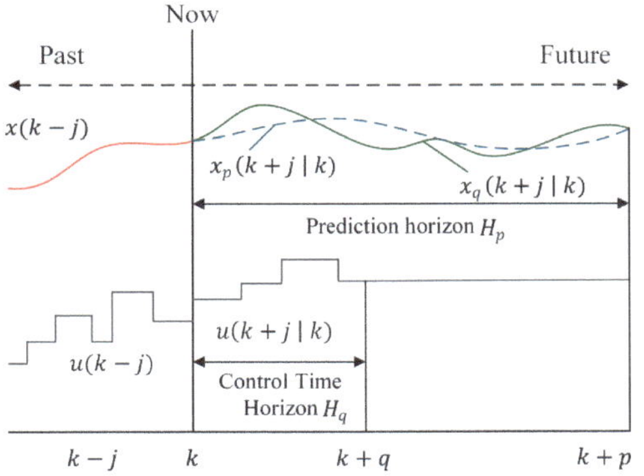

Figure 10. Model Predictive Control (MPC) principle.

In Equations (18) and (19), $x(t)$ represents the state variable at time t,$u(t)$ denotes the control variable at time t, L represents the performance index at time k, and J_k denotes the performance index at time instant k over the prediction horizon $k-k+p$.

From Equation (18), it can be observed that the performance index is calculated solely within the prediction horizon. Similarly, the performance index at time instant $k+1$ can be expressed as:

$$minJ_{k+1} = \sum_{t=k+1}^{k+p+1} L(x(t), u(t)) \tag{20}$$

3.3. System Constraints

At any given time instant, the total grid output power can be expressed as follows:

$$P_{total} = P_{fc1} + P_{fc2} + P_{Batt} + P_{sc} \tag{21}$$

In the design of system operational constraints, trade-offs must be considered among performance, lifetime, and safety. The power constraints of the fuel cell are intended to meet load demands while preventing overload operation, thereby extending its service life. The power constraints and SOC range of the lithium battery are employed to prevent excessive charging and discharging cycles, thus enhancing cycle life and ensuring thermal management stability. The power constraints and SOC range of the supercapacitor are established to provide rapid transient power response while minimizing efficiency losses and thermal issues. These constraints comprehensively incorporate the performance characteristics of the fuel cell, battery, and supercapacitor, aiming to meet system load demands while achieving an optimal balance between energy efficiency and component longevity. The operational constraints for the fuel cell, lithium battery, and supercapacitor are defined as follows:

$$\begin{cases} 0\text{kW} \le P_{fc} \le 120\text{kW} \\ -100\text{kW} \le P_{batt} \le 100\text{kW} \\ -120\text{kW} \le P_{sc} \le 120\text{kW} \\ 20\% \le SOC_{batt} \le 80\% \\ 10\% \le SOC_{sc} \le 90\% \end{cases} \tag{22}$$

3.4. First Optimization Stage

Due to the distinct characteristics of supercapacitors and batteries, the ship's load power is decomposed according to frequency components. The high-frequency components of the load power should be absorbed by the supercapacitor. In this study, low-pass filtering is employed to decompose the high-frequency power signals. The low-pass filtering control strategy is illustrated in Figure 11.

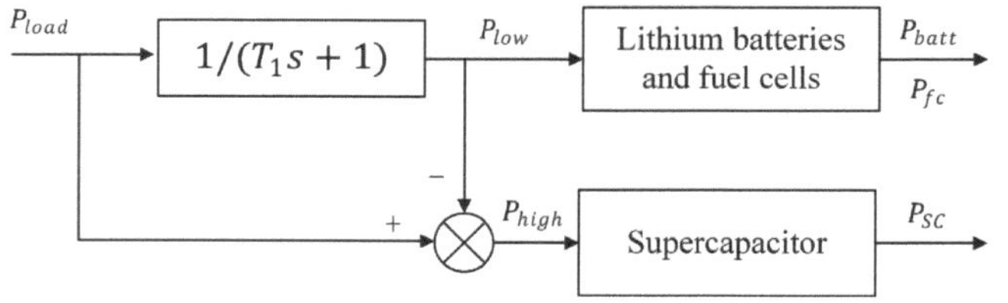

Figure 11. Low-pass filtering control strategy.

In the figure, T_1 represents the filtering time constant, P_{low} and P_{high} denote the low-frequency and high-frequency components of the load power, respectively, while P_{batt}, P_{fc}, and P_{sc} represent the output powers of the lithium battery, fuel cell, and supercapacitor, respectively. A larger T_1 value results in smoother low-frequency power components, consequently directing more high-frequency power components to the supercapacitor for absorption. The T_1 value should be synchronized with the control period of the MPC in the primary optimization phase, where larger MPC control periods, corresponding to extended time scales, necessitate larger τ values to achieve smoother low-frequency power profiles.

In this phase, the MPC controller focuses on optimizing low-frequency power distribution over the prediction horizon. It determines reference values for battery charge–discharge power, supercapacitor charge–discharge power, and fuel cell power output. The optimization objective function for this phase aims to minimize overall hydrogen consumption while accounting for the performance degradation of both fuel cells and lithium batteries. In this primary optimization phase, the total fuel cell power is equally distributed between two fuel cells using a balanced allocation method, with the controller sampling time set to 2 s and the prediction horizon equal to the control horizon. The MPC controller objective function is formulated as:

$$\begin{gathered}\min J_1 = \sum_{i=1}^{N} C_{fc1} + \sum_{i=1}^{N} C_{fc2} + \sum_{i=1}^{N} C_{batt} + \sum_{i=1}^{N} k_{fc\_deg} C_{fc\_deg} + \sum_{i=1}^{N} k_{batt\_deg} C_{batt\_deg} \\ \text{s.t.}(21),(22)\end{gathered} \tag{23}$$

where C_{fc1} and C_{fc2} represent the hydrogen consumption rates of fuel cell 1 and fuel cell 2, respectively; C_{batt} denotes the equivalent hydrogen consumption rate of the lithium battery; k_{fc_deg} and k_{Batt_deg} are degradation weights that indicate the optimization priorities for fuel cell and lithium battery degradation within the overall system performance. Based on the data presented in Figure 2, the hydrogen consumption rate function is obtained through quadratic polynomial fitting:

$$C_{fc} = a \cdot P_{fc}^2 + b \cdot P_{fc} + c \tag{24}$$

where a, b, and c are the fitting coefficients of the fuel cell. To minimize hydrogen consumption in the hybrid power system, the electrical energy consumed by the lithium battery can be equivalently expressed in terms of the chemical energy consumed by the fuel cell [34], calculated as:

$$C_{batt} = \begin{cases} \frac{kP_{batt}}{\eta_{dis}\eta_{fc\_avg}LHV_{H_2}}, P_{batt} \geq 0 \\ \frac{kP_{batt}\eta_{chg}}{\eta_{fc\_avg}LHV_{H_2}}, P_{batt} \leq 0 \end{cases} \tag{25}$$

where η_{dis} and η_{chg} represent the charging and discharging efficiencies of the lithium battery, respectively; c denotes the average efficiency of the fuel cell. To ensure the lithium battery operates within an appropriate SOC range and prevent damage from excessive charging or discharging, a correction function k is introduced. The magnitude of k determines the rate of convergence, where larger values indicate faster convergence and smaller values indicate slower convergence. This can be calculated using the following expression:

$$k = 1 - 2\mu \frac{(SOC - 0.5(SOC_{max} + SOC_{min}))}{SOC_{max} - SOC_{min}} \tag{26}$$

where SOC_{max} and SOC_{min} denote the upper and lower SOC thresholds for the lithium battery, established at 0.8 and 0.2, respectively, in this investigation, and μ represents the balance coefficient. Recognizing lifetime degradation as a critical determinant of fuel cell performance and system economics, this study implements an equivalence methodology derived from the ECMS (Equivalent Consumption Minimization Strategy) to quantify fuel cell degradation in terms of equivalent hydrogen consumption. This approach achieves dual objectives: minimizing dynamic load fluctuations in fuel cell operation while simultaneously decelerating the degradation process through the explicit consideration of performance deterioration mechanisms, ultimately enhancing overall system durability. The equivalent hydrogen consumption $C_{\deg}$ attributable to fuel cell degradation is formulated as:

$$C_{\deg} = \frac{D_{fc}c_{fc}}{c_{H_2}} \tag{27}$$

where c_{fc} and c_{H_2} represent the unit costs of the fuel cell and hydrogen, respectively, while D_{fc} denotes the fuel cell degradation rate as derived from Equation (2). Since the exclusive consideration of fuel cell lifetime degradation could result in accelerated deterioration and premature end-of-life of the lithium battery, the objective function incorporates battery degradation metrics. The associated degradation cost of the lithium battery is formulated as:

$$C_{batt\_deg} = D_{batt}Q_{batt} \tag{28}$$

The objective function (Equation (23)) of the MPC controller is solved using the Sequential Dynamic Programming (SDP) algorithm.

3.5. Second Optimization Stage

The initial solution set obtained from the first optimization phase serves as the reference power for the second phase:

$$P^{ref} = \begin{bmatrix} P_{fc1}^{ref} \\ P_{fc2}^{ref} \\ P_{batt}^{ref} \\ P_{sc}^{ref} \end{bmatrix} \tag{29}$$

The real-time power optimization controller in the secondary phase performs power allocation based on instantaneous load demands. This phase's optimization objective is to achieve optimal power distribution by comprehensively considering the characteristics of energy storage devices and high-frequency components of power demands while ensuring fuel economy. Given the differential degradation rates among units in the dual fuel cell system, the uniform distribution method employed in the primary phase cannot achieve optimal efficiency. Therefore, this phase necessitates further consideration of maximizing the overall efficiency of the fuel cell system. The objective function for the secondary optimization phase is expressed as:

$$\begin{gathered} \min J_2 = -\alpha_1\left(\frac{P_{fc1}+P_{fc2}}{\frac{P_{fc1}}{\eta_1}+\frac{P_{fc2}}{\eta_2}}\right) + \alpha_2(C_{fc\_\deg} + C_{batt\_\deg}) + \alpha_3\left(P_{batt} - P_{batt}^{ref}\right)^2 \\ +\alpha_4\left(\left(P_{fc1} + P_{fc2}\right) - \left(P_{fc1}^{ref} + P_{fc2}^{ref}\right)\right)^2 + \alpha_5\left(P_{uc} - P_{uc}^{ref}\right)^2 \\ s.t.(21),(22) \end{gathered} \tag{30}$$

where coefficients α_1, α_2, α_3, α_4, and α_5 represent penalty factors. The first term aims to maximize the efficiency of the dual fuel cell system; the second term reflects the degradation costs of both fuel cells and the lithium battery; and the third, fourth, and fifth terms represent the tracking requirements for reference values generated by the upper-level controller, including fuel cell power references and charge–discharge power references for the battery pack and supercapacitor. These tracking terms ensure the maintenance of fuel economy optimization performance established at the macro time scale in the first phase.

The nonlinear programming problem described by Equation (30) is solved using the GUROBI commercial optimization solver.

4. Simulation Results

To comprehensively evaluate the optimization effectiveness of the proposed strategy, this study conducts a multi-dimensional comparative analysis. The proposed strategy is initially benchmarked against both the conventional single-level Model Predictive Control (MPC) strategy and the Equivalent Consumption Minimization Strategy (ECMS). Furthermore, to validate the global optimality of the proposed strategy over the entire operational cycle, comparisons are made with the Dynamic Programming (DP) strategy. This multi-tiered comparative validation approach effectively demonstrates the optimization performance of the proposed strategy.

4.1. Evaluation Criterion

To objectively evaluate the economic performance of the power system, this study defines the total equivalent hydrogen consumption $C_{H_2\_equ}$, which not only encompasses the direct hydrogen consumption of fuel cells but also incorporates the equivalent hydrogen consumption derived from SOC deviations of both the lithium battery and the supercapacitor from their initial states, as expressed in Equation (23):

$$C_{H_2\_equ} = C_{fc} + \frac{(SOC_{batt\_init} - SOC_{batt\_end})Q_{batt} + (SOC_{sc\_init} - SOC_{sc\_end})Q_{sc}}{\eta_{fc\_avg} LHV_{H_2}} \quad (31)$$

where SOC_{batt_end} and SOC_{sc_end} represent the final SOC of the lithium battery and supercapacitor, respectively; η_{fc_avg} denotes the average efficiency of the fuel cell system. To objectively evaluate the optimization performance of various control strategies, this study establishes a single-voyage cost assessment index for the vessel's power system. This index comprehensively incorporates three critical dimensions: hydrogen fuel consumption cost, fuel cell system degradation cost, and lithium battery degradation cost, enabling a holistic characterization of the power system's economic performance, as shown in Equation (24).

$$C_{total} = C_{H_2\_equ} \cdot c_{H_2} + D_{fc} \cdot c_{fc} \cdot P_{fc\_rated} + D_{batt} \cdot c_{batt} \cdot Q_{batt} \quad (32)$$

where c_{batt} represents the unit cost coefficient, which is determined through market research to be 35 CNY/kg for hydrogen, 3500 CNY/kW for fuel cells, and 2000 CNY/kWh for lithium batteries; P_{fc_rated} and Q_{batt} denote the rated power of the fuel cell system and the nominal capacity of the lithium battery system, respectively.

4.2. Power Distribution

Figure 12a–c illustrates the operational characteristics of the vessel's hybrid power system under four different strategies, depicting the power profiles of the fuel cell system, lithium battery, and supercapacitor. The simulation results demonstrate that the proposed strategy achieves optimal real-time power control based on power demand forecasting, enabling coordinated optimization across both long and short time scales. The fuel cell system predominantly operates in a power-following mode with enhanced operational efficiency. This approach not only ensures sufficient power redundancy in the hybrid system to accommodate load variations but also effectively minimizes the dynamic load fluctuations of the fuel cell, maximizing the peak-shaving and valley-filling capabilities of the energy storage system. The results indicate that the lithium battery power profile exhibits relatively smooth characteristics, while the high-frequency power components of the load demand are primarily absorbed by the supercapacitor. This effectively prevents high-frequency power fluctuations in battery output, achieving the intended control objectives. In contrast, the conventional single-layer MPC strategy, which must solve for both real-time and predicted power simultaneously, demonstrates suboptimal power distribution due to prediction accuracy limitations and system constraints. This results in underutilization of the energy storage system's auxiliary function and increases fuel cell power output fluctuations. The traditional ECMS strategy, which only addresses real-time power distribution without future power information, tends to converge to local optima during optimization. To maintain the SOC of the energy storage system, this approach leads to higher fuel cell power output, resulting in increased power variation rates and hydrogen consumption. Consequently, it fails to ensure fuel economy and yields a suboptimal overall performance.

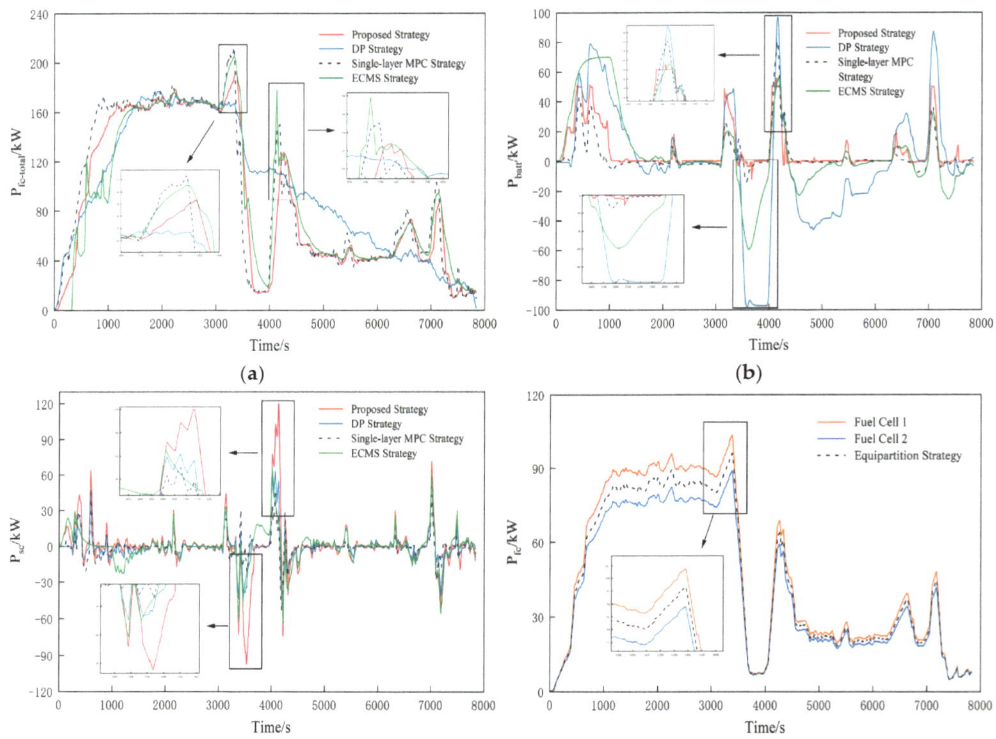

Figure 12. (**a**) Total fuel cell power; (**b**) lithium battery power; (**c**) supercapacitor power; (**d**) fuel cell power distribution.

The proposed strategy accounts for the differential degradation states of fuel cells, with the power distribution between two fuel cells illustrated in Figure 12d. The fuel cell with inferior performance (SOH = 0.7) exhibits a significantly lower power fluctuation amplitude compared to the one with superior performance (SOH = 1). This demonstrates that the proposed strategy effectively mitigates the aging rate of the Multi Fuel Cell System (MFCS), thereby reducing the frequency and costs associated with fuel cell replacement. Furthermore, according to calculations using Equation (33), the overall efficiency of the fuel cell system reaches 58.5% when using the strategy proposed in this paper, while it is 57.1% when using the equal distribution strategy. The comparison results show that the method in this paper improved the overall efficiency of the fuel cell system by 1.4%, effectively enhancing the system's energy utilization rate.

$$\eta_{total} = \frac{\sum(P_{FC1} + P_{FC2})}{\sum(\frac{P_{FC1}}{\eta_{FC1}} + \frac{P_{FC2}}{\eta_{FC2}})} \tag{33}$$

4.3. Energy Storage Unit State of Charge

Figure 13 illustrate the State of Charge (SOC) profiles of the lithium battery and supercapacitor under four different control strategies. The simulation results demonstrate that all strategies maintain the SOC of both energy storage units within their constraint boundaries. Specifically, the lithium battery SOC fluctuation ranges are 46.6–60% for the single-layer MPC strategy, 26.9–60% for the ECMS strategy, and 27.1–71.2% for the DP strategy, which exhibits the widest variation. The proposed strategy maintains the battery SOC within 34.2–60%. As a locally optimizing approach, the proposed strategy deliberately restricts SOC deviations from median values to reserve sufficient capacity for subsequent power demands. The single-layer MPC strategy demonstrates a narrower battery SOC fluctuation range compared to the proposed strategy, primarily due to increased battery power output in the first optimization stage of our strategy, which aims to limit fuel cell power variations. From the perspective of energy storage system performance evaluation, the proposed strategy demonstrates favorable balanced characteristics: Regarding lithium battery degradation, the capacity loss is only marginally higher than the single-layer MPC strategy, indicating effective capacity loss control while maximizing battery performance utilization. For the supercapacitor, this strategy exhibits the relatively largest SOC fluctuation range, an inherent consequence of the dual-layer optimization structure. This effectively accomplishes power smoothing for both the lithium battery and fuel cell system. It should be noted that due to the strategy's emphasis on mitigating fuel cell and lithium battery degradation, there exists a significant deviation between the final and initial SOC values of the energy storage elements. This represents a deliberate trade-off within the system's global optimization process.

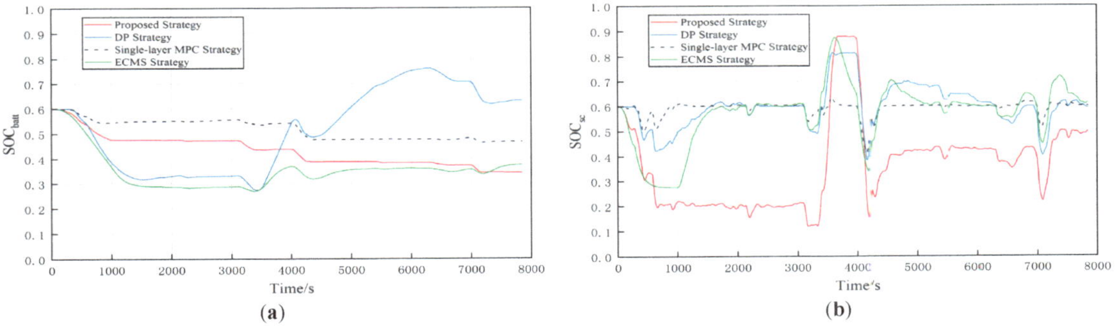

Figure 13. (**a**) Lithium battery SOC; (**b**) supercapacitor SOC.

4.4. Comparative Analysis of Coordinated Degradation of Fuel Cell and Lithium Battery Systems

Based on Equations (2) and (12), the degradation rates of the fuel cell system and lithium battery system were calculated. Figure 14 shows the degradation rate curves of the fuel cell and lithium battery under four different energy management strategies. For the DP, single-layer MPC, and ECMS strategies, fuel cell total power distribution follows an equal allocation approach. At the end of the operating cycle, the differential degradation rates between the dual power sources under the proposed strategy, DP, single-layer MPC, and ECMS are 0.0019%, 0.101%, 0.0218%, and 0.0545%, respectively.

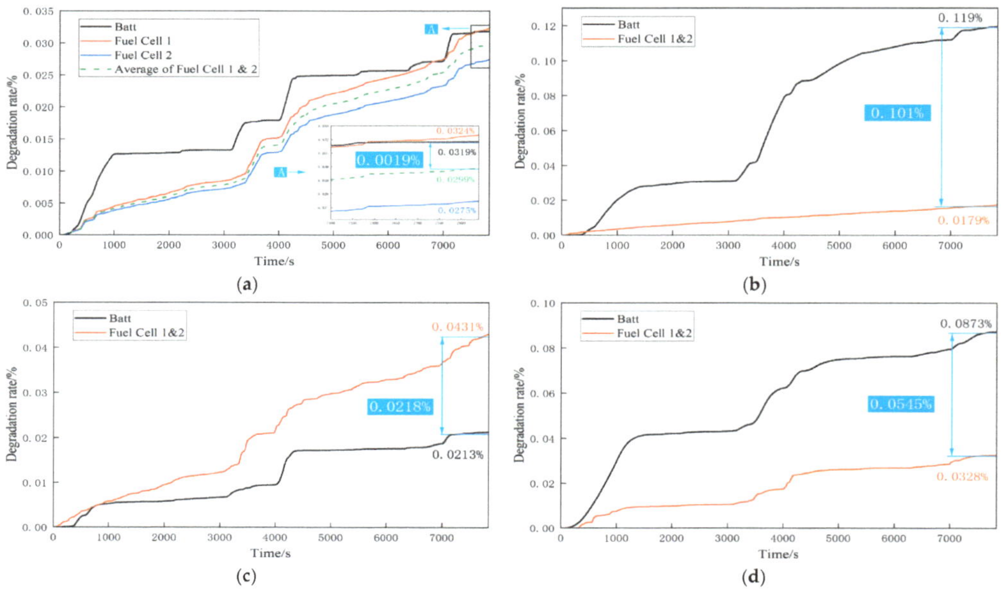

Figure 14. Comparison of life decay rates of dual power sources: (**a**) Proposed strategy; (**b**) DP strategy; (**c**) Single-layer MPC strategy; (**d**) ECMS strategy.

Compared to the DP, single-layer MPC, and ECMS strategies, the proposed strategy achieves remarkable reductions in the differential degradation rates between dual power sources of 98.12%, 91.36%, and 96.48%, respectively. This superior optimization performance can be attributed to several key factors: First, although the DP strategy achieves the lowest total voyage cost, it excessively relies on the lithium battery to minimize fuel cell degradation. This leads to premature battery life depletion, failing to achieve balanced optimization between voyage economics and coordinated power source degradation. Second, the single-layer MPC strategy's insufficient utilization of the lithium battery's regulatory capabilities results in a fuel cell degradation rate exceeding that of the battery, thereby exacerbating the degradation inconsistency between the two systems. Third, the ECMS strategy, in its pursuit of minimizing equivalent hydrogen consumption, is constrained by its tendency to converge to local optima. This limitation increases power fluctuations in both fuel cell and lithium battery systems, consequently amplifying the disparity in degradation rates between the two power sources.

This strategy achieves the synchronized degradation control of fuel cell and lithium battery systems by organically integrating fuel economy with coordinated power degradation management. From the perspectives of long-term system economics and maintenance, this coordinated control strategy has significant practical implications: First, by achieving synchronized degradation of lithium batteries and fuel cells, this strategy can significantly extend the overall service life of the ship's power system. When fuel cells and lithium battery systems degrade at similar rates, it prevents the need for complete system replacement due to premature failure of a single component, thereby reducing equipment renewal costs. Second, coordinated degradation control significantly reduces maintenance requirements. By avoiding uneven aging between components, the frequency of maintenance inspections can be reduced. This not only directly saves maintenance costs but also reduces vessel downtime for maintenance, improving operational efficiency. These research findings demonstrate that while achieving coordinated power source degradation, this strategy brings significant economic benefits to vessel operators and has practical guiding significance for advancing ship power systems toward more sustainable and economical development.

To further validate the effectiveness of the proposed strategy, the simulation results were compared with existing experimental studies. Lu et al. [35] built a fuel cell stack durability test platform to experimentally study the coordinated life degradation of dual power sources. Their experimental results showed that under the CWC1 combined working conditions, after adopting the PEMS strategy, the difference in degradation rates between the fuel cell and battery decreased from 0.00538% to 0.00021% compared to the ECMS strategy. The strategy proposed in this paper reduced the degradation rate difference from 0.0545% to 0.0019%, with a relative improvement (96.48%) basically consistent with the experimental results (96.1%). Through comparisons with experimental studies, it can be found that the performance improvement demonstrated by the proposed strategy in simulation and HIL testing shows good consistency with actual test results.

4.5. Comparative Analysis of Power Source Stress

Power source stress analysis primarily examines the high-frequency power output characteristics of both fuel cell and lithium battery systems, serving as a critical factor influencing their respective service life [36]. Lower power source operational stress, characterized by the reduced frequency and amplitude of power output fluctuations, is conducive to maintaining optimal source performance. The power source stress can be effectively quantified through the standard deviation σ of high-frequency components obtained via Haar wavelet transformation. Figure 15 illustrates the comparative analysis of power source stress under different control strategies.

Under the proposed strategy, the average stress per fuel cell unit is 0.1537, while the power source stress indices for DP, single-layer MPC, and ECMS strategies are 0.0782, 0.242, and 0.2462, respectively. Compared to real-time optimization strategies (single-layer MPC and ECMS), the proposed strategy achieves significant reductions in individual fuel cell stress of 36.5% and 37.5%, respectively. While the DP strategy exhibits the lowest fuel cell stress, this is achieved at the expense of lithium battery longevity. Regarding lithium battery stress performance, the proposed strategy demonstrates a characteristic pattern of "superior macroscopic optimization with elevated microscopic stress levels". This distinctive feature primarily stems from the strategy's optimization mechanism: during the first optimization phase, the lithium battery assumes the critical role of compensating for fuel cell power fluctuations. Although this design reduces the system's overall energy demand, it results in a relatively aggressive battery power allocation strategy, manifested as more frequent charge–discharge transitions, leading to higher power source stress compared to single-layer MPC and ECMS strategies.

However, considering the online real-time optimization nature of the proposed strategy, this stress distribution characteristic actually facilitates synchronized degradation control between fuel cells and lithium batteries. Through the rational allocation of power source stress, the strategy achieves the comprehensive optimization of voyage economics and system durability while maintaining real-time response capabilities. These results further validate the superiority of the proposed strategy in coordinated power source management.

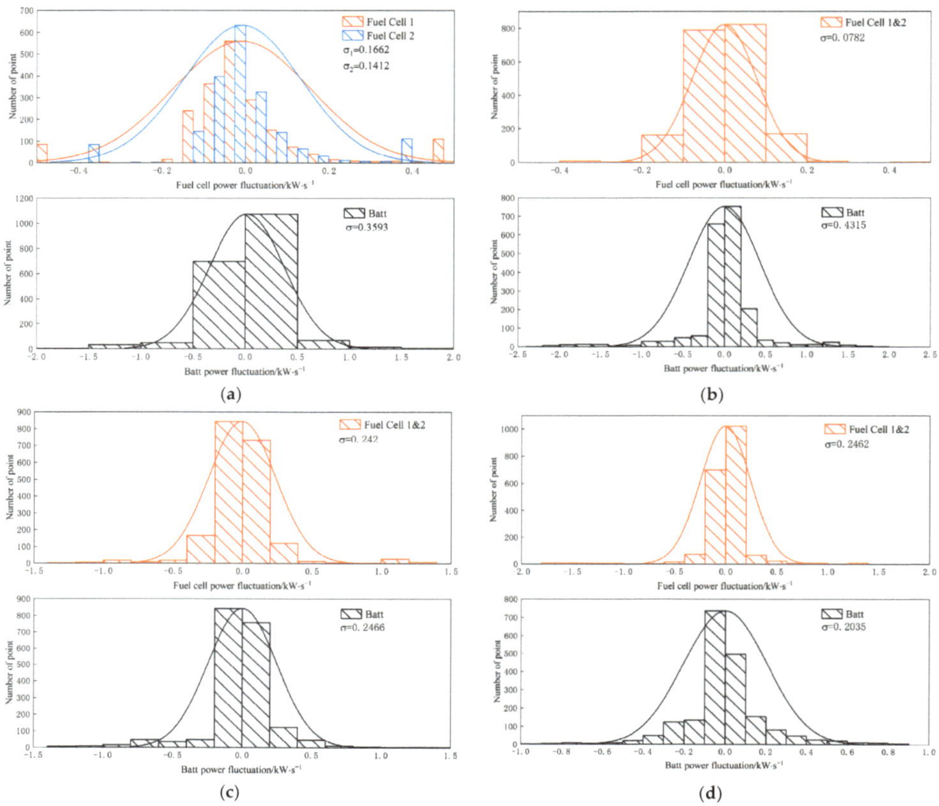

Figure 15. Stress analysis and comparison of power source under different strategies: (**a**) Proposed strategy; (**b**) DP strategy; (**c**) Single-layer MPC strategy; (**d**) ECMS strategy.

4.6. Comprehensive Performance Comparison

Table 4 presents a comprehensive comparison of key performance metrics across different strategies. The proposed strategy achieves comparable equivalent hydrogen consumption to single-layer MPC and ECMS strategies. Although the DP strategy demonstrates the lowest hydrogen consumption, its high computational complexity and poor real-time performance make it impractical for real-world applications. In this trade-off, the proposed strategy opts to restrict fuel cell operation within high-efficiency variable load ranges to suppress degradation, sacrificing some instantaneous efficiency while laying the foundation for overall performance improvement. The proposed strategy significantly reduces fuel cell degradation costs by 30.4% and 8.9% compared to single-layer MPC and ECMS strategies, respectively. However, due to the reduced fuel cell degradation leading to a relatively increased battery load, the battery degradation cost of the proposed strategy is higher than that of single-layer MPC but notably lower than DP and ECMS strategies. From an economic perspective, the proposed strategy demonstrates significant advantages, saving 12.6% and 11.7% in costs compared to single-layer MPC and ECMS strategies, respectively. Although the total cost is 1.6% higher than the DP strategy, the proposed strategy overcomes the limitations of high computational complexity and poor real-time performance associated with DP, providing a more practical solution for real-world applications. The proposed strategy shows far lower power source degradation rate differences compared to other strategies, confirming its more balanced and rational energy distribution. Additionally, the fuel cell high-frequency stress is significantly lower than single-layer MPC and ECMS strategies, contributing to the extended fuel cell life; although the lithium battery high-frequency stress is relatively higher, this trade-off is a necessary compromise to achieve overall low degradation rates, as battery systems possess superior dynamic response capabilities and better tolerance compared to fuel cells.

Table 4. Comparison of the results of each strategy.

Result	Proposed Strategy	DP Strategy	Single-Layer MPC	ECMS Strategy
Equivalent hydrogen consumption (kg)	10.9	10.5	10.9	10.9
Cost of FC degradation (CNY)	251.5	149.9	361.5	275.6
Battery degradation cost (CNY)	38.2	143.5	25.5	104.7
Total cost (CNY)	672.5	661.8	769.4	761.2
Difference in attenuation rate between FC and BAT (%)	0.0019%	0.101%	0.022%	0.054%
FC High frequency stress (σ)	0.1537	0.0782	0.242	0.2462
BAT High frequency stress (σ)	0.3593	0.4315	0.2466	0.2035

Comprehensive analysis indicates that the proposed strategy achieves multi-objective coordinated optimization through its control algorithm: while maintaining energy efficiency, it significantly improves the system's overall economic performance through the reasonable distribution of dynamic loads and balanced component degradation rates while maintaining near-globally optimal control effects.

5. Conclusions

This paper presents a multi-timescale energy management strategy that incorporates coordinated power source degradation. The strategy features an innovative dual-layer optimization architecture that synergistically integrates Model Predictive Control (MPC)'s predictive optimization capabilities with real-time control responsiveness. An attention-enhanced CNN-LSTM prediction model was developed, achieving high-precision forecasting with a power prediction RMSE of merely 3.69 kW, thus providing robust decision support for strategy implementation. The simulation results demonstrate that compared to single-layer MPC strategy and the ECMS strategy, the proposed strategy reduced total voyage costs by 12.6% and 11.7%, respectively, while mitigating fuel cell high-frequency stress by 36.5% and 37.5%, with a performance approximating the globally optimal DP strategy solution. By incorporating differentiated fuel cell degradation and lithium battery aging characteristics, the strategy significantly enhanced dual power source lifetime synchronization, reducing degradation rate disparities by 98.12%, 91.36%, and 96.48% compared to DP, single-layer MPC, and ECMS strategies, respectively. Furthermore, effective power smoothing was achieved through coordinated energy storage system control. This research contributes a solution for fuel cell hybrid vessels that combines optimization performance with real-time operability, demonstrating substantial potential for engineering implementation.

To address the current limitations of this study, future research directions may focus on the following aspects:

- Development of adaptive multi-mode energy management strategies tailored to different voyage phases (departure, cruising, port entry, and berthing), enhancing system operational efficiency throughout the entire voyage.

- Development of adaptive power distribution strategies for multi-stack fuel cell systems that consider differential degradation rates among individual stacks, incorporating real-time performance monitoring and fault diagnosis capabilities to optimize both system efficiency and longevity.
- Development of load power prediction models integrating multidimensional information (vessel speed, heading, and sea conditions), improving prediction accuracy and robustness.

Author Contributions: Conceptualization, X.Z., X.Y., and M.Z.; methodology, X.Z.; software, Y.W. and L.L.; validation, X.Z. and X.Y.; formal analysis, S.N.; investigation, X.Z.; resources, X.Y.; data curation, C.Z.; writing—original draft preparation, X.Z. and X.Y.; writing—review and editing, X.Y. and M.Z.; visualization, L.L. and S.N.; supervision, X.Y. and M.Z.; project administration, X.Y. and M.Z.; funding acquisition, X.Y. and M.Z. All authors have read and agreed to the published version of the manuscript.

Funding: This work is supported by the China National Key Research and Development Project (Grant No: 2023YFB4301704) and the China National Key Laboratory of Electromagnetic Energy Technology Open Fund (Grant No: 61422172220403).

Institutional Review Board Statement: Not applicable.

Informed Consent Statement: Not applicable.

Data Availability Statement: Data will be made available on request.

Conflicts of Interest: The authors declare no conflicts of interest.

References

1. Wang, K.; Yan, X.; Yuan, Y.; Jiang, X.; Lin, X.; Negenborn, R.R. Dynamic Optimization of Ship Energy Efficiency Considering Time-Varying Environmental Factors. *Transp. Res. Part D Transp. Environ.* **2018**, *62*, 685–698. [CrossRef]
2. Pan, P.; Sun, Y.; Yuan, C.; Yan, X.; Tang, X. Research Progress on Ship Power Systems Integrated with New Energy Sources: A Review. *Renew. Sustain. Energy Rev.* **2021**, *144*, 111048. [CrossRef]
3. Qu, J.; Wang, H.; Zhou, J.; Zhang, B. Research on Hybrid Energy Management Strategy for Zero-Carbon Ships. In Proceedings of the 2023 IEEE/IAS Industrial and Commercial Power System Asia (I&CPS Asia), Chongqing, China, 7 July 2023; pp. 837–843.
4. Becherif, M.; Claude, F.; Hervier, T.; Boulon, L. Multi-Stack Fuel Cells Powering a Vehicle. *Energy Procedia* **2015**, *74*, 308–319. [CrossRef]
5. Belmonte, N.; Luetto, C.; Staulo, S.; Rizzi, P.; Baricco, M. Case Studies of Energy Storage with Fuel Cells and Batteries for Stationary and Mobile Applications. *Challenges* **2017**, *8*, 9. [CrossRef]
6. Roh, G.; Kim, H.; Jeon, H.; Yoon, K. Fuel Consumption and CO2 Emission Reductions of Ships Powered by a Fuel-Cell-Based Hybrid Power Source. *JMSE* **2019**, *7*, 230. [CrossRef]
7. Fan, A.; Li, Y.; Liu, H.; Yang, L.; Tian, Z.; Li, Y.; Vladimir, N. Development Trend and Hotspot Analysis of Ship Energy Management. *J. Clean. Prod.* **2023**, *389*, 135899. [CrossRef]
8. Roslan, S.B.; Tay, Z.Y.; Konovessis, D.; Ang, J.H.; Menon, N.V. Rule-Based Control Studies of LNG–Battery Hybrid Tugboat. *JMSE* **2023**, *11*, 1307. [CrossRef]
9. Liu, C.; Wang, Y.; Wang, L.; Chen, Z. Load-Adaptive Real-Time Energy Management Strategy for Battery/Ultracapacitor Hybrid Energy Storage System Using Dynamic Programming Optimization. *J. Power Sources* **2019**, *438*, 227024. [CrossRef]
10. Sun, C.; Sun, F.; He, H. Investigating Adaptive-ECMS with Velocity Forecast Ability for Hybrid Electric Vehicles. *Appl. Energy* **2017**, *185*, 1644–1653. [CrossRef]
11. Zeng, Y.; Cai, Y.; Kou, G.; Gao, W.; Qin, D. Energy Management for Plug-In Hybrid Electric Vehicle Based on Adaptive Simplified-ECMS. *Sustainability* **2018**, *10*, 2060. [CrossRef]
12. Ge, Y.; Zhang, J.; Zhou, K.; Zhu, J.; Wang, Y. Research on Energy Management for Ship Hybrid Power System Based on Adaptive Equivalent Consumption Minimization Strategy. *JMSE* **2023**, *11*, 1271. [CrossRef]
13. Hou, J.; Sun, J.; Hofmann, H. Mitigating Power Fluctuations in Electrical Ship Propulsion Using Model Predictive Control with Hybrid Energy Storage System. In Proceedings of the 2014 American Control Conference, Portland, OR, USA, 4–6 June 2014; pp. 4366–4371.

14. Chen, L.; Gao, D.; Xue, Q. Energy Management Strategy of Hybrid Ships Using Nonlinear Model Predictive Control via a Chaotic Grey Wolf Optimization Algorithm. *JMSE* **2023**, *11*, 1834. [CrossRef]
15. Moghadari, M.; Kandidayeni, M.; Boulon, L.; Chaoui, H. Operating Cost Comparison of a Single-Stack and a Multi-Stack Hybrid Fuel Cell Vehicle Through an Online Hierarchical Strategy. *IEEE Trans. Veh. Technol.* **2023**, *72*, 267–279. [CrossRef]
16. Macias Fernandez, A.; Kandidayeni, M.; Boulon, L.; Chaoui, H. An Adaptive State Machine Based Energy Management Strategy for a Multi-Stack Fuel Cell Hybrid Electric Vehicle. *IEEE Trans. Veh. Technol.* **2020**, *69*, 220–234. [CrossRef]
17. Xie, P.; Asgharian, H.; Guerrero, J.M.; Vasquez, J.C.; Araya, S.S.; Liso, V. A Two-Layer Energy Management System for a Hybrid Electrical Passenger Ship with Multi-PEM Fuel Cell Stack. *Int. J. Hydrogen Energy* **2023**, *50*, 1005–1019. [CrossRef]
18. Kandidayeni, M.; Kelouwani, S.; Boulon, L.; Trovão, J.P. Designing a Hierarchical Energy Management Strategy for a Hybrid Multi-Stack Fuel Cell System. In Proceedings of the 2023 IEEE Vehicle Power and Propulsion Conference (VPPC), Milan, Italy, 24 October 2023; pp. 1–5.
19. Ahmadi, P.; Torabi, S.H.; Afsaneh, H.; Sadegheih, Y.; Ganjehsarabi, H.; Ashjaee, M. The Effects of Driving Patterns and PEM Fuel Cell Degradation on the Lifecycle Assessment of Hydrogen Fuel Cell Vehicles. *Int. J. Hydrogen Energy* **2020**, *45*, 3595–3608. [CrossRef]
20. Gao, J.; Lan, H.; Cheng, P.; Hong, Y.-Y.; Yin, H. Optimal Scheduling of an Electric Propulsion Tugboat Considering Various Operating Conditions and Navigation Uncertainties. *JMSE* **2022**, *10*, 1973. [CrossRef]
21. Fletcher, T.; Thring, R.; Watkinson, M. An Energy Management Strategy to Concurrently Optimise Fuel Consumption & PEM Fuel Cell Lifetime in a Hybrid Vehicle. *Int. J. Hydrogen Energy* **2016**, *41*, 21503–21515. [CrossRef]
22. Tang, L.; Rizzoni, G.; Onori, S. Energy Management Strategy for HEVs Including Battery Life Optimization. *IEEE Trans. Transp. Electrific.* **2015**, *1*, 211–222. [CrossRef]
23. Xu, B.; Oudalov, A.; Ulbig, A.; Andersson, G.; Kirschen, D.S. Modeling of Lithium-Ion Battery Degradation for Cell Life Assessment. *IEEE Trans. Smart Grid* **2018**, *9*, 1131–1140. [CrossRef]
24. Han, X.; Lu, L.; Zheng, Y.; Feng, X.; Li, Z.; Li, J.; Ouyang, M. A Review on the Key Issues of the Lithium Ion Battery Degradation among the Whole Life Cycle. *eTransportation* **2019**, *1*, 100005. [CrossRef]
25. Şahin, M.; Blaabjerg, F.; Sangwongwanich, A. A Comprehensive Review on Supercapacitor Applications and Developments. *Energies* **2022**, *15*, 674. [CrossRef]
26. Ren, P.; Pei, P.; Li, Y.; Wu, Z.; Chen, D.; Huang, S. Degradation Mechanisms of Proton Exchange Membrane Fuel Cell under Typical Automotive Operating Conditions. *Prog. Energy Combust. Sci.* **2020**, *80*, 100859. [CrossRef]
27. Schmittinger, W.; Vahidi, A. A Review of the Main Parameters Influencing Long-Term Performance and Durability of PEM Fuel Cells. *J. Power Sources* **2008**, *180*, 1–14. [CrossRef]
28. Wang, X.; Li, Q.; Wang, T.; Han, Y.; Chen, W. Optimized Energy Management Strategy Based on SQP Algorithm for PEMFC Hybrid Locomotive. In Proceedings of the 2019 IEEE Transportation Electrification Conference and Expo, Asia-Pacific (ITEC Asia-Pacific), Seogwipo-si, Republic of Korea, 8–10 May 2019; pp. 1–5.
29. Song, K.; Ding, Y.; Hu, X.; Xu, H.; Wang, Y.; Cao, J. Degradation Adaptive Energy Management Strategy Using Fuel Cell State-of-Health for Fuel Economy Improvement of Hybrid Electric Vehicle. *Appl. Energy* **2021**, *285*, 116413. [CrossRef]
30. Ghaderi, R.; Kandidayeni, M.; Soleymani, M.; Boulon, L.; Trovao, J.P.F. Online Health-Conscious Energy Management Strategy for a Hybrid Multi-Stack Fuel Cell Vehicle Based on Game Theory. *IEEE Trans. Veh. Technol.* **2022**, *71*, 5704–5714. [CrossRef]
31. Wan, A.; Chang, Q.; AL-Bukhaiti, K.; He, J. Short-Term Power Load Forecasting for Combined Heat and Power Using CNN-LSTM Enhanced by Attention Mechanism. *Energy* **2023**, *282*, 128274. [CrossRef]
32. Darby, M.L.; Nikolaou, M. MPC: Current Practice and Challenges. *Control Eng. Pract.* **2012**, *20*, 328–342. [CrossRef]
33. Schwenzer, M.; Ay, M.; Bergs, T.; Abel, D. Review on Model Predictive Control: An Engineering Perspective. *Int. J. Adv. Manuf. Technol.* **2021**, *117*, 1327–1349. [CrossRef]
34. Truong, H.V.A.; Do, T.C.; Dang, T.D. Enhancing Efficiency in Hybrid Marine Vessels through a Multi-Layer Optimization Energy Management System. *JMSE* **2024**, *12*, 1295. [CrossRef]
35. Lu, D. Online Optimization of Energy Management Strategy for FCV Control Parameters Considering Dual Power Source Lifespan Decay Synergy. *Appl. Energy* **2023**, *348*, 121516. [CrossRef]
36. Wang, T.; Li, Q.; Wang, X.; Chen, W.; Breaz, E.; Gao, F. A Power Allocation Method for Multistack PEMFC System Considering Fuel Cell Performance Consistency. *IEEE Trans. Ind. Appl.* **2020**, *56*, 5340–5351. [CrossRef]

Journal of
*Marine Science
and Engineering*

MDPI

Article

Prior Knowledge-Based Two-Layer Energy Management Strategy for Fuel Cell Ship Hybrid Power System

Lin Liu, Xiangguo Yang, Xin Li *, Xingwei Zhou, Yufan Wang, Telu Tang, Qijia Song and Yifan Liu

The School of Naval Architecture, Ocean and Energy Power Engineering, Wuhan University of Technology, Wuhan 430063, China; 299437@whut.edu.cn (L.L.); yxglyr@whut.edu.cn (X.Y.); 345550@whut.edu.cn (X.Z.); wangyuf@whut.edu.cn (Y.W.); 331552@whut.edu.cn (T.T.); 299486@whut.edu.cn (Q.S.); l2023201239@whut.edu.cn (Y.L.)

* Correspondence: xinli1005@whut.edu.cn

Abstract: Implementing energy management is crucial in the fuel cell and battery or supercapacitor hybrid energy systems of ships. Traditional real-time energy management strategies often struggle to adapt to complex operating conditions; to address this issue and mitigate fuel cell fluctuations during real-time operations while extending the lifespan of lithium-ion batteries, this paper proposes a two-layer energy management system (EMS) based on prior knowledge of ship operation. In the first layer of the EMS, which operates offline, dynamic programming (DP) and low-pass filtering (LPF) are used to allocate power optimally for different typical ship operating conditions. Distribution results are then used to train an SSA-BP neural network, creating an offline strategy library. In the second layer, operating in real-time, the current load power is input into a support vector machine (SVM) to classify the current operating condition. The corresponding strategy from the offline library is then selected and used to provide energy distribution recommendations based on the real-time load and the state of charge (SOC) of the lithium-ion batteries and supercapacitors. The proposed EMS was validated using different ship load cycles. The results demonstrate that, compared to second-order filtering-based real-time energy management strategies, the proposed method reduces fuel cell power fluctuations by 44% and decreases lithium-ion battery degradation by 28%. Furthermore, the simulation results closely align with the offline optimization results, indicating that the proposed strategy achieves near-optimal energy management in real-time ship operations with minimal computational overhead.

Keywords: prior knowledge; hybrid power system; energy management strategy; fuel cell; optimal analysis

1. Introduction

With the rapid growth of global ocean development, the continuous increases in the types, numbers, and sizes of ships has resulted in severe marine pollution. Consequently, international organizations such as the International Maritime Organization have implemented low-carbon strategies and greenhouse gas (GHG) reduction initiatives [1–4]. Shipboard power grids face several challenges, including a limited capacity, complex operating conditions, significant load fluctuations, and instability caused by shore power connections [5–9]. Due to the greenhouse gas emissions and environmental pollution associated with traditional fossil-fueled ships, the development of hydrogen fuel cell systems has garnered significant attention. However, fuel cells exhibit slow dynamic responses to load fluctuations, making it difficult to adapt to the complex operating conditions of ships. As a result, hybrid energy storage systems (HESSs) incorporating batteries and supercapacitors are required to stabilize the power grid and enhance grid stability [10–12]. An efficient EMS is critical for coordinating power flows, reducing operational costs, preventing stability issues, and ensuring safe operation [13–16].

Currently, EMS research can be categorized into three main types: rule-based control (RBC) strategies, optimization-based control strategies, and artificial intelligence-based strategies. For rule-based control strategies, the rules rely on expert experience or mathematical calculations and include methods such as rule-based control [17,18], fuzzy logic control [19], low-pass filter-based control [20,21], and wavelet transform-based control [22], among others. These strategies are computationally efficient, intuitive, easy to implement, robust, and effective in deterministic scenarios. However, they lack self-learning and adaptive capabilities, making them unsuitable for complex operating conditions.

The second category comprises optimization-based strategies, which are more complex but also more effective. They use objective functions to evaluate hybrid power system performance, such as minimizing costs or maximizing lifespan. Examples include dynamic programming [23], model predictive control (MPC) [24], and particle swarm optimization (PSO) [25]. DP is an algorithm used to find the global optimal solution to a problem. Given known operating conditions, it can solve for the optimal result of the objective function. For instance, W. Tang et al. [26] propose a hierarchical EMS based on adaptive dynamic programming (ADP); the proposed EMS ensures real-time planning speed and good following performance while reducing energy consumption. Similarly, N. Xu et al. [27] propose a global optimization framework of "information layer–physical layer–energy layer–dynamic programming" (IPE-DP), which can realize the unity of different information scenarios, different vehicle configurations, and energy conversions. Although DP can achieve offline optimization, its computation time is relatively long, and when the discretized grid is too fine, it can lead to the "curse of dimensionality" [28]. Therefore, it is necessary to apply the offline optimal strategy in real time by combining other methods.

The third category is artificial intelligence (AI)-based strategies, which leverage advances in AI technology for energy management control. These include methods like neural networks [29], reinforcement learning [30], and support vector machines [31]. AI algorithms have been widely applied to vehicle energy management and are increasingly used in hybrid power systems [32,33]. For example, C. Qi et al. [34] propose a multi-agent reinforcement learning algorithm that incorporates vehicle operation features, enabling generalization across different vehicle models. Reference [35] developed a deep reinforcement learning-based framework for plug-in hybrid electric vehicles (PHEVs), combining offline training with online control. By addressing continuous state and action spaces through Lagrange relaxation, the system effectively learned from real driving data, resolving safety and overestimation issues while achieving an optimal energy management system.

To effectively adapt to highly dynamic conditions and ensure fuel-efficient operations while meeting instantaneous load demands, many researchers have combined multiple approaches. WU et al. [36] propose a predictive EMS using multiple neural networks, where DP results trained a pattern recognition network, and a recurrent neural network was applied for online dynamic estimation. Yang, N. et al. [37] propose a real-time EMS for hybrid electric vehicles by integrating reinforcement learning into an MPC framework, addressing RL's drawbacks of a long learning time and poor adaptability, and significantly improving MPC's real-time performance. Waqar Waheed and Qingshan Xu [38] demonstrate and incorporate a deep learning neural network model with time series analysis and feature selection to forecast the complex and variable hourly load demand and involve a comprehensive comparison between the DNN approach and other alternatives for short-term load forecasting.

After reviewing the current efforts on the development of onboard energy management systems, it becomes evident that it is inevitable to make trade-offs among multiple factors. Considering the unique characteristics of the onboard loads, long-term regularity and short-term unpredictability, both aspects play pivotal roles in optimizing power-sharing decisions. The dynamic nature of real-time power demands has remained a challenge.

To solve the above problems, a two-layer EMS based on prior knowledge of the ship is proposed. This strategy demonstrates strong adaptability to ship loads and low computational costs, enabling optimal power distribution in real time. The main contributions of this paper are as follows:

- Proposing a two-layer EMS integrating historical operating conditions—this EMS comprises an offline optimization layer and a real-time energy management layer.
- Leveraging historical operating characteristics—using neural networks to extract optimal allocation results, the strategy provides near-optimal management strategies in real time with low computational costs.
- Eliminating the need for a manual analysis of offline optimal strategies—the proposed EMS only requires typical ship operating conditions as input for training, eliminating the need for a manual analysis of the offline optimal strategy characteristics. This allows the system to derive corresponding control strategies for different cruising missions and complex conditions, making it easily applicable to the energy management systems of other ships.

The remainder of the paper is organized as follows: Section 2 presents the configuration of the ship's integrated power system and provides a model for the hybrid energy storage system (HESS). Section 3 elaborate on the offline optimization strategy of the first layer and the real-time energy management method of the second layer, respectively. Section 4 validates the proposed strategy by generating random load conditions and compares the performance of TLPF, LPF-DP, and the proposed two-layer energy management model (TLM). Finally, Section 5 provides the conclusions.

2. System Modeling and Configuration

The ship studied in this paper is a hydrogen-powered vessel designed for transportation, patrolling, and emergency tasks in a specific reservoir area. This vessel uses hydrogen fuel cells as its primary power source and adopts a twin rudder–propeller propulsion system. It is characterized by zero emissions, silent operation, and comfort. Due to its relatively fixed navigation routes and operational tasks, the ship's load exhibits a consistent pattern. The following subsections provide details of the corresponding systems.

2.1. Integrated DC Power System for Ships

The partially simplified structure of the DC grid of a hydrogen-powered ship after modification is shown in Figure 1. The main components include the fuel cell stack, hybrid energy storage system, power converters, thrusters, various auxiliary systems, and regional power supply networks.

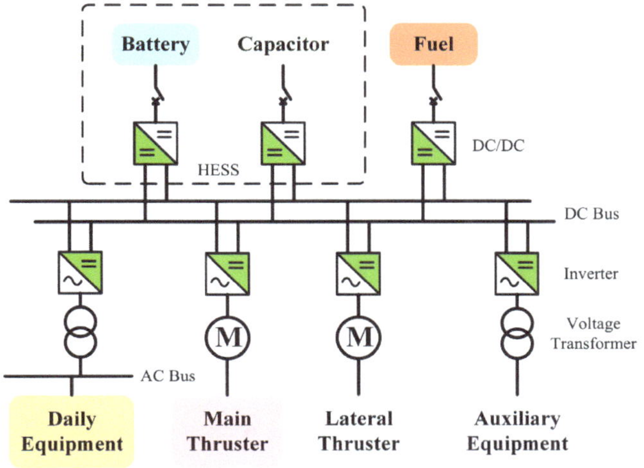

Figure 1. Partial structural diagram of ship DC grid.

2.2. System Modeling and Parameters

The hydrogen fuel cell system mainly consists of the fuel cell power generation module, hydrogen supply system, auxiliary devices, control and energy management devices, and emergency shutdown device (ESD), among others. The ship is equipped with 12 sets of 70 kW fuel cell power generation modules, which convert the electrochemical reaction of hydrogen and oxygen into electrical and thermal energy through the fuel cell stack inside the module. The main technical parameters of the fuel cell stack are shown in Table 1.

Table 1. Fuel cell parameters.

Parameter	Value
Quantity (sets)	12
Rated power (kW)	70
Size (mm)	1120 × 580 × 625
Weight (kg)	≤300

The ship's power is provided by the fuel cell system, battery, and supercapacitor. The power expression can be described as follows:

$$P_{load} = P_{FC,DC} + P_{BAT,DC} + P_{SC,DC} \tag{1}$$

The $LiFePO_4$ battery, due to its high voltage, high specific capacity, and long cycle life, is a good choice for the ship's battery pack. The main parameters of the $LiFePO_4$ battery used in this study are shown in Table 2 ($V_{bat,cell}$ is the nominal voltage). The Rint model is an equivalent circuit model with a simple structure and fewer parameters that can effectively describe the basic situation of the battery's terminal voltage change with current due to internal resistance during charging and discharging, and is widely used in research. The battery pack capacity C_{bat} (Ah), open circuit voltage $V_{OC,bat}$, and internal resistance R_{bat} can be calculated as follows, where $N_{bat,s}$ and $N_{bat,p}$ represent the number of series and parallel cells in the battery system, and $R_{cell,bat}$ is the internal resistance of the battery cell:

$$\begin{cases} C_{bat} = N_{bat,p} C_{bat,cell} \\ V_{OC,bat} = N_{bat,s} V_{bat,cell} \\ R_{bat} = N_{bat,s} R_{cell,bat} / N_{bat,p} \end{cases} \quad (2)$$

Table 2. Parameters of lithium–iron–phosphate battery cells.

Parameter	Value
$V_{bat,cell}$ (V)	3.2
$C_{bat,cell}$ (Ah)	60
Cell stored energy (kW∗h)	0.192
Cell mass (kg)	~2

Similar to lithium-ion batteries, the capacitance C_{sc}, open-circuit voltage $V_{OC,sc}$, and internal resistance R_{sc} of the supercapacitor pack can be calculated as follows, where $N_{sc,s}$ and $N_{sc,p}$ represent the numbers of series and parallel connections, and R_{cell} is the internal resistance of the supercapacitor cell:

$$\begin{cases} C_{sc} = N_{sc,p} C_{sc,cell} \\ V_{OC,sc} = N_{sc,s} V_{OC,sc,cell} \\ R_{sc} = N_{sc,s} R_{cell} / N_{sc,p} \end{cases} \quad (3)$$

The SOC of the supercapacitor is defined as follows:

$$SOC_{SC} = \frac{V_{OC,SC}}{V_{SC,MAX}} \quad (4)$$

where $V_{OC,SC}$ is the current voltage of the supercapacitor and $V_{SC,MAX}$ is the rated voltage of the supercapacitor.

The size of the composite energy storage system in this study is based on reference [39] and adjusted according to the configuration of the hydrogen-powered ship in this research. The numbers of series and parallel connections for the battery are 160 and 2, respectively, while for the supercapacitor, the numbers of series and parallel connections are 25 and 5, respectively. The charge and discharge efficiency is set at 98%.

3. Methods

3.1. Typical Operating Conditions of the Ship

During different operational phases and mission scenarios, a ship's load exhibits diverse characteristics. Accurately extracting the typical operating conditions of a ship is crucial for performance evaluation, operational optimization, and equipment maintenance. By analyzing historical load data from the ship's operations, the typical operating conditions can be effectively summarized. Before extracting these typical conditions, performing essential preprocessing of the raw data is necessary to remove anomalies and missing values caused by sensor failures, communication interference, or other factors, ensuring data quality and usability.

Key features such as the mean, variance, and variance-to-mean ratio of power fluctuations are selected to characterize the ship's operating conditions. The mean represents the average power level, the variance measures the dispersion relative to the mean, and the variance-to-mean ratio reflects the temporal variation of power, calculated as the difference between each second's power value and that of the previous second.

The K-Means clustering algorithm, a commonly used partition-based clustering method, is employed due to its simplicity, efficiency, and ease of implementation. Using K-Means clustering, the ship's operating load data are classified, extracting five typical operating conditions. Figure 2a–e illustrates these conditions: complex operating condition, stepwise power increase condition, stepwise power decrease condition, high-power steady-state operation, and low-power steady-state operation.

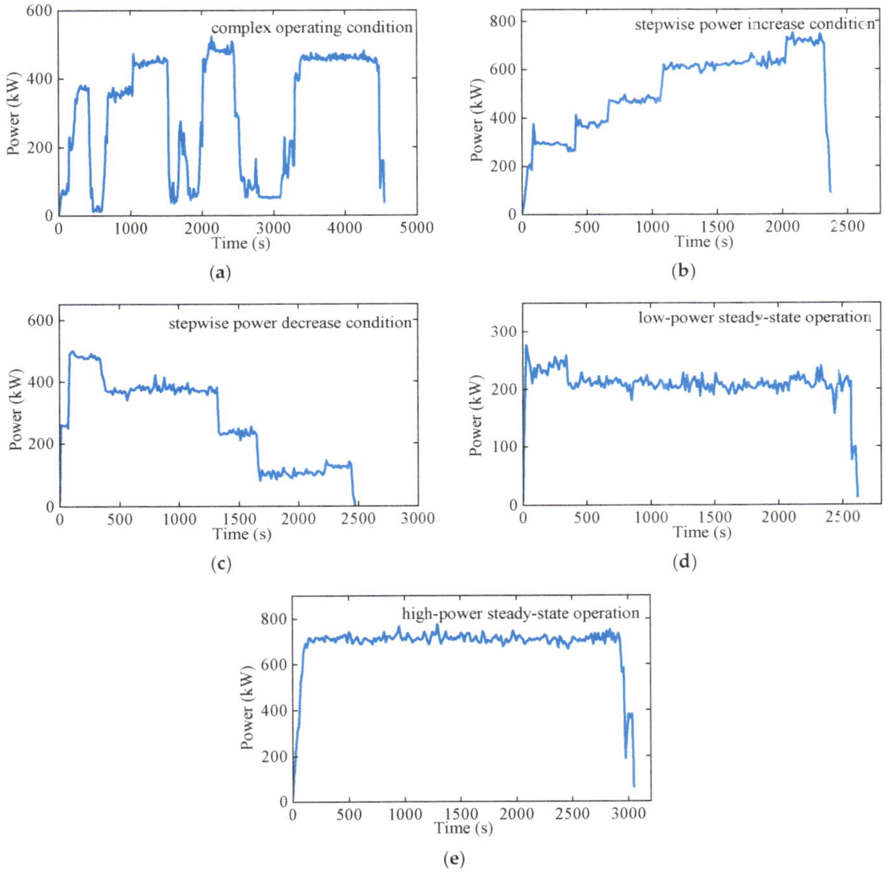

Figure 2. Typical operating conditions of the ship: (**a**) complex operating condition; (**b**) stepwise power increase condition; (**c**) stepwise power decrease condition; (**d**) high-power steady-state operation; and (**e**) low-power steady-state operation.

The vessel studied in this paper is the "Three Gorges Hydrogen Vessel No. 1" in Chi-na. The load data shown in Figure 2, along with the corresponding ship configuration, are sourced from this vessel.

3.2. *Offline Energy Management Strategy*

3.2.1. Optimization Problem

The objective of the energy management system optimization in this study is to minimize the average fluctuations of the fuel cell and the degradation of the lithium-ion battery pack. Additionally, a penalty function is introduced to limit the ripple of the lithium battery pack, thereby improving the power supply quality of the grid. Severe fluctuations in load power transmitted to the fuel cell stack can directly cause output voltage instability, accelerate fuel cell degradation, and reduce the efficiency of the fuel cell stack. Power fluctuations in the fuel cell stack are often accompanied by uneven changes in heat generation rates, leading to thermal management issues. Therefore, the power fluctuation of the fuel cell stack is set as one of the objective functions. The expression for the average fuel cell fluctuation can be described as follows [40]:

$$f_1 = \frac{\sum_{t=0}^{T} (P_{FC,DC}(t+1) - P_{FC,DC}(t))^2}{T} \tag{5}$$

where $P_{FC,DC}(t)$ is the output power of the fuel cell stack at time t, and T is the running time.

The degradation of lithium battery life is reflected in its actual usable capacity being less than the initial capacity. When the actual capacity drops to 80% of the initial capacity, the battery is considered to have reached the end of its life. Improper cycling conditions, such as frequent charging and discharging or high-current charging and discharging, can accelerate battery degradation and even lead to failure. Therefore, the design of energy management strategies must consider the cost of lithium battery life degradation. The cost model for lithium battery life degradation can be expressed as follows [41]:

$$f_2 = \frac{C_B \cdot E_{bat}}{2L(D) \cdot D \cdot E_{bat} \cdot \eta_{ch} \cdot \eta_{dis}} \cdot P_{dis} \cdot \Delta t \tag{6}$$

In the above equation, C_B represents the average cost of batteries (USD/kWh), the factor of 2 in the denominator represents one complete charge–discharge cycle, and E_{bat} is the energy released by the battery. $L(D)$ is the average cycle number of the lithium battery as a function of the depth of discharge, D represents the depth of discharge (DOD), and η_{ch} and η_{dis} are the charging and discharging efficiencies of the lithium battery, respectively. P_{dis} is the discharge power, and Δt is the discharge time. The charge and discharge efficiency is set at 98% and C_B is 500 (USD/kW·h).

The relationship between the typical average cycle number of a lithium battery and its depth of discharge can be modeled as follows [41]:

$$L(D) = a \cdot D^{-b} \tag{7}$$

In the above equation, a and b are the parameters obtained from fitting the life cycle curve.

When the battery operates within the rated current range, the cycle number is primarily related to the DOD. Here, the DOD is defined as the proportion of energy released during a single discharge event relative to the total capacity of the battery, expressed as follows [42]:

$$D = \frac{E_{dis}}{E_{bat}} = \frac{P_{dis}(t) \cdot \Delta t}{E_{bat}} \tag{8}$$

Substituting Equations (7) and (8) into Equation (6) and simplifying yields the following:

$$f_2 = \frac{C_B}{2a \cdot E_{bat}^{b-1} \cdot \eta_{ch} \cdot \eta_{dis}} \cdot (P_{dis} \cdot \Delta t)^b \tag{9}$$

To ensure that the fuel cell stack and the EMS operate under normal conditions, the corresponding constraints need to be applied:

$$\begin{cases} SOC_{bat} \in [SOC_{bat,\min}, SOC_{bat,\max}] \\ SOC_{sc} \in [SOC_{sc,\min}, SOC_{sc,\max}] \\ P_{FC,DC} \in [P_{FC,\min}, P_{FC,\max}] \\ P_{sc,DC} \in [P_{sc,\min}, P_{sc,\max}] \\ P_{bat} \in [P_{bat,\min}, P_{bat,\max}] \end{cases} \quad (10)$$

3.2.2. DP-LPF-Based Power Splitting Strategy

When a ship operates under different conditions, its load exhibits varying fluctuation characteristics. Under stable conditions, the load typically consists of low-frequency fluctuations with smaller variations, while under complex conditions, the power changes rapidly and exhibits larger high-frequency fluctuations. Therefore, a power allocation method based on the nature of power fluctuations is more consistent with the characteristics of shiploads. Among these methods, the low-pass filtering-based power allocation method is the most studied and widely used. This approach divides the load into high-frequency and low-frequency components through a low-pass filter, where the high-frequency power is handled by the Hybrid Energy Storage System (HESS), and the low-frequency power is managed by the fuel cell. This reduces power fluctuations in the fuel cell stack. The control strategy is illustrated in Figure 3.

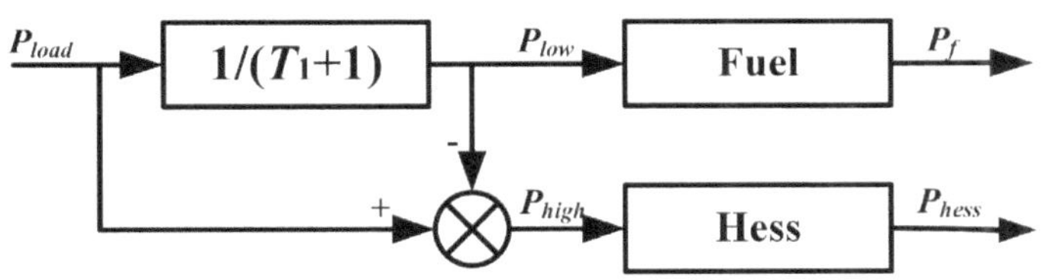

Figure 3. Low−pass filtering power allocation strategy.

Dynamic programming can determine the optimal control input by designing an optimization objective function. For the battery/supercapacitor HESS, the optimization objective in this work is to minimize battery capacity degradation while limiting the power ripple of the lithium battery. The load power of the hydrogen-powered ship is discretized with a 1-s step size, and the state variables and control variables are discretized into discrete arrays. The DP algorithm is then used to solve the objective function over the time steps. The state variables in this paper are $[SOC_{sc}, SOC_{bat}]$, and the control variables are $[P_{sc,DC}, P_{bat,DC}]$. The initial values are set to $SOC_{sc} = 0.7$ and $SOC_{bat} = 0.7$. Based on the typical operating load of the ship, the DP algorithm adopts a backward optimization approach. After discretization, the algorithm starts from the last step's state variables and searches for the optimal solution at each step through the control variables, thereby determining the optimal decision for each step. A forward solution is then performed to obtain the overall optimal strategy for the optimization process. The goal is to minimize lithium battery capacity degradation while incorporating a penalty function to limit the power ripple of the lithium battery.

Based on the above strategy, the optimal power allocation modes are provided for five typical operating conditions of the ship. The detailed allocation results are shown in Figure 4. These results provide effective guidance strategies for the operation of the power system.

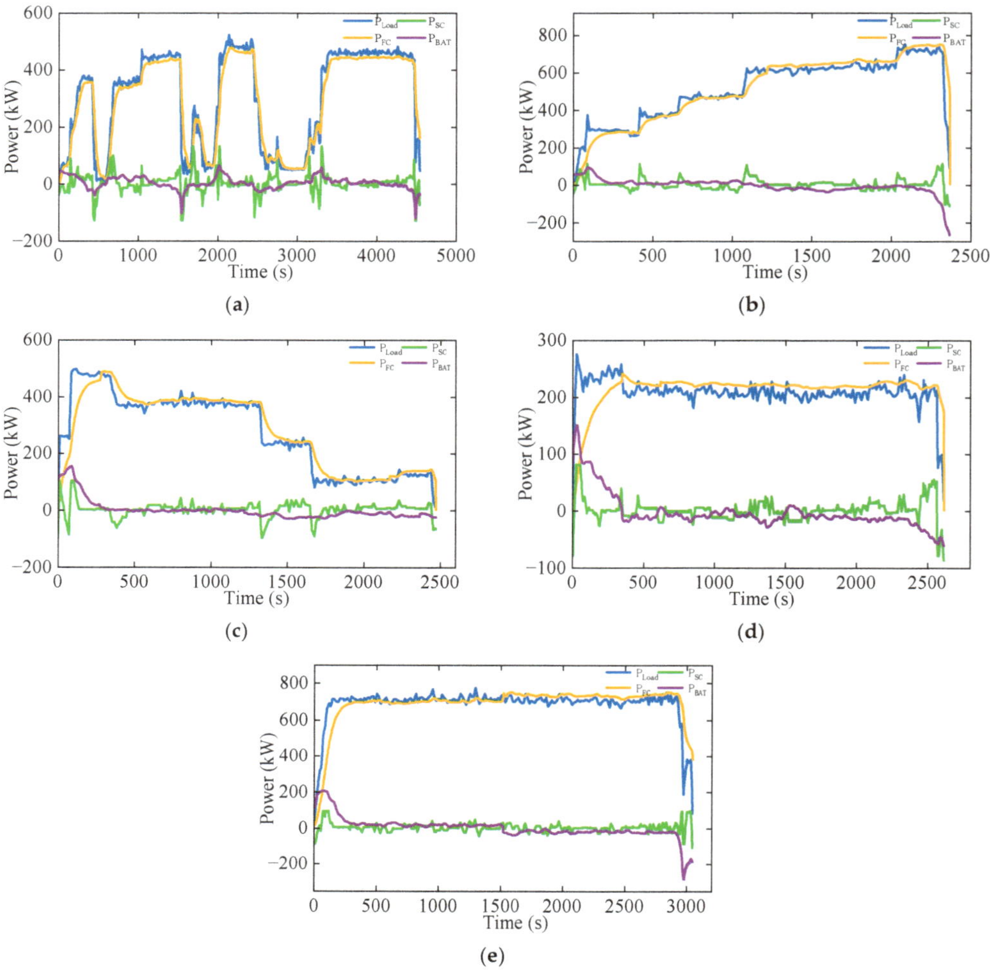

Figure 4. Offline power allocation under five typical operating conditions: (**a**) power allocation results under operating condition 1; (**b**) power allocation results under operating condition 2; (**c**) power allocation results under operating condition 3; (**d**) power allocation results under operating condition 4; and (**e**) power allocation results under operating condition 5.

1. The fuel cell handles the main power variations and strives to reduce power fluctuations to minimize grid instability, thereby improving the power supply quality and safety.
2. The supercapacitor manages the high-frequency components of the remaining power filtered by the low-pass filter, handling short-term high-frequency power fluctuations.
3. The lithium battery provides power to compensate for higher low-frequency power demands while ensuring that the SOC of the lithium-ion battery returns to its initial level as much as possible before and after use.

3.3. Proposed Real-Time Two-Layer Energy Management Strategy

In this paper, a two-layer energy management system based on prior knowledge is proposed to reduce power fluctuations in the fuel cell stack, enhance the stability and safety of the ship's power grid, reduce the probability of power outages, and consider the service life of the lithium-ion battery pack. The overall structure of the proposed EMS is shown in Figure 5. This figure serves as the core representation of the research methodology employed in this study, encapsulating the key elements and their interrelationships within the EMS framework.

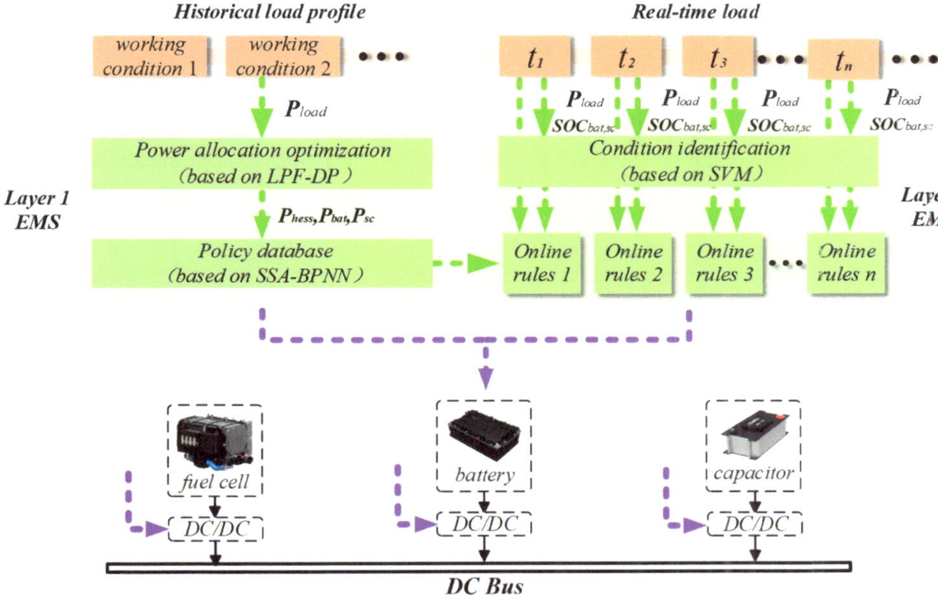

Figure 5. The EMS framework proposed in this article.

The EMS of this study focuses on a hybrid power system integrating fuel cells with a hybrid energy storage system, including lithium batteries and supercapacitors. The fuel cell stack serves as the primary power source, supplying the majority of the energy to the ship's power grid. Meanwhile, lithium-ion batteries and supercapacitors function as auxiliary power sources to stabilize power grid performance, reduce the load on the primary power source, and enhance energy recovery and utilization. The proposed method is designed for application in small- to medium-sized ships equipped with electric propulsion systems.

The first layer leverages prior knowledge of the ship's operating conditions. It begins with the offline computation of the power distribution results for the fuel cell stack and the hybrid energy storage system under typical operating scenarios. An improved BP neural network is employed to learn the optimal power distribution strategies corresponding to various typical conditions, thereby constructing an offline-trained strategy repository.

The second layer employs a support vector machine model trained using the characteristic parameters of typical ship operating conditions. Real-time load data are processed through the SVM model to identify the current operating condition of the ship. Based on the identified operating condition, the corresponding optimal power distribution strategy is retrieved from the offline strategy repository. Additionally, the SOC of the lithium battery and supercapacitor is updated in real time to ensure the reliable operation of the energy storage system.

3.3.1. First-Layer Energy Management System

Since the DP algorithm requires prior knowledge of the entire driving condition's speed profile and its solving process suffers from the curse of dimensionality, which requires significant computation time, it is challenging to apply this method to real-time ship controllers. Therefore, extracting the optimal energy management strategy from the offline power allocation results is necessary.

The backpropagation neural network, also known as the error backpropagation neural network, is a multilayer feedforward neural network trained using the error backpropagation algorithm. It is one of the most widely used neural network models and possesses powerful nonlinear mapping capabilities, adaptability, and generalization abilities. It can be applied to various fields, such as function approximation, pattern recognition, classification, and data compression. Therefore, the BPNN can be trained in advance using offline power allocation results, allowing it to provide timely power distribution outcomes during real-time control based on the current ship operating conditions.

The working principle of the BPNN involves inputting data into the network through the input layer, processing it sequentially in the hidden layer based on the connection weights between neurons, and finally reaching the output layer to obtain the output result. Assume the network has n inputs, m outputs, and s neurons in the hidden layer. The output of the hidden layer is b_j, the threshold of the hidden layer is θ_j, and the threshold of the output layer is θ_k. The transfer function of the hidden layer is F_1, and that of the output layer is F_2. The weight between the input layer and the hidden layer is w_{ij}, and the weight between the hidden layer and the output layer is w_{jk}. The output of the grid y_k is then obtained, and the desired output is t_k. The output of the jth neuron in the hidden layer can be expressed as follows [43]:

$$b_j = F_1\left(\sum_{i=1}^{n} w_{ij}x_i - \theta_j\right) \qquad (i = 1,2,\ldots,n; j = 1,2,\ldots,s) \tag{11}$$

The output layer can be expressed as follows:

$$y_k = F_2\left(\sum_{j=1}^{s} w_{jk}b_j - \theta_k\right) \qquad (j = 1,2,\ldots,n; k = 1,2,\ldots,m) \tag{12}$$

The error function can be expressed as follows:

$$e = \sum_{k=1}^{m} (t_k - y_k)^2 \tag{13}$$

The BP neural network reduces network errors by continuously adjusting the weights. Weight training generally uses the gradient descent method, which is prone to falling into local optima and unable to achieve global optimization. The Sparrow Search Algorithm (SSA) has strong global search capabilities and does not rely on gradient information. It simulates the foraging and anti-predation behaviors of sparrow populations to explore the entire solution space.

By introducing SSA, the globally minimal weights for BP neural network errors can be obtained, improving the overall network performance. The SSA is set with an initial population of 30, a maximum number of 50 evolution iterations, a safety value of 0.6, and a discoverer ratio of 0.7.

To apply the optimization results of DP, the SSA-BP neural network is used to learn the optimal power allocation data obtained from DP. The optimization data of five typical operating conditions are used as samples to train corresponding neural network models, resulting in control strategies suitable for the five typical conditions. The initial, SOC_{sc} and SOC_{bat}, and P_{load} of the ship's typical operating conditions are taken as inputs for the neural network, while the outputs are the supercapacitor power and lithium battery power. The number of hidden neurons is determined through iterative optimization.

To verify the effectiveness of the SSA-BP neural network proposed in this paper, the BP neural network and NFN [29] were used to learn the optimal power allocation data obtained from DP, and the power allocation results output by different neural networks were compared.

Figure 6 shows the offline optimization results and three different neural network outputs under complex ship operating conditions. From Figure 6a, b, it can be concluded that compared to other neural networks, the proposed SSA-BP neural network has the highest similarity with the results obtained from offline computation. Figure 6c, d indicates that the average absolute error and root mean square error between the output results of SSA-BP and offline results are the smallest. Table 3 compares the two algorithms. The computation time for SSA-BP is only 49.8 s, while that for LPF-DP is 21,092 s, indicating that SSA-BP has a response time of just 10.9 ms, which meets the requirements for real-time power allocation on ships. It significantly improves computation speed, with the fuel cell fluctuation rate f_1 increasing by only 11% and the lithium battery degradation cost f_2 increasing by just 9.9%.

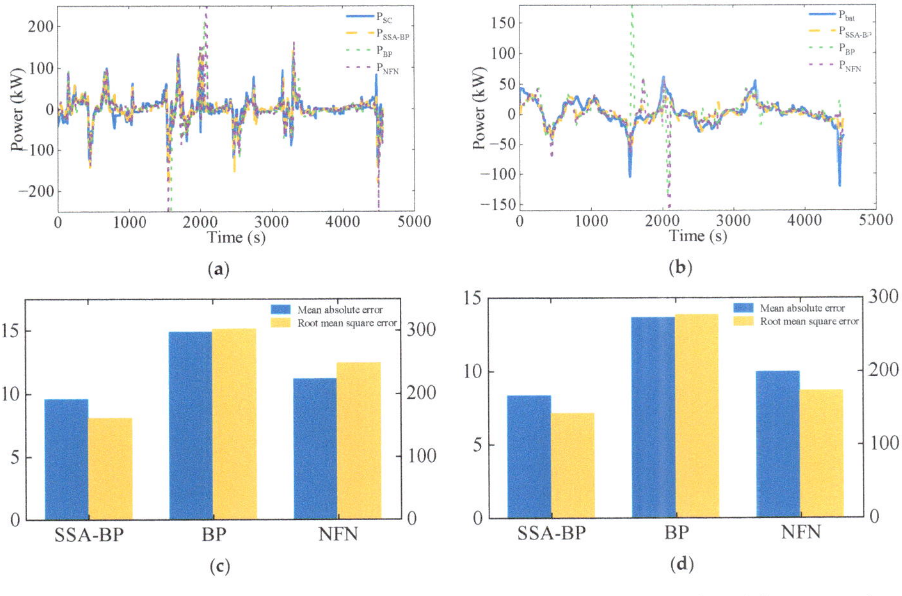

Figure 6. Comparison of offline optimization results and training results for three different neural networks. (**a**) Comparison of the supercapacitor power; (**b**) comparison of the power of lithium-ion batteries; (**c**) comparison of the average absolute error and root mean square error of the supercapacitor output from different neural networks; (**d**) comparison of the average absolute error and root mean square error of lithium-ion battery power output from different neural networks.

Table 3. Comparison of the results for the parameters.

Parameter	LPF-DP	SSA-BPNN
f_1 $(\Delta P^2/s)$	2.7554	3.0584
f_2 (USD)	1.084	1.192
Time (S)	21,092	49.8

The regression relationship between the actual power and estimated power is shown in Figure 7. The R values of different datasets are all above 0.93, indicating a fitting accuracy of over 93%. This demonstrates that the neural network optimized by SSA has a high degree of fitting to the offline power allocation results, allowing it to derive near-optimal strategies during real-time power allocation that closely match the offline calculations.

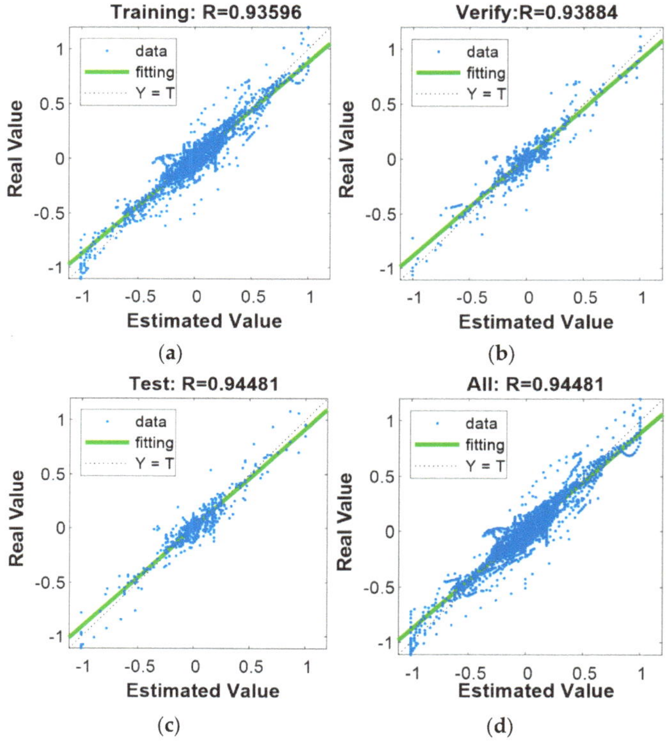

Figure 7. Neural network fitting regression diagrams. (**a**) Fits of the training set and dataset; (**b**) fits of the verification set and dataset; (**c**) fits of the test set and dataset; and (**d**) fits of all sets and the dataset.

3.3.2. Second-Layer Energy Management System

The second-layer energy management system is built upon the offline strategy repository established in the first layer. First, the characteristic parameters of typical operating conditions are input into the SVM model for training. During the real-time operation of the ship, the real-time load is processed through the SVM model to identify the current operating condition of the ship. Based on the identified condition, the optimal power distribution strategy is retrieved from the offline strategy repository created in the first layer. Furthermore, the SOC of the lithium battery and supercapacitor is updated in real time to ensure the stable and efficient operation of the energy storage system

The SVM model is a widely used algorithm in the field of machine learning due to its simplicity and broad applicability. Based on the fundamental strategy of supervised learning, it enables the learning machine to pursue structural risk minimization, thereby improving the generalization ability while achieving a minimal confidence range and empirical risk. With its excellent performance in handling classification and regression problems, SVM has shown significant potential in the field of ship operating condition identification.

Its core idea is to find an optimal hyperplane in the feature space that separates samples from different classes as accurately as possible while maximizing the margin between the two classes and the hyperplane. For a given sample S, the hyperplane is defined as follows [44]:

$$g(x) = w^T S + b = 0 \tag{14}$$

where w is the coefficient and b is the intercept. Using the above formula, the distance from each point to the classification hyperplane can be calculated. Suppose there is a point $O(X_1, X_2, \ldots, X_N)$ in the dataset, where X_i is the i variable, and n is the dimension. The distance calculation formula is as follows:

$$d = \frac{|w^T S + b|}{||w||} \tag{15}$$

where $S = (X_1, X_2, \ldots, X_N)^T$ and $||w||$ is the norm of the hyperplane. All support vectors are searched, and the distances are calculated. The hyperplanes corresponding to all eligible distances are compared, and the hyperplane corresponding to the maximum distance, i.e., when $2/||w||$ is maximized, is retained for S and b.

For linearly inseparable datasets, support vector machines introduce a kernel function to map the data from the original feature space to a higher-dimensional feature space, making the data linearly separable in this new space. Commonly used kernel functions include the linear kernel, polynomial kernel, and Gaussian Radial Basis Function (RBF) kernel.

Based on the characteristics of the ship's operating condition identification problem, an extended method of multi-class SVM is adopted. The load data from five typical operating conditions of the ship are input into the SVM for training, with an 8:2 split between the training set and the test set.

During actual ship operation, the load exhibits extremely complex variation patterns. On one hand, the load on the power system varies significantly across different operational phases, such as starting, accelerating, cruising at a constant speed, decelerating, and docking. Moreover, external factors such as environmental changes frequently impact the load, making the variations even more intricate. On the other hand, sensor interference cannot be ignored. The various sensors installed on the ship may experience signal transmission instability and data collection deviations due to prolonged operation and the challenging marine environment.

Given the aforementioned complexities, traditional static data processing methods struggle to effectively handle these diverse and variable interfering factors, allowing data noise to infiltrate and affect the accuracy of operating condition identification. Using a rolling time window for condition identification is an effective strategy. The rolling time window dynamically segments and analyzes the ship's operational data based on a predefined time range. By continuously updating the data within the time window, it can better adapt to load variations and filter out data noise caused by various interfering factors, thereby improving the precision and reliability of condition identification.

The ship condition recognition in this study adopts a rolling time window, where the time window $T_l = 20$ s, t is the discretized time step (set to 1 s in this study). This means that the condition identification at the current moment is based on the common features of data from the current moment and the preceding 19 s to determine the ship's current operating condition category. The SVM-based strategy achieves an accuracy of 90.2%. This indicates that SVM can accurately identify the ship's current operating condition during operation.

4. Analysis of the Simulation Results

4.1. Dynamic Load Cycles of Ship

To validate the effectiveness of the proposed two-layer energy management model control strategy, tests under dynamic load cycles are necessary. In this study, two types of composite load cycles are provided. The first is the Random Load Cycle (RLC), constructed by randomly splicing portions of five typical operating conditions. The second is the Combined Load Cycle (CLC), formed by sequentially connecting the five typical operating conditions. These two dynamic ship load cycles are employed to simulate the real-time operational loads of ships. Figure 8 illustrates the two random dynamic load patterns used to verify the proposed strategies in this paper. The detailed explanation of the extraction of these five typical operating conditions is provided in Section 3.1, and the data for these typical operating conditions comes from China's "Three Gorges Hydrogen Boat One".

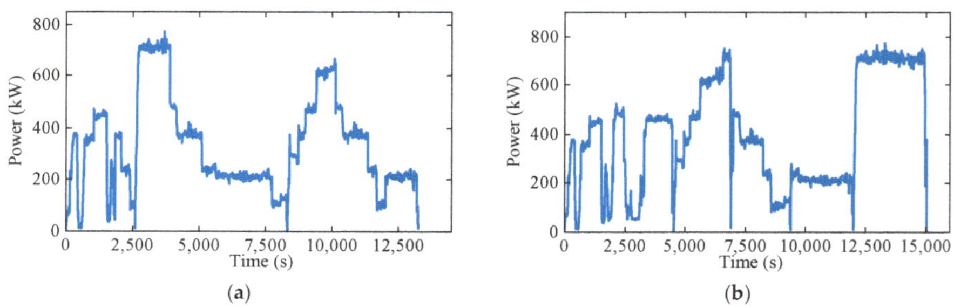

Figure 8. Dynamic load cycles of the ship: (**a**) Random Load Cycle (RLC) and (**b**) Combined Load Cycle (CLC).

4.2. Verification

To verify the effectiveness of the proposed two-layer energy management model (TLM) for the real-time operation of ships, this study compares it with the adaptive second-order filtering optimization method (TLFP) and the offline optimization strategy (LPF-DP) proposed in this paper.

In hybrid power systems, ships operating under different conditions exhibit different load fluctuation characteristics. Under stable conditions, the load primarily consists of low-frequency power with small fluctuations, while under complex conditions, the load is dominated by high-frequency power with rapid changes and large fluctuations.

Therefore, a power allocation method based on the nature of power fluctuations better aligns with the characteristics of shiploads and is more suitable for the joint operation of different energy storage components. Low-pass filtering can effectively smooth the control process of energy flow, allocating power signals of different frequencies to the corresponding components based on their characteristics. Compared to first-order low-pass filters, second-order low-pass filters have stronger attenuation capabilities for high-frequency signals and can effectively extract low-frequency signals, facilitating dynamic coordination. Additionally, the adaptive second-order filtering strategy achieves power allocation for complex conditions by updating the filter coefficients in real-time.

Figure 9 shows the results of operating condition classification for dynamic load profiles, achieving accuracy rates of 90.2% (Figure 9a) and 92.7% (Figure 9b). During transitions between operating conditions, there is a temporary decrease in classification accuracy. However, due to the similar characteristics shared by some operating conditions, the offline strategy repository is still able to provide effective power distribution rules. These results indicate that the proposed two-layer energy management strategy is capable of accurately identifying the current operating condition during ship operation.

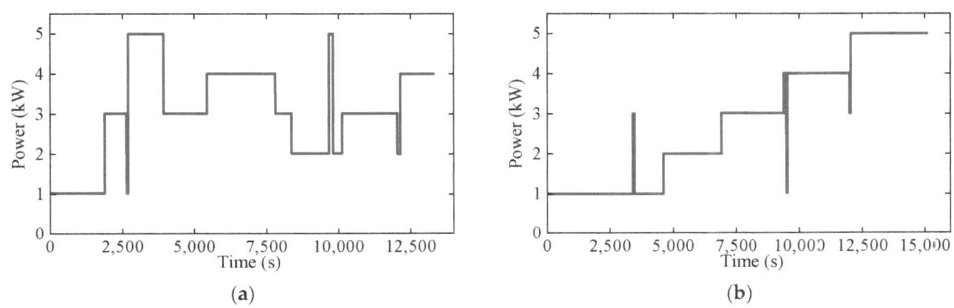

Figure 9. Real-time recognition results for the operating conditions. (**a**) Recognition results for RLC and (**b**) recognition results for CLC.

The detailed power splitting condition of RLC is plotted in Figure 10a–e, and the detailed power splitting condition of CLC is plotted in Figure 10f–j. From Figure 10a,f, it can be observed that the LPF-DP achieves the smallest power fluctuation for the fuel cell system, demonstrating the best smoothing performance. In contrast, the TLFP results in the largest power fluctuations. The proposed TLM achieves power fluctuations in the fuel cell system that are slightly higher than those of the offline optimization strategy but still outperform the TLFP energy management strategy. The performance of different options in the RLC and CLC is given in Figure 11. Under the RLC and CLC, the proposed two-layer energy management model reduces the fuel cell power fluctuations by 44.06% and 31.32%, respectively, compared to the TLFP strategy. However, compared to the LPF-DP strategy, the power fluctuations increased by 9.58% and 12.9%, respectively.

From Figure 10b,c,g,h, it can be observed that the power fluctuation frequency of the lithium battery pack under the LPF-DP and TLM is significantly lower than that of the TLFP. Conversely, the power fluctuation frequency of the supercapacitor pack is higher than that of the TLFP strategy. This allocation strategy aligns well with the power characteristics of lithium battery packs and supercapacitor packs. Under the RLC and CLC, the capacity degradation cost of the lithium battery pack in the TLM is reduced by 28.9% and 21.49%, respectively, compared to the TLFP strategy, while increasing by 8.91% and 13.36%, respectively, compared to the LPF-DP strategy.

The computation times for different energy management strategies during the ship's random dynamic load tests are presented in Table 4. According to the table, the LPF-DP requires 44,003 s and 60,783 s for the RLC and CLC conditions, corresponding to response times of 3309.2 ms and 4026.6 ms, respectively. In contrast, the response times for the two-layer energy management model are 9.1 ms and 9.6 ms, respectively. The TLFP requires 4.06 ms and 4.01 ms. While the DP algorithm can typically achieve optimal results, it demands a significant amount of computational time, rendering it unsuitable for practical applications. The TLM substantially reduces the computational time compared to the offline optimization strategy while achieving similar optimization results.

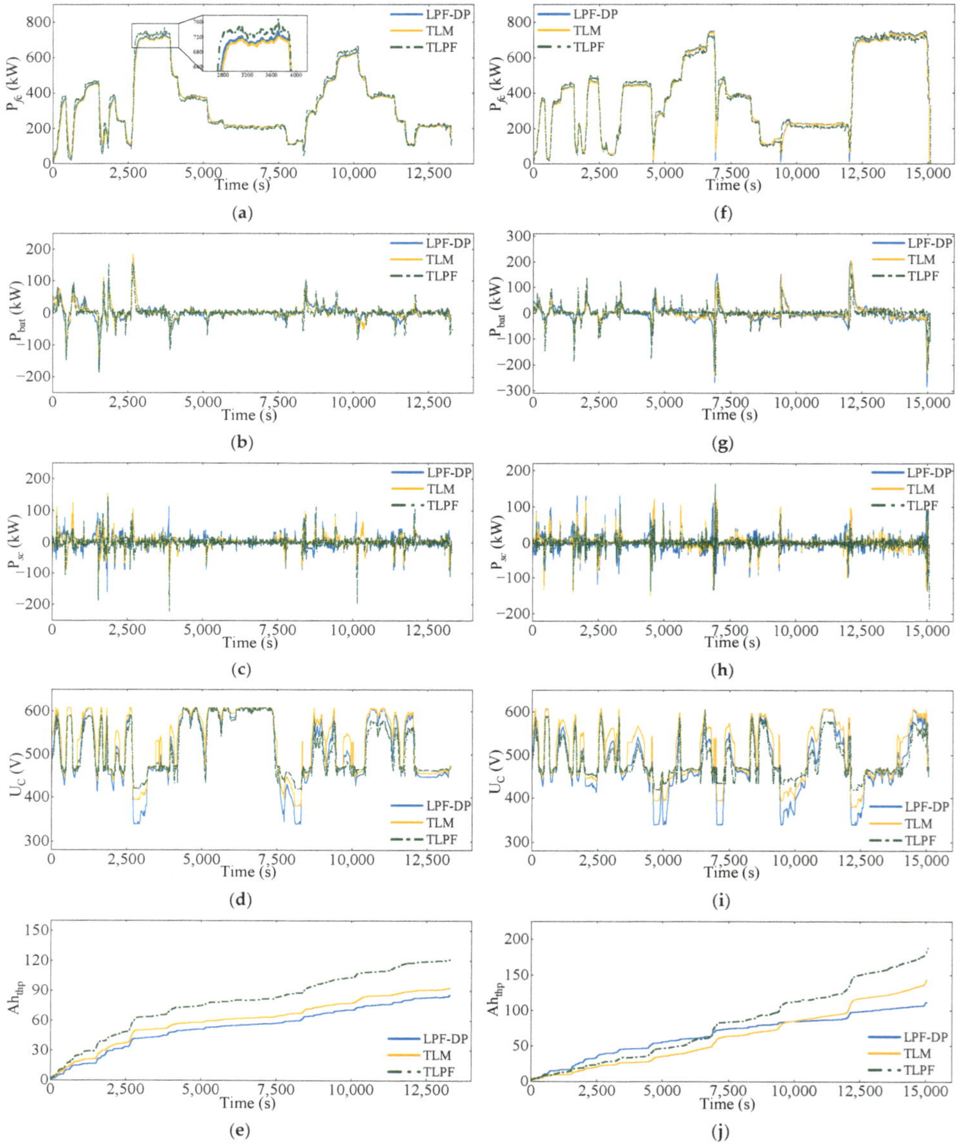

Figure 10. Power allocation results under different strategies. (**a**) Power allocation results for fuel cells under the RLC; (**b**) power allocation results for lithium batteries under the RLC; (**c**) power allocation results for supercapacitors under the RLC; (**d**) UC voltage comparison of the RLC; (**e**) Ah throughput comparison of the RLC; (**f**) power allocation results for fuel cells under the CLC; (**g**) power allocation results for lithium batteries under the CLC; (**h**) power allocation results for supercapacitors under the CLC; (**i**) UC voltage comparison of the CLC; and (**j**) Ah throughput comparison of the CLC.

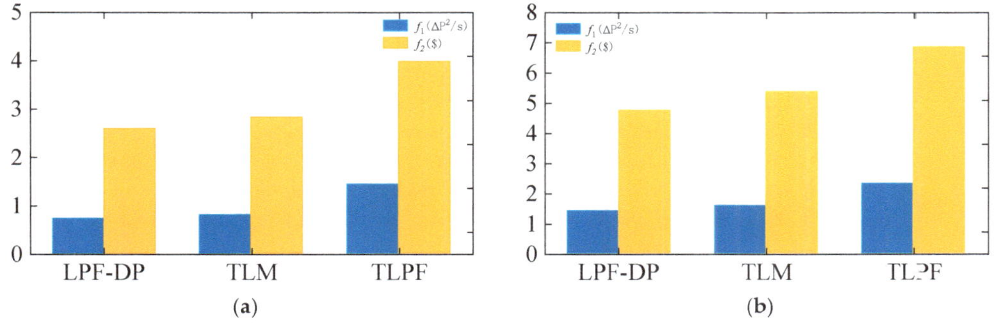

Figure 11. Performance of different energy management strategies. (**a**) Optimization results for different strategies under the RLC; (**b**) optimization results for different strategies under the CLC.

Table 4. Response times for a single step.

Parameters	LPF-DP	TLM	TLFP
$RLC - SponseTime$ (ms)	3309.2	9.1	4.06
$CLC - SponseTime$ (ms)	4026.6	9.6	4.01

5. Conclusions and Discussion

In response to the complex operating conditions of ships and by leveraging historical operational data, this paper proposes a two-layer energy management system for hydrogen-powered hybrid ship systems. The first layer of the EMS is based on offline optimization, utilizing DP and LPF to determine globally optimal strategies. Typical operating conditions of the ship are identified, and the optimal power distribution for each condition is obtained using optimization algorithms. These results are then used to train an improved BP neural network, thereby constructing an offline strategy library. The second layer of the EMS operates in real time to manage load variations. It first identifies the ship's current operating condition and then provides real-time power distribution strategies based on the offline strategy library.

To validate the effectiveness of the proposed approach, a simulated load representing complex ship operating conditions was used for testing. Compared with the LPF, the proposed method reduces the fuel cell fluctuation rate by 44.06% and decreases the lithium battery capacity degradation cost by 28.9%, while achieving results that are closely aligned with those of offline optimization. Additionally, the real-time response time of the system is 9.1 ms, meeting the real-time requirements of a ship EMS. These results demonstrate that the proposed approach can provide efficient energy management in real-time applications, offering a novel energy management system for the real-time operation of the maritime industry.

A traditional real-time EMS relies on analyzing historical operational data and deriving allocation parameters through complex calculations, leading to rule-based solutions that are heavily dependent on these parameters. Such strategies struggle to adapt to diverse ship types and complex operating conditions. In contrast, the two-layer EMS proposed in this study bypasses the need for precise rule extraction from offline optimization and achieves outcomes closer to offline results. While designed for hydrogen-powered ships, the proposed EMS is adaptable to various vessels by utilizing their specific historical data. As more operational data are incorporated, the offline strategy library can expand, ensuring stable and reliable energy management under diverse cruising and environmental conditions.

This study has certain limitations that warrant further investigation. First, the proposed EMS does not fully address emergencies caused by environmental or human factors, such as sudden power fluctuations or unexpected equipment failures during ship operations. These challenges may impact the strategy's effectiveness and stability in practical applications. Future research should focus on enhancing the adaptability of energy management strategies by simulating extreme scenarios and developing robust response mechanisms to ensure safe and reliable operation under complex and dynamic conditions.

Second, the EMS in this study is based on a single fuel cell and a hybrid energy storage system. However, modern ship power grids typically include multiple fuel cells and lithium-ion battery groups, which experience varying degrees of degradation over time. This non-uniform degradation can significantly affect the overall grid performance. Therefore, future work should explore coordination mechanisms for multiple fuel cell and battery groups while considering the impact of degradation on energy management strategies.

Author Contributions: Conceptualization, X.Y. and L.L.; methodology, L.L., X.Y., and Y.W.; software, L.L.; validation, L.L., X.Y., and X.L.; resources, T.T. and Q.S.; data curation, L.L.; investigation, Y.L. and X.Z.; writing—original draft, L.L.; writing—review and editing, L.L. and X.Y.; supervision, X.Y. and X.L.; project administration, X.Y.; funding acquisition, X.Y. All authors have read and agreed to the published version of the manuscript.

Funding: This work was supported by the China National Key Research and Development Project (Grant No. 2023YFB4301704) and the China National Key Laboratory of Electromagnetic Energy Technology Open Fund (Grant No. 61422172220403).

Institutional Review Board Statement: Not applicable.

Informed Consent Statement: Not applicable.

Data Availability Statement: Data will be made available upon request.

Conflicts of Interest: The authors declare no conflicts of interest.

References

1. Van Roy, W.; Merveille, J.-B.; Van Nieuwenhove, A.; Scheldeman, K.; Maes, F. Policy Recommendations for International Regulations Addressing Air Pollution from Ships. *Mar. Policy* **2024**, *159*, 105913. [CrossRef]
2. Fan, A.; Wang, J.; He, Y.; Perčić, M.; Vladimir, N.; Yang, L. Decarbonising Inland Ship Power System: Alternative Solution and Assessment Method. *Energy* **2021**, *226*, 120266. [CrossRef]
3. Van Roy, W.; Merveille, J.-B.; Scheldeman, K.; Van Nieuwenhove, A.; Schallier, R.; Van Roozendael, B.; Maes, F. Assessment of the Effect of International Maritime Regulations on Air Quality in the Southern North Sea. *Atmosphere* **2023**, *14*, 969. [CrossRef]
4. Joung, T.-H.; Kang, S.-G.; Lee, J.-K.; Ahn, J. The IMO Initial Strategy for Reducing Greenhouse Gas(GHG) Emissions, and Its Follow-up Actions towards 2050. *J. Int. Marit. Saf. Environ. Aff. Shipp.* **2020**, *4*, 1–7. [CrossRef]
5. Badakhshan, S.; Rahman, J.; Zhang, J. Black Start of Coastline Power Networks From Grid-Forming Ship-to-Grid Services. *IEEE Trans. Smart Grid* **2024**, *15*, 1670–1679. [CrossRef]
6. Zhang, D.; Chu, X.; Liu, C.; He, Z.; Zhang, P.; Wu, W. A Review on Motion Prediction for Intelligent Ship Navigation. *J. Mar. Sci. Eng.* **2024**, *12*, 107. [CrossRef]
7. Wang, Z.; Wu, Y.; Chu, X.; Liu, C.; Zheng, M. Risk Identification Method for Ship Navigation in the Complex Waterways via Consideration of Ship Domain. *J. Mar. Sci. Eng.* **2023**, *11*, 2265. [CrossRef]
8. Daniel, H. Unlocking Shore Power in St. Lawrence and Great Lakes for Cargo Ships. *Transp. Res. Part D Transp. Environ.* **2024**, *131*, 104230. [CrossRef]
9. Bullock, S.; Hoolohan, C.; Larkin, A. Accelerating Shipping Decarbonisation: A Case Study on UK Shore Power. *Heliyon* **2023**, *9*, e17475. [CrossRef] [PubMed]
10. Wang, Y.; Sun, Z.; Chen, Z. Energy Management Strategy for Battery/Supercapacitor/Fuel Cell Hybrid Source Vehicles Based on Finite State Machine. *Appl. Energy* **2019**, *254*, 113707. [CrossRef]
11. Tobajas, J. Resilience-Oriented Schedule of Microgrids with Hybrid Energy Storage System Using Model Predictive Control. *Appl. Energy* **2022**, *306*, 118092. [CrossRef]

12. Wang, Z.; Li, J.; Hu, C.; Li, X.; Zhu, Y. Hybrid Energy Storage System and Management Strategy for Motor Drive with High Torque Overload. *J. Energy Storage* **2024**, *75*, 109432. [CrossRef]
13. Thirunavukkarasu, G.S.; Seyedmahmoudian, M.; Jamei, E.; Horan, B.; Mekhilef, S.; Stojcevski, A. Role of Optimization Techniques in Microgrid Energy Management Systems—A Review. *Energy Strategy Rev.* **2022**, *43*, 100899. [CrossRef]
14. Ganesh, A.H.; Xu, B. A Review of Reinforcement Learning Based Energy Management Systems for Electrified Powertrains: Progress, Challenge, and Potential Solution. *Renew. Sustain. Energy Rev.* **2022**, *154*, 111833. [CrossRef]
15. Yi, F.; Lu, D.; Wang, X.; Pan, C.; Tao, Y.; Zhou, J.; Zhao, C. Energy Management Strategy for Hybrid Energy Storage Electric Vehicles Based on Pontryagin's Minimum Principle Considering Battery Degradation. *Sustainability* **2022**, *14*, 1214. [CrossRef]
16. Yu, X.; Lin, C.; Tian, Y.; Zhao, M.; Liu, H.; Xie, P.; Zhang, J. Real-Time and Hierarchical Energy Management-Control Framework for Electric Vehicles with Dual-Motor Powertrain System. *Energy* **2023**, *272*, 127112. [CrossRef]
17. Ding, N.; Prasad, K.; Lie, T.T. Design of a Hybrid Energy Management System Using Designed Rule-Based Control Strategy and Genetic Algorithm for the Series-parallel Plug-in Hybrid Electric Vehicle. *Int. J. Energy Res.* **2021**, *45*, 1627–1644. [CrossRef]
18. Restrepo, M.; Cañizares, C.A.; Simpson-Porco, J.W.; Su, P.; Taruc, J. Optimization- and Rule-Based Energy Management Systems at the Canadian Renewable Energy Laboratory Microgrid Facility. *Appl. Energy* **2021**, *290*, 116760. [CrossRef]
19. Rodriguez, R.; Trovão, J.P.F.; Solano, J. Fuzzy Logic-Model Predictive Control Energy Management Strategy for a Dual-Mode Locomotive. *Energy Convers. Manag.* **2022**, *253*, 115111. [CrossRef]
20. Maghfiroh, H.; Wahyunggoro, O.; Cahyadi, A.I. Low Pass Filter as Energy Management for Hybrid Energy Storage of Electric Vehicle: A Survey. *Automot. Exp.* **2023**, *6*, 466–484. [CrossRef]
21. Ramos, G.A.; Costa-Castelló, R. Energy Management Strategies for Hybrid Energy Storage Systems Based on Filter Control: Analysis and Comparison. *Electronics* **2022**, *11*, 1631. [CrossRef]
22. Wang, C.; Yang, R.; Yu, Q. Wavelet Transform Based Energy Management Strategies for Plug-in Hybrid Electric Vehicles Considering Temperature Uncertainty. *Appl. Energy* **2019**, *256*, 113928. [CrossRef]
23. Du, C.; Huang, S.; Jiang, Y.; Wu, D.; Li, Y. Optimization of Energy Management Strategy for Fuel Cell Hybrid Electric Vehicles Based on Dynamic Programming. *Energies* **2022**, *15*, 4325. [CrossRef]
24. Quan, S.; Wang, Y.-X.; Xiao, X.; He, H.; Sun, F. Real-Time Energy Management for Fuel Cell Electric Vehicle Using Speed Prediction-Based Model Predictive Control Considering Performance Degradation. *Appl. Energy* **2021**, *304*, 117845. [CrossRef]
25. Aguila-Leon, J.; Vargas-Salgado, C.; Chiñas-Palacios, C.; Díaz-Bello, D. Energy Management Model for a Standalone Hybrid Microgrid through a Particle Swarm Optimization and Artificial Neural Networks Approach. *Energy Convers. Manag.* **2022**, *267*, 115920. [CrossRef]
26. Tang, W.; Wang, Y.; Jiao, X.; Ren, L. Hierarchical Energy Management Strategy Based on Adaptive Dynamic Programming for Hybrid Electric Vehicles in Car-Following Scenarios. *Energy* **2023**, *265*, 126264. [CrossRef]
27. Xu, N.; Kong, Y.; Yan, J.; Zhang, Y.; Sui, Y.; Ju, H.; Liu, H.; Xu, Z. Global Optimization Energy Management for Multi-Energy Source Vehicles Based on "Information Layer—Physical Layer—Energy Layer—Dynamic Programming" (IPE-DP). *Appl. Energy* **2022**, *312*, 118668. [CrossRef]
28. Liu, C.; Wang, Y.; Wang, L.; Chen, Z. Load-Adaptive Real-Time Energy Management Strategy for Battery/Ultracapacitor Hybrid Energy Storage System Using Dynamic Programming Optimization. *J. Power Sources* **2019**, *438*, 227024. [CrossRef]
29. Wang, S.; Tan, Q.; Ding, X.; Li, J. Efficient Microgrid Energy Management with Neural-Fuzzy Optimization. *Int. J. Hydrog. Energy* **2024**, *64*, 269–281. [CrossRef]
30. Wu, J.; Huang, C.; He, H.; Huang, H. Confidence-Aware Reinforcement Learning for Energy Management of Electrified Vehicles. *Renew. Sustain. Energy Rev.* **2024**, *191*, 114154. [CrossRef]
31. Li, Y.; Zhou, S.; Liu, J.; Tong, J.; Dang, J.; Yang, F.; Ouyang, M. Multi-Objective Optimization of the Atkinson Cycle Gasoline Engine Using NSGA III Coupled with Support Vector Machine and Back-Propagation Algorithm. *Energy* **2023**, *262*, 125262. [CrossRef]
32. Kurucan, M.; Özbaltan, M.; Yetgin, Z.; Alkaya, A. Applications of Artificial Neural Network Based Battery Management Systems: A Literature Review. *Renew. Sustain. Energy Rev.* **2024**, *192*, 114262. [CrossRef]
33. Elsisi, M.; Amer, M.; Dababat, A.; Su, C.-L. A Comprehensive Review of Machine Learning and IoT Solutions for Demand Side Energy Management, Conservation, and Resilient Operation. *Energy* **2023**, *281*, 128256. [CrossRef]
34. Qi, C.; Song, C.; Xiao, F.; Song, S. Generalization Ability of Hybrid Electric Vehicle Energy Management Strategy Based on Reinforcement Learning Method. *Energy* **2022**, *250*, 123826. [CrossRef]
35. Zhang, H.; Peng, J.; Tan, H.; Dong, H.; Ding, F. A Deep Reinforcement Learning-Based Energy Management Framework with Lagrangian Relaxation for Plug-In Hybrid Electric Vehicle. *IEEE Trans. Transp. Electrific.* **2021**, *7*, 1146–1160 [CrossRef]
36. Wu, Y.; Zhang, Y.; Li, G.; Shen, J.; Chen, Z.; Liu, Y. A Predictive Energy Management Strategy for Multi-Mode Plug-in Hybrid Electric Vehicles Based on Multi Neural Networks. *Energy* **2020**, *208*, 118366. [CrossRef]

37. Yang, N.; Ruan, S.; Han, L.; Liu, H.; Guo, L.; Xiang, C. Reinforcement Learning-Based Real-Time Intelligent Energy Management for Hybrid Electric Vehicles in a Model Predictive Control Framework. *Energy* **2023**, *270*, 126971. [CrossRef]
38. Waheed, W.; Xu, Q. Data-Driven Short Term Load Forecasting with Deep Neural Networks: Unlocking Insights for Sustainable Energy Management. *Electr. Power Syst. Res.* **2024**, *232*, 110376. [CrossRef]
39. Jiang, H.; Xu, L.; Li, J.; Hu, Z.; Ouyang, M. Energy Management and Component Sizing for a Fuel Cell/Battery/Supercapacitor Hybrid Powertrain Based on Two-Dimensional Optimization Algorithms. *Energy* **2019**, *177*, 386–396. [CrossRef]
40. Lee, Y.; Lee, H.; Gim, J.; Seo, I.; Lee, G. Technical Measures to Mitigate Load Fluctuation for Large-Scale Customers to Improve Power System Energy Efficiency. *Energies* **2020**, *13*, 4812. [CrossRef]
41. Han, S.; Han, S.; Aki, H. A Practical Battery Wear Model for Electric Vehicle Charging Applications. *Appl. Energy* **2014**, *113*, 1100–1108. [CrossRef]
42. Fanoro, M.; Božanić, M.; Sinha, S. A Review of the Impact of Battery Degradation on Energy Management Systems with a Special Emphasis on Electric Vehicles. *Energies* **2022**, *15*, 5889. [CrossRef]
43. Zheng, Y.; Lv, X.; Qian, L.; Liu, X. An Optimal BP Neural Network Track Prediction Method Based on a GA–ACO Hybrid Algorithm. *J. Mar. Sci. Eng.* **2022**, *10*, 1399. [CrossRef]
44. Akram, A.S.; Abbas, S.; Khan, M.A.; Athar, A.; Ghazal, T.M.; Hamadi, H.A. Smart Energy Management System Using Machine Learning. *Comput. Mater. Contin.* **2024**, *78*, 959–973. [CrossRef]

Journal of
Marine Science and Engineering

MDPI

Article

Parameter Identification of Maritime Vessel Rudder PMSM Based on Extended Kalman Particle Filter Algorithm

Tianqing Yuan [1,2], Tianli Wang [1,2], Jing Bai [3,*] and Jingwen Fan [3]

1 Key Laboratory of Modern Power System Simulation and Control & Renewable Energy Technology, Ministry of Education (Northeast Electric Power University), Jilin 132012, China; 20192929@neepu.edu.cn (T.Y.); 2202200254@neepu.edu.cn (T.W.)

2 Department of Electrical Engineering, Northeast Electric Power University, Jilin 132012, China

3 Yuda Engineering (Jilin) Co., Ltd., Siping 136000, China; tskfjw@163.com

* Correspondence: baijingjane@163.com

Abstract: To address the issue of system parameter variations during the operation of a maritime light vessel rudder permanent magnet synchronous motor (PMSM), an extended Kalman particle filter (EKPF) algorithm that combines a particle filter (PF) with an extended Kalman filter (EKF) is proposed in this paper. This approach enables the online identification of motor resistance and inductance. For highly nonlinear problems that are challenging for traditional methods such as Kalman filtering, this algorithm is typically a statistical and effective estimation method that usually yields good results. Firstly, a standard linear discrete parameter identification model is established for a PMSM. Secondly, the PF algorithm based on Bayesian state estimation as a foundation for subsequent research is derived. Thirdly, the advantages and limitations of the PF algorithm are analyzed, addressing issues such as sample degeneracy, by integrating it with the Kalman filtering algorithm. Specifically, the EKPF algorithm for online parameter identification is employed. Finally, the identification model within MATLAB/Simulink is constructed and the simulation studies are executed to ascertain the viability of our suggested algorithm. The outcomes from these simulations indicate that the proposed EKPF algorithm identifies resistance and inductance values both swiftly and precisely, markedly boosting the robustness and enhancing the control efficacy of the PMSM.

Keywords: PMSM; particle filtering algorithm; extended Kalman particle filtering algorithm; parameter identification algorithm

1. Introduction

In modern maritime light vessel navigation systems, the accuracy and reliability of the autopilot system are crucial for navigational safety. As the core component of the autopilot system, the steering gear motor is required to possess high dynamic response, high efficiency, and precise control characteristics. Low-power permanent magnet synchronous motors (PMSMs) have been identified as the ideal choice for steering gear motors due to their excellent power density, efficiency, and control performance [1,2]. Over the years, with the advancement in PMSM drive control technology, scholars both domestically and internationally have proposed a plethora of control techniques to achieve the high-performance control of permanent magnet motors [3–6]. Whether it is speed control, position control, or torque control, achieving the high dynamic response and precise control goals requires accurate motor parameters. However, these parameters are significantly influenced by factors such as temperature, stator current, and magnetic flux saturation, leading to substantial variations. Consequently, methods such as vector control and direct torque control often fall short of achieving the results predicted by theoretical analysis. In the complex marine environment, the steering gear motor must maintain high-performance operation under various conditions. To realize efficient PMSM control, accurate parameter identification becomes crucial [7–11], leading to the emergence of various motor parameter identification methods. This is of great practical significance for enhancing the performance and reliability of the autopilot system. Depending on the operating conditions of the PMSM, the parameter identification can be categorized into offline parameter identification or online parameter identification. Offline parameter identification, due to its non-interference with the online operational status of the motor and conservation of the online control resources, has garnered widespread attention in the industrial sector. On the other hand, online parameter identification methods can be synchronized with the motor operation, enabling the real-time acquisition of the motor parameters. They meet the need for accurate parameter tracking under different motor operating conditions and reflect the changes in the motor status over time. Compared to offline parameter identification, the online methods offer better real-time performance and flexibility [12]. The traditional online identification methods include least squares estimation [13], model reference adaptive control [14,15], and Kalman filtering [16,17].

Considering that the inductance parameters are influenced by the PMSM operation status, a motor model utilizing the transient voltage equation is developed and a forgetting factor is incorporated to enhance the least squares identification method in [18], which can improve the algorithm's tracking capabilities. When the forgetting factor is set to a certain value, the least squares method may struggle to ensure robustness. Ref. [8] treats the error between the theoretical and actual outputs as a variable and dynamically adjusts the forgetting factor during the identification process, thereby accelerating the convergence speed of the algorithm while ensuring good robustness. Assuming that the stator resistance and permanent magnet flux linkage are known and constant, Ref. [19] achieves the online identification of the d-axis and q-axis inductances through a model reference adaptive control algorithm. Ref. [20] proposes a disturbance compensation-based model reference adaptive system, designing a disturbance estimator to estimate the external disturbances in real time and updating the adaptation rate based on the disturbance, thus reducing the impact of parameter uncertainty and disturbances on the system and expanding the application scenarios of the model reference adaptive control algorithm. The Kalman filtering algorithm provides real-time updates of estimated values for linear systems, achieving optimal parameter estimation. However, in cases where the noise is unknown or the modeling error is significant, the Kalman filtering algorithm often fails to achieve the desired effect. Since the motor system is nonlinear, many scholars have proposed improvements to the Kalman filtering algorithm, such as extended Kalman filtering [21,22], unscented Kalman filtering [23], and cubature Kalman filtering [24], to extend its application scope. The prerequisite for using Kalman filtering is that the measurement noise follows a Gaussian distribution. However, in practice, the statistical properties of the system noise are often unknown or time-varying. In [25], an adaptive Kalman filtering algorithm based on variational Bayesian inference is presented, which selects appropriate conjugate distributions to estimate the covariance matrix of noise with minimal changes, to some extent addressing the aforementioned issues. However, it is limited by the linear Gaussian state model. The H_∞ filtering algorithm has good robustness to unknown noise but at the cost of accuracy. The H_∞ filtering algorithm is only suitable for linear systems, and, when applied to nonlinear systems, it still has many drawbacks. Therefore, modifications of the H_∞ filtering algorithm have been proposed. Xia proposes a fitted H_∞ filtering algorithm based on fitted transformation to address the problems in nonlinear uncertain systems, which shows good robustness [26,27]. Furthermore, outliers in the system can also reduce the estimation accuracy of the H_∞ filtering algorithm. Therefore, based on the H_∞ filtering algorithm, Zhao studies a series of derivative algorithms [28–30], among which the Krein space robust unscented Kalman filter based on the generalized maximum likelihood can effectively handle outliers [30].

In addition to the aforementioned common traditional identification methods, there are some identification methods that are highly effective for handling highly nonlinear systems but have not been widely applied due to certain limitations. For example, the PF algorithm, which suffers from low computational efficiency [31], has only been applied with the continuous improvement in computer computational capabilities in recent years. While the PF algorithm is relatively simple and can accurately estimate target parameters, its limitations include high computational complexity, particle degeneracy, and potential delays in responding to dynamic system changes. To address these issues, this paper employs the EKPF algorithm, which combines the extended Kalman filtering algorithm with the particle filtering algorithm, for parameter identification. This approach mitigates problems such as particle degeneracy to some extent and improves the overall performance of the particle filter.

2. Establishing the Parameter Identification Model for PMSM

The mathematical model of the PMSM is typically formulated in the synchronous rotating reference frame d-q. The stator voltage equation is provided by

$$\begin{cases} u_{\mathrm{d}} = R_{\mathrm{s}} i_{\mathrm{d}} + \frac{\mathrm{d}\psi_{\mathrm{d}}}{\mathrm{d}t} - \omega_{\mathrm{e}}\psi_{\mathrm{q}} \\ u_{\mathrm{q}} = R_{\mathrm{s}} i_{\mathrm{q}} + \frac{\mathrm{d}\psi_{\mathrm{q}}}{\mathrm{d}t} + \omega_{\mathrm{e}}\psi_{\mathrm{d}} \end{cases} \tag{1}$$

The stator flux linkage equation is

$$\begin{cases} \psi_{\mathrm{d}} = L_{\mathrm{d}} i_{\mathrm{d}} + \psi_f \\ \psi_{\mathrm{q}} = L_{\mathrm{q}} i_{\mathrm{q}} \end{cases} \tag{2}$$

Substituting Equation (2) into Equation (1):

$$\begin{cases} u_{\mathrm{d}} = R_{\mathrm{s}} i_{\mathrm{d}} + L_{\mathrm{d}}\frac{\mathrm{d}}{\mathrm{d}t} i_{\mathrm{d}} - \omega_{\mathrm{e}} L_{\mathrm{q}} i_{\mathrm{q}} \\ u_{\mathrm{q}} = R_{\mathrm{s}} i_{\mathrm{q}} + L_{\mathrm{q}}\frac{\mathrm{d}}{\mathrm{d}t} i_{\mathrm{q}} + \omega_{\mathrm{e}}(L_{\mathrm{d}} i_{\mathrm{d}} + \psi_f) \end{cases} \tag{3}$$

In the equations, u_{d} and u_{q} are the voltages along the d and q axes, respectively; i_{d} and i_{q} are the currents along the d and q axes, respectively; R_{s} is the stator resistance; ψ_{d} and ψ_{q} are the components of stator flux linkage along the d and q axes, respectively; ω_{e} is the electrical angular speed of the rotor; L_{d} and L_{q} are the d- and q-axis inductances, respectively; ψ_{f} is the flux linkage of the permanent magnet.

Selecting the components of current along the d and q axes, i_{d} and i_{q}, and the parameters to be identified, L_{d}, L_{q}, and R_{s}, as state variables, we establish the state-space equations for the permanent magnet synchronous motor. This paper primarily focuses on surface-mounted permanent magnet synchronous motors, where $L_{\mathrm{d}} = L_{\mathrm{q}} = L_{\mathrm{s}}$. Therefore, after rearrangement, Equation (3) becomes

$$\frac{\mathrm{d}}{\mathrm{d}t}\begin{bmatrix} i_{\mathrm{d}} \\ i_{\mathrm{q}} \\ R_{\mathrm{s}} \\ L_{\mathrm{s}} \end{bmatrix} = \begin{bmatrix} -\frac{R_{\mathrm{s}}}{L_{\mathrm{s}}} & \omega_{\mathrm{e}} & 0 & 0 \\ -\omega_{\mathrm{e}} & -\frac{R_{\mathrm{s}}}{L_{\mathrm{s}}} & 0 & 0 \\ 0 & 0 & 0 & 0 \\ 0 & 0 & 0 & 0 \end{bmatrix}\begin{bmatrix} i_{\mathrm{d}} \\ i_{\mathrm{q}} \\ R_{\mathrm{s}} \\ L_{\mathrm{s}} \end{bmatrix} + \begin{bmatrix} \frac{1}{L_{\mathrm{s}}} & 0 \\ 0 & \frac{1}{L_{\mathrm{s}}} \\ 0 & 0 \\ 0 & 0 \end{bmatrix}\begin{bmatrix} u_{\mathrm{d}} \\ u_{\mathrm{q}} - \omega_{\mathrm{e}}\psi_f \end{bmatrix} \tag{4}$$

Due to the presence of coupling terms in the coefficient matrix of state Equation (4), direct identification of R_{s} and L_{s} is rather complex. Therefore, we introduce intermediate variables a and b to simplify the identification equation. Let $a = R_{\mathrm{s}}/L_{\mathrm{s}}$ and $b = 1/L_{\mathrm{s}}$, Equation (4) becomes

$$\frac{\mathrm{d}}{\mathrm{d}t}\begin{bmatrix} i_{\mathrm{d}} \\ i_{\mathrm{q}} \\ a \\ b \end{bmatrix} = \begin{bmatrix} 0 & \omega_{\mathrm{e}} & -i_{\mathrm{d}} & u_{\mathrm{d}} \\ -\omega_{\mathrm{e}} & 0 & -i_{\mathrm{q}} & u_{\mathrm{q}} - \omega_{\mathrm{e}}\psi_f \\ 0 & 0 & 0 & 0 \\ 0 & 0 & 0 & 0 \end{bmatrix}\begin{bmatrix} i_{\mathrm{d}} \\ i_{\mathrm{q}} \\ a \\ b \end{bmatrix} + w \tag{5}$$

The output y can be expressed as

$$y = \begin{bmatrix} 1 & 0 & 0 & 0 \\ 0 & 1 & 0 & 0 \end{bmatrix} \begin{bmatrix} i_d \\ i_q \\ a \\ b \end{bmatrix} + v \tag{6}$$

In the above two equations, w represents the process noise of the system, while v represents the measurement noise of the system. $x = [i_d\ i_q\ a\ b]^T$ denotes the state variable matrix of the system, and $y = [i_d\ i_q]$ represents the output variable matrix of the system. By discretizing Equations (5) and (6) with a sampling period T_s, we obtain the following equations:

$$\begin{cases} x_{k+1} = F_k x_k + w_k \\ y_k = H_k x_k + v_k \end{cases} \tag{7}$$

In the equations, w_k and v_k are mutually independent white noise processes. F_k and H_k are coefficient matrices defined as follows:

$$\begin{cases} F_k = \begin{bmatrix} 1 & \omega_e T_s & -i_d T_s & u_d T_s \\ -\omega_e T_s & 1 & -i_q T_s & \left(u_q - \omega_e \psi_f\right) T_s \\ 0 & 0 & 1 & 0 \\ 0 & 0 & 0 & 1 \end{bmatrix} \\ H_k = \begin{bmatrix} 1 & 0 & 0 & 0 \\ 0 & 1 & 0 & 0 \end{bmatrix} \end{cases} \tag{8}$$

Using the filtering algorithm, parameters a and b can be identified through the above model. Then, by utilizing the relationship between a, b, R_s, and L_s, the identification values of resistance R_s and inductance L_s can be obtained. The specific identification process is illustrated in Figure 1, where the s-Function module represents the part where the filtering algorithm program is implemented.

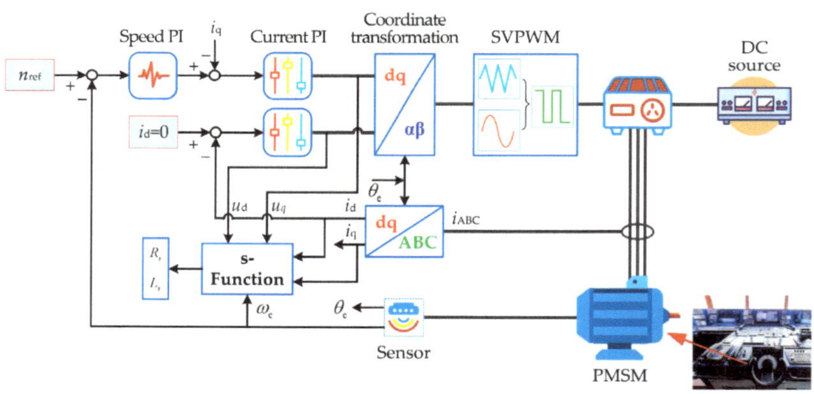

Figure 1. Identification process flowchart.

3. Particle Filter

3.1. Bayesian State Estimation

Assume the equation describing a nonlinear system is as follows:

$$\begin{aligned} x_{k+1} &= f_k(x_k, w_k) \\ y_k &= h_k(x_k, v_k) \end{aligned} \tag{9}$$

where k is the time index, x_k is the state variable, w_k is the process noise, y_k is the measurement value, and v_k is the measurement noise. The functions $f_k(\cdot)$ and $h_k(\cdot)$ are the process equation and measurement equation of the time-varying nonlinear system, respectively. The noises w_k and v_k are independent white noises with known probability density functions. The purpose of the Bayesian estimator is to estimate the probability density of the state x_k conditioned on the measurement values $y_1, y_2, \ldots, y_k$. This conditional probability density is represented as

$$p(x_k \mid Y_k) = \text{The probability density of the state } x_k \text{ conditioned on the measurement values } y_1, y_2, \ldots, y_k. \quad (10)$$

At k = 1, the first measurement value is obtained. Therefore, the initial condition of the estimator is the probability density function of x_0 conditioned on the absence of measurement values, denoted as Y_0, expressed as

$$p(x_0) = p(x_0|Y_0) \quad (11)$$

Once $p(x_k \mid Y_k)$ is computed, the most suitable estimate of x_k can be determined based on the specific problem. When the conditional probability density function $p(x_k \mid Y_k)$ is multimodal, the mean of x_k cannot be used as an estimate.

To compute the conditional probability density function $p(x_k \mid Y_k)$, a recursive method is required. Prior to this, it is essential to ascertain $p(x_k \mid Y_{k-1})$, representing the probability density of x_k conditioned on all measurements up to time k. This step is fundamental in deriving a recursive solution for evaluating $p(x_k \mid Y_k)$. This can be expressed utilizing the Bayes' theorem and the properties of joint probability density functions as follows:

$$p(x_k \mid Y_{k-1}) = \int p[(x_k, x_{k-1}) \mid Y_{k-1}]dx_{k-1} = \int p[x_k \mid (x_{k-1}, Y_{k-1})]p(x_{k-1} \mid Y_{k-1})dx_{k-1} \quad (12)$$

From Equation (9), it can be observed that x_k is fully determined by x_{k-1} and w_{k-1}. Therefore, $p[x_k \mid (x_{k-1}, Y_{k-1})] = p(x_k \mid x_{k-1})$, and we have

$$p(x_k \mid Y_{k-1}) = \int p(x_k \mid x_{k-1})p(x_{k-1} \mid Y_{k-1})dx_{k-1} \quad (13)$$

In the equation, the second probability density function on the right-hand side, $p(x_k \mid x_{k-1})$, is initially unknown but becomes known at the start (refer to Equation (11)) This function $p(x_k \mid x_{k-1})$ denotes the probability density of the state at time k, given the state at time (k − 1), which is established. Given our knowledge of the system equation $f_k(\cdot)$ and the noise w_k, this probability density function is ascertainable.

Now, consider the posterior conditional probability density function of x_k. Express this probability density function utilizing Bayes' theorem and the properties of joint probability density functions as follows:

$$\begin{aligned} p(x_k \mid Y_k) &= \frac{p(Y_k|x_k)}{p(Y_k)}p(x_k) = \frac{p[(y_k, Y_{k-1})|x_k]}{p(y_k, Y_{k-1})} \underbrace{\frac{p(x_k \mid Y_{k-1})p(Y_{k-1})}{p(Y_{k-1} \mid x_k)}}_{p(x_k)} \\ &= \frac{p(x_k, y_k, Y_{k-1})}{p(x_k)p(y_k, Y_{k-1})} \frac{p(x_k, Y_{k-1})p(Y_{k-1})}{p(Y_{k-1})p(Y_{k-1}|x_k)} \end{aligned} \quad (14)$$

Multiply the numerator and denominator simultaneously by $p(x_k, y_k)$:

$$p(x_k \mid Y_k) = \frac{p(x_k, y_k, Y_{k-1})p(x_k, Y_{k-1})p(Y_{k-1})}{p(x_k)p(y_k, Y_{k-1})p(Y_{k-1})p(Y_{k-1} \mid x_k)} \frac{p(x_k, y_k)}{p(x_k, y_k)} \quad (15)$$

In Equation (15), the conditional probability density function is derived several times by utilizing the ratio of the joint probability density function to the marginal probability density function. This method effectively captures the relationship between the conditioned and conditioning events; we obtain

$$p(x_k \mid Y_k) = \frac{p[Y_{k-1} \mid (x_k, y_k)]p(y_k \mid x_k)p(x_k \mid Y_{k-1})}{p(y_k \mid Y_{k-1})p(Y_{k-1} \mid x_k)} \tag{16}$$

Note that y_k is a function of x_k, so $p[Y_{k-1} \mid (x_k, y_k)] = p(Y_{k-1} \mid x_k)$. After rearranging, we obtain

$$p(x_k \mid Y_k) = \frac{p(y_k \mid x_k)p(x_k \mid Y_{k-1})}{p(y_k \mid Y_{k-1})} \tag{17}$$

All probability density functions on the right-hand side of the equation are known. From the measurement equation $h_k(\cdot)$ and the probability density function of the measurement noise v_k, we can obtain $p(y_k \mid x_k)$. From Equation (13), we know $p(x_k \mid Y_{k-1})$. The probability density function of $p(y_k \mid Y_{k-1})$ can be obtained.

$$\begin{aligned} p(y_k \mid Y_{k-1}) &= \int p[(y_k, x_k) \mid Y_{k-1}]dx_k \\ &= \int p[y_k \mid (x_k, Y_{k-1})]p(x_k \mid Y_{k-1})dx_k \end{aligned} \tag{18}$$

y_k is completely determined by x_k and v_k, so $p[y_k \mid (x_k, Y_{k-1})] = p(y_k \mid x_k)$, and

$$p(y_k \mid Y_{k-1}) = \int p[y_k \mid x_k]p(x_k \mid Y_{k-1})dx_k \tag{19}$$

From the above discussion, we can conclude that $P(y_k \mid x_k)$ can be obtained from the probability density function of the measurement equation $h_k(\cdot)$ and the measurement noise v_k, and $p(x_k \mid Y_{k-1})$ can be obtained from Equation (13).

In summary, the recursive equations for the Bayesian state estimator can be summarized as follows:

1. The system equation and measurement equation are as follows:

$$\begin{aligned} x_{k+1} &= f_k(x_k, w_k) \\ y_k &= h_k(x_k, v_k) \end{aligned} \tag{20}$$

where w_k and v_k are independent white noise processes with known probability density functions.

2. Assuming the probability density function $p(x_0)$ of the initial state is known, the estimator is initialized as follows:

$$p(x_0|Y_0) = p(x_0) \tag{21}$$

3. For k = 1, 2, . . . , execute the following equations:

(a) Obtain the prior probability density function from Equation (13) as follows:

$$p(x_k \mid Y_{k-1}) = \int p(x_k \mid x_{k-1})p(x_{k-1} \mid Y_{k-1})dx_{k-1} \tag{22}$$

(b) Obtain the posterior probability density function from Equations (17) and (19) as follows:

$$p(x_k \mid Y_k) = \frac{p(y_k \mid x_k)p(x_k \mid Y_{k-1})}{\int p(y_k \mid x_k)p(x_k \mid Y_{k-1})dx_k} \tag{23}$$

3.2. Establishment of the Particle Filter

The PF is derived from Bayesian state estimation and is used for numerically implementing Bayesian filters. The concept is straightforward and intuitive. At the outset of the estimation process, we randomly generate a specific number N of state vectors according to the initial probability density function $p(x_0)$. These state vectors are termed particles and are labeled as $x^+_{0,i}$ (where $i = 1, \ldots, N$). For each subsequent time step k = 1, 2, ..., the next batch of particles is produced using state equation $f(\cdot)$.

$$x^-_{k,i} = f_{k-1}\left(x^+_{k-1,i}, w^i_{k-1}\right) \ (i = 1, \cdots, N) \tag{24}$$

where each noise vector w^i_{k-1} is randomly generated based on the known probability density function of w_{k-1}. After obtaining the measurement at time k, the likelihood probability density of each particle $x^-_{k,i}$, denoted as $p(y_k \mid x^-_{k,i})$, is computed. For example, if given an m-dimensional measurement equation $y_k = h(x_k) + v_k$, and $v_k \sim N(0, R)$, the likelihood probability density q_i for the measurement y^* under the condition that x_k equals the particle $x^-_{k,i}$ can be computed as follows:

$$\begin{aligned} q_i &= p\left[(y_k = y^*) \mid (x_k = x^-_{k,i})\right] = p\left[v_k = y^* - h_k\left(x^-_{k,i}\right)\right] \\ &\sim \frac{1}{(2\pi)^{m/2}|R|^{1/2}} \exp\left(\frac{-\left[y^* - h\left(x^-_{k,i}\right)\right]^T R^{-1}\left[y^* - h\left(x^-_{k,i}\right)\right]}{2}\right) \end{aligned} \tag{25}$$

In the above equation, "~" indicates that this probability density function is proportional to the expression on the right-hand side. If this equation is applied to all particles $x^-_{0,i}$ ($i = 1, \ldots, N$), then the "relative" likelihood probability density of the state being equal to each particle will be correct. Normalize the likelihood probability density obtained from Equation (25).

$$q_i = \frac{q_i}{\sum_{j=1}^{N} q_j} \tag{26}$$

This guarantees that the total of all likelihood probability densities sums to 1.

Next, we perform resampling. The resampling process entails discarding particles with minimal weights and concentrating on those with larger weights. By resampling, we amplify the presence of particles with substantial weights, thereby better representing the posterior distribution using particles and their associated weights. Resample particles from the computed likelihood probability density functions; i.e., randomly generate a new set of particles $x^+_{k,i}$ based on the likelihood probability density function q_i. There can be several different resampling methods; one of the most direct methods is as follows. For 1, ..., N, perform the following two steps:

1. Generate a random number r from a uniform distribution in the interval [0, 1].

2. Accumulate the likelihood function q_i until the cumulative sum is greater than r. That is, for $\sum_{m=1}^{j-1} q_m < r$, but $\sum_{m=1}^{j} q_m \geq r$, the new particle $x^+_{k,i}$ is set equal to the old particle $x^-_{k,j}$.

As the sample size N tends to ∞, the set probability density function of the new particles $x^+_{k,i}$ approximates $p(x_k \mid y_k)$.

The steps of resampling can be summarized as follows:

$$x^+_{k,i} = \text{Probability of } x^-_{k,j} \text{ in terms of } q_j \tag{27}$$

The resampling process diagram is as Figure 2.

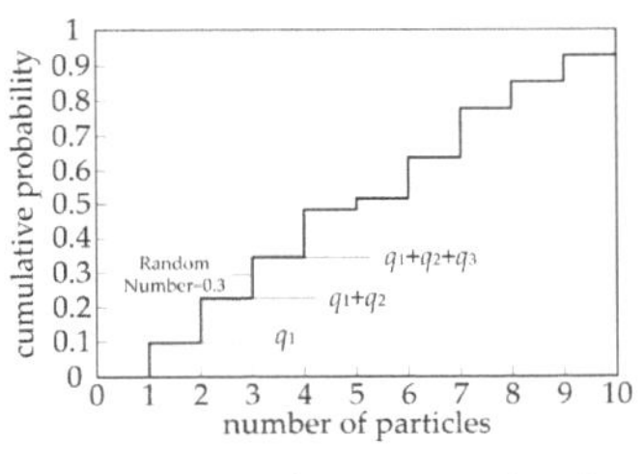

Figure 2. Resampling process flowchart.

The primary bottleneck of the PF is often computational complexity. To address this, more efficient resampling techniques can be used. With particles $x^{+}_{k,i}$ now distributed according to $p(x_k \mid y_k)$, arbitrary statistical properties of this function can be computed.

In summary, the steps of particle filter can be summarized as follows:

1. The system equation and measurement equation are as follows:

$$\begin{aligned} x_{k+1} &= f_k(x_k, w_k) \\ y_k &= h_k(x_k, v_k) \end{aligned} \tag{28}$$

where w_k and v_k are independent white noise processes with known probability density functions.

2. Assuming the probability density function $p(x_0)$ of the initial state is known, N initial particles $x^{+}_{0,i}(i = 1, \ldots, N)$ are randomly generated based on $p(x_0)$. The parameter N serves as a trade-off between computational load and estimation accuracy.

3. For k = 1, 2, . . . , execute the following steps:

(a) Execute time update using the known process equation and the probability density function of the process noise to obtain the prior particles $x^{-}_{k,i}$:

$$x^{-}_{k,i} = f_{k-1}\left(x^{+}_{k-1,i}, w^{i}_{k-1}\right) \ (i = 1, \cdots, N) \tag{29}$$

where each noise vector w^{i}_{k-1} is randomly generated according to the known probability density function of w_{k-1}.

(b) Compute the likelihood probability density q_i for each particle $x^{-}_{k,i}$ given the measurement y_k. This can be obtained by estimating $p(y_k \mid x^{-}_{k,i})$ using the probability density function of the nonlinear measurement equation and the measurement noise.

(c) Normalize the obtained likelihood probability density as follows:

$$q_i = \frac{q_i}{\sum_{j=1}^{N} q_j} \tag{30}$$

Now, the sum of all likelihood probability densities equals 1.

(d) Perform resampling, i.e., randomly generate a set of posterior particles $x^{+}_{k,i}$ based on the likelihood probability densities q_i.

(e) Now that we have a set of particles $x^{+}_{k,i}$ distributed according to the probability density function $p(x_k \mid y_k)$, arbitrary statistical properties of this probability density function can be computed.

4. Extended Kalman Particle Filter

PF has many advantages, such as being a statistical and effective estimation method for highly nonlinear systems, often yielding good results. However, achieving good performance with PF comes with significantly increased computational complexity. Additionally, when the regions of high probability density for the state-space probability density functions $p(y_k \mid x_k)$ and $p(x_k \mid y_{k-1})$ do not overlap, sample degeneracy occurs, causing all particles to converge to a single value. This issue can be mitigated by increasing the number of particles N, but this rapidly leads to a large computational burden and typically only delays sample degeneration. Sample degeneration and other factors may prevent PF from promptly reflecting system dynamics.

Improving the performance of the particle filter can be achieved through combination with other filters. In this study, we adopt the EKPF, which integrates the EKF with the PF to enhance its performance and mitigate the issue of sample degeneration. In this approach, at each measurement instant, the EKF iterates over the particles, and these measurements are used to resample the particles. This is akin to running N Kalman filters concurrently and performing one resampling step after each measurement. After obtaining $x^{-}_{k,i}$ as per Equation (24), $x^{+}_{k,i}$ can be iteratively obtained from $x^{-}_{k,i}$ using the iteration equations of the EKF [32].

$$\begin{aligned} P_{k,i}^{-} &= F_{k-1,i}P_{k-1,i}^{+}F_{k-1,i}^{T} + Q_{k-1} \\ K_{k,i} &= P_{k,i}^{-}H_{k,i}^{T}\left(H_{k,i}P_{k,i}^{-}H_{k,i}^{T} + R_{k}\right)^{-1} \\ x_{k,i}^{+} &= x_{k,i}^{-} + K_{k,i}\left[y_{k} - h\left(x_{k,i}^{-}\right)\right] \\ P_{k,i}^{+} &= (I - K_{k,i}H_{k,i})P_{k,i}^{-} \end{aligned} \tag{31}$$

$K_{k,i}$ represents the Kalman gain for the i-th particle, and $P^{-}{}_{k,i}$ denotes the a priori estimate error covariance for the i-th particle. The matrices F and H, representing the partial derivatives, are defined as follows:

$$\begin{aligned} F_{k-1,i} &= \left.\frac{\partial f}{\partial x}\right|_{x=x_{k-1,i}^{+}} \\ H_{k,i} &= \left.\frac{\partial h}{\partial x}\right|_{x=x_{k,i}^{-}} \end{aligned} \tag{32}$$

Next, following the procedure in Section 3.2, resample $x^{+}{}_{k,i}$ and its corresponding covariance matrix $P^{+}{}_{k,i}$. This method, which updates prior particles $x^{-}{}_{k,i}$ based on the measurement at time k before resampling, helps to mitigate particle degeneracy.

In summary, the EKPF is as follows.

1. The system equation and measurement equation are as follows:

$$\begin{aligned} x_{k+1} &= f_{k}(x_{k}, w_{k}) \\ y_{k} &= h_{k}(x_{k}, v_{k}) \end{aligned} \tag{33}$$

where w_k and v_k are independent white noise processes with known probability density functions.

2. Assuming the probability density function $p(x_0)$ of the initial state is known, N initial particles $x^{+}{}_{0,i}$ are randomly generated based on $p(x_0)$, with corresponding covariance matrices $P^{+}{}_{0,i} = P^{+}{}_{0}(i = 1, \ldots, N)$. The choice of N should balance computational complexity and estimation accuracy.

3. For sampling time k = 1, 2, . . . , follow these steps:

(a) Based on the process equation and the probability density function of the process noise, obtain the prior particles $x^{-}{}_{k,i}$ and covariance $P^{-}{}_{k,i}$ through the time update equation:

$$\begin{aligned} x_{k,i}^{-} &= f_{k-1}\left(x_{k-1,i}^{+}, w_{k-1}^{i}\right) \\ P_{k,i}^{-} &= F_{k-1,i}P_{k-1,i}^{+}F_{k-1,i}^{T} + Q_{k-1} \\ F_{k-1,i} &= \left.\frac{\partial f}{\partial x}\right|_{x=x_{k-1,i}^{+}} \end{aligned} \tag{34}$$

Here, each noise vector $w^{i}{}_{k-1}$ is randomly generated based on the known probability density function of w_{k-1}.

(b) Update the posterior particles and their covariance matrices based on the prior particles and their covariance matrices:

$$\begin{aligned} H_{k,i} &= \left.\frac{\partial h}{\partial x}\right|_{x=x_{k,i}^{-}} \\ K_{k,i} &= P_{k,i}^{-}H_{k,i}^{T}\left(H_{k,i}P_{k,i}^{-}H_{k,i}^{T} + R_{k}\right)^{-1} \\ x_{k,i}^{+} &= x_{k,i}^{-} + K_{k,i}\left[y_{k} - h\left(x_{k,i}^{-}\right)\right] \\ P_{k,i}^{+} &= (I - K_{k,i}H_{k,i})P_{k,i}^{-} \end{aligned} \tag{35}$$

(c) Compute the likelihood probability density q_i for each particle $x^{+}{}_{k,i}$ conditioned on the measurement y_k. The likelihood probability density $p(y_k \mid x^{+}{}_{k,i})$ can be estimated based on the nonlinear measurement equation and the probability density function of the measurement noise.

(d) Normalize the probabilities obtained in the previous step:

$$q_i = \frac{q_i}{\sum_{j=1}^{N} q_j} \tag{36}$$

At this point, all likelihood probabilities sum to 1.

(e) Correct the posterior particles $x^+{}_{k,i}$ and their corresponding covariance matrices $P^+{}_{k,i}$ based on the probabilities q_i. This constitutes the resampling process.

(f) Now that we have a set of posterior particles $x^+{}_{k,i}$ and covariance matrices $P^+{}_{k,i}$, we can compute statistical quantities based on these.

5. Simulation Analysis and Comparison

To confirm the feasibility of the proposed parameter identification algorithm, this chapter carries out relevant simulation verification. In the simulation process, a simulation model is built using MATLAB/Simulink, and the two methods mentioned above are implemented in the s-Function module of Simulink. The flowchart of the s-Function module is shown in Figure 3.

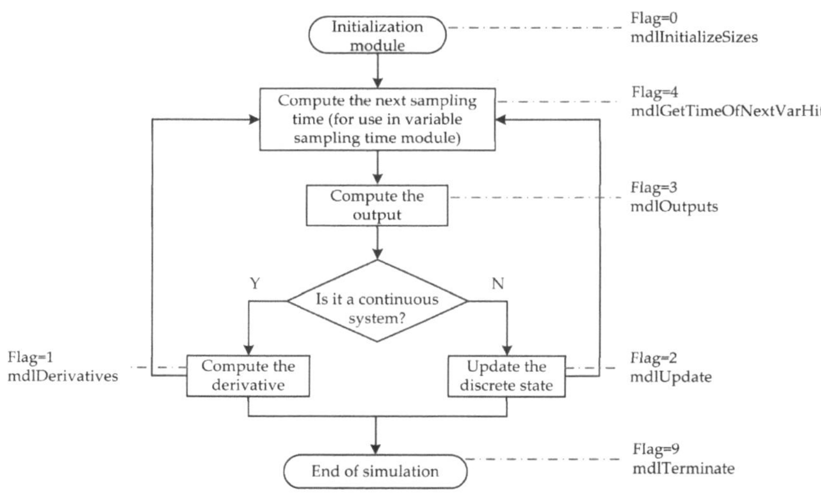

Figure 3. s-Function flowchart.

The s-Function module is implemented using a non-graphical method to create a dynamic system. In this module, the input variables needed for the motor control system are obtained in real time, the state variables are updated in real time, and, thus, the real-time identification of the motor parameters in the control model is achieved. The selected input variables are $\mathbf{u} = [i_\mathrm{d}\ i_\mathrm{q}\ \omega_\mathrm{e}\ u_\mathrm{d}\ u_\mathrm{q}]$, state variables are $\mathbf{x} = [i_\mathrm{d}\ i_\mathrm{q}\ a\ b]$ (where $a = R_\mathrm{s}/L_\mathrm{s}$ and $b = 1/L_\mathrm{s}$), and output variables are $\mathbf{y} = [i_\mathrm{d}\ i_\mathrm{q}]$. The chosen number of particles N is 2000. The particle cloud is generated by providing an initial particle and standard deviation of the initial particle cloud using the "randn" function. The motor parameters used are shown in Table 1.

Table 1. Motor parameters used in the present study.

Electromagnetic Parameters	Value	Unit
DC voltage	24	V
Stator resistance	0.84	Ω
d-axis inductance	3	mH
q-axis inductance	3	mH
Flux linkage	0.01	Wb
Number of pole pairs	4	-

In the first part of this section, simulation analysis is conducted on the PF algorithm to verify its performance and identify any limitations. In the second part, simulation analysis is performed on the EKPF algorithm. Firstly, steady-state condition parameter identification simulation is conducted to validate the effectiveness of the method. Subsequently, to assess the robustness of the EKPF algorithm, simulation analyses are performed under three different operating conditions: variations in motor load, alterations in stator resistance, and changes in stator inductance.

5.1. PF Simulation Analysis

In this section, the PF algorithm is simulated with the following parameter settings: initial particles are set to [0; 5; 208.696; 434.78], initial standard deviation of particle cloud is set to [2; 8; 30; 60], process noise covariance matrix Q is diag([0 0 0.9 1.18]), measurement noise covariance matrix R is diag([1 1]), motor load is set to 0.3 N·m, and motor speed is set to 900 rpm, as shown in Figure 4. Three scenarios are simulated: (1) resistance and inductance remain stable; (2) resistance changes abruptly from 0.84 Ω to 1.2 Ω; (3) inductance changes abruptly from 3 mH to 4 mH.

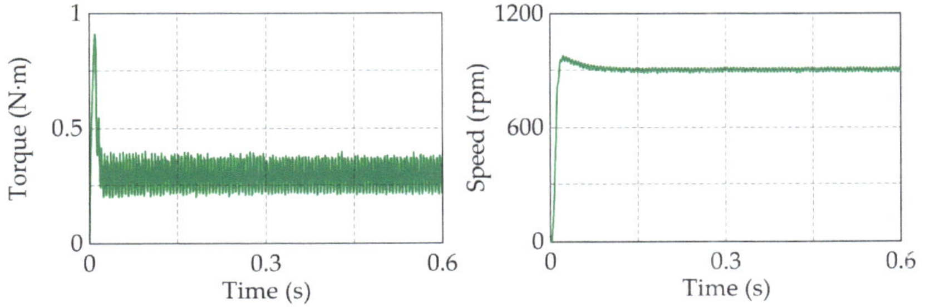

Figure 4. Load torque and speed.

(1) Resistance and inductance remain stable. The simulation results are shown in Figure 5.

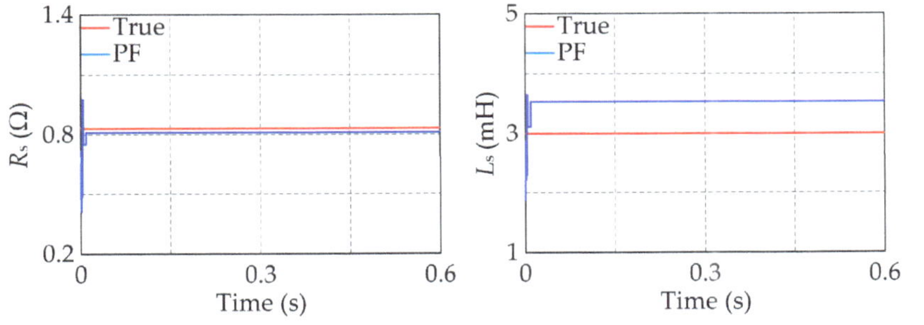

Figure 5. Identification of *Rs* and *Ls* under steady state.

From Figure 5, it can be observed that, for stable parameters, the PF algorithm achieves fast and accurate identification of R_s and L_s.

(2) When the resistance changes abruptly from 0.84 Ω to 1.2 Ω, the simulation results are shown in Figure 6.

(3) The inductance abruptly changes from 3 mH to 4 mH, as shown in the simulation results in Figure 7.

From Figures 6 and 7, it can be observed that, when the stator resistance or stator inductance undergoes a sudden change, the identification process fails to promptly reflect the dynamic changes in the system.

In summary, the PF algorithm can perform identification relatively quickly and accurately to some extent. However, due to the issue of particle degeneracy in the PF algorithm, it may fail to respond promptly to dynamic changes in the system.

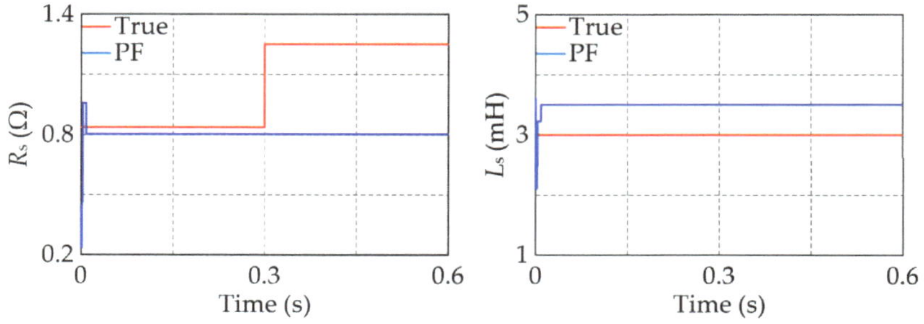

Figure 6. Illustrates the parameter identification of *Rs* and *Ls* under the sudden change in stator resistance.

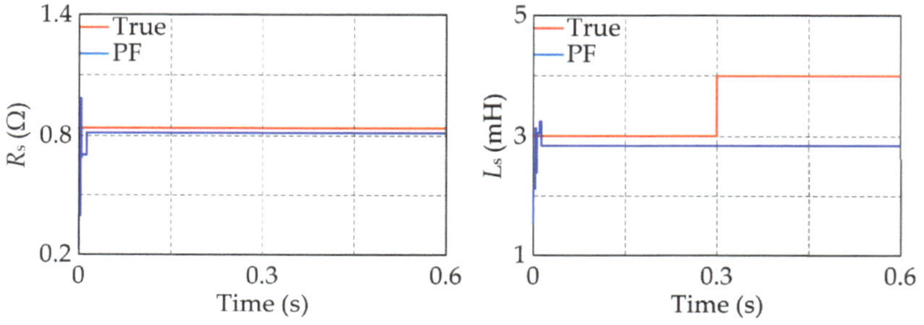

Figure 7. Depicts the parameter identification of *Rs* and *Ls* under the sudden change in inductance from 3 mH to 4 mH.

5.2. EKPF Simulation Analysis

In this section, simulations are conducted to analyze the EKPF algorithm. The parameter settings for the identification algorithm are as follows: the initial particles are set to [0; 5; 208.696; 434.78], the standard deviation of the initial particle cloud is [2; 8; 30; 60], the process noise covariance matrix Q is diag([0 0 0.9 1.18]), the measurement noise covariance matrix R is diag([1 1]), and the estimation error covariance matrix P_0 is diag([0.01 0.1 3 3]). The performance and robustness of the algorithm are validated separately for steady-state and dynamic conditions.

5.2.1. Steady-State Performance

The motor operates under steady-state conditions with a load torque of 0.3 N·m and a speed of 900 rpm, as shown in Figure 8, while the stator resistance and stator inductance remain constant at 0.84 Ω and 3 mH, respectively. The simulation results are shown in Figure 9.

From Figure 9, it can be observed that the EKPF algorithm achieves parameter identification of resistance and inductance in a short time, with the final identification results deviating by around 3% from the actual values. This demonstrates the effectiveness of the parameter identification algorithm.

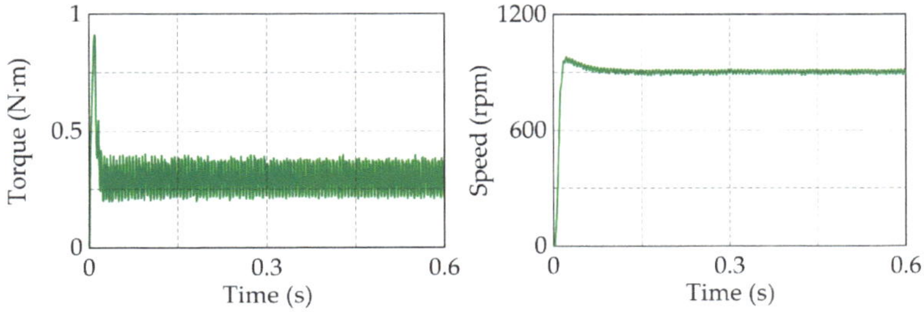

Figure 8. Load torque and speed when torque remains constant.

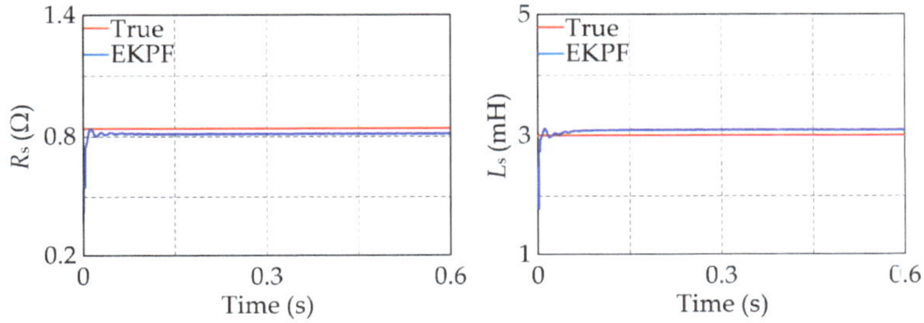

Figure 9. Parameter identification of R_s and L_s.

5.2.2. Robustness Verification

(a) Load Torque

The motor speed is maintained at 900 rpm, while the torque is adjusted from 0.2 N·m to 0.4 N·m within a span of 0.5 s. The outcomes of this simulation are displayed in Figure 10.

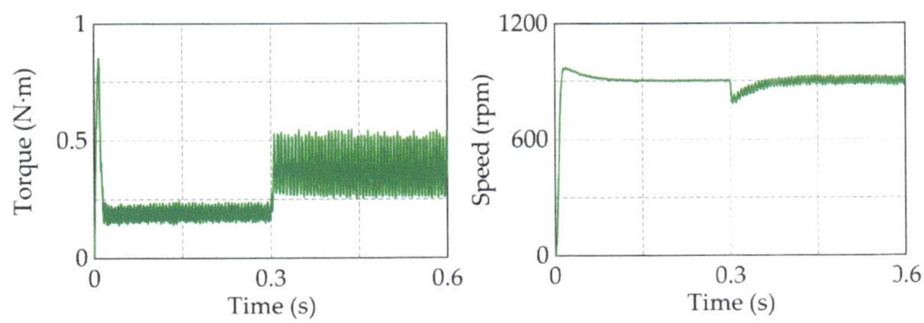

Figure 10. Load torque and speed when torque changes.

From Figure 11, it can be observed that the EKPF algorithm ensures parameter identification even when the torque changes (doubles). The difference between the identification results before and after the change is within 2%, demonstrating the robustness of the EKPF parameter identification algorithm to torque variations.

(b) Stator Resistance

In this section, simulations were performed under two conditions: abrupt changes in resistance due to motor faults and gradual increases in resistance from factors such as temperature rise. The stator resistance shifts suddenly from 0.84 Ω to 1.2 Ω and then increases gradually to 1.2 Ω. The motor speed is fixed at 900 rpm, with the motor load set at 0.3 N·m. The outcomes of these simulations are presented in Figure 12.

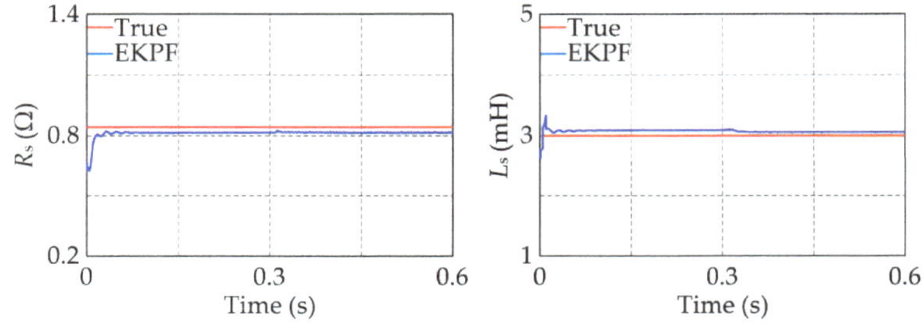

Figure 11. Parameter identification of *Rs* and *Ls* under load torque variation.

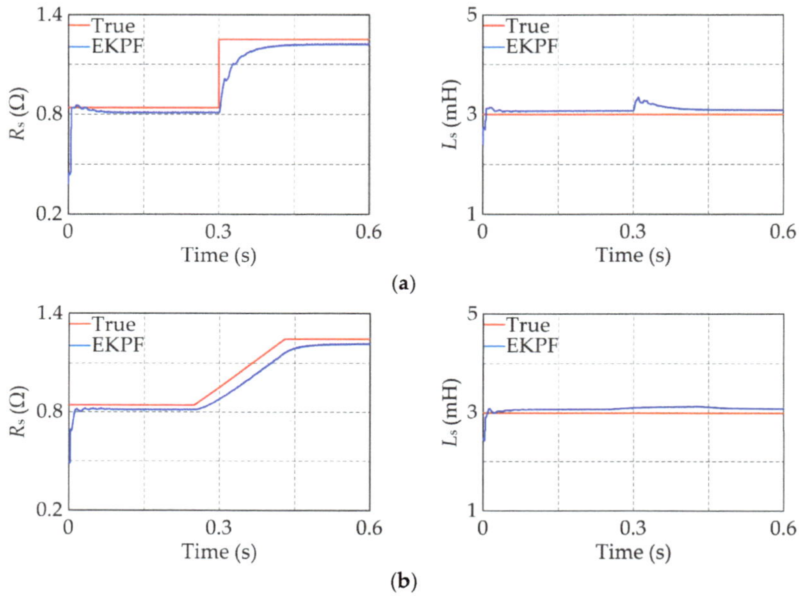

Figure 12. Parameter identification of *Rs* and *Ls* under stator resistance variation: (**a**) sudden resistance change due to motor faults; (**b**) slow increase in resistance due to factors such as temperature rise.

From Figure 12, it can be observed that the EKPF algorithm ensures parameter identification during resistance changes, and it maintains a satisfactory response speed during sudden resistance changes, demonstrating the robustness of the EKPF parameter identification algorithm to resistance variations.

(c) Stator inductance

In this section, simulations were conducted for the inductance change conditions corresponding to the previous section: sudden inductance change due to motor failure and gradual increase in inductance due to improper design. The stator inductance changed abruptly from 3 mH to 4 mH, and gradually increased to 4 mH. The motor speed was established at 900 rpm, and the motor load was configured at 0.3 N·m. The results of the simulation are depicted in Figure 13.

From Figure 13, it can be observed that the EKPF algorithm ensures parameter identification during inductance changes, and it maintains a satisfactory response speed during sudden inductance changes, demonstrating the robustness of the EKPF parameter identification algorithm to inductance variations.

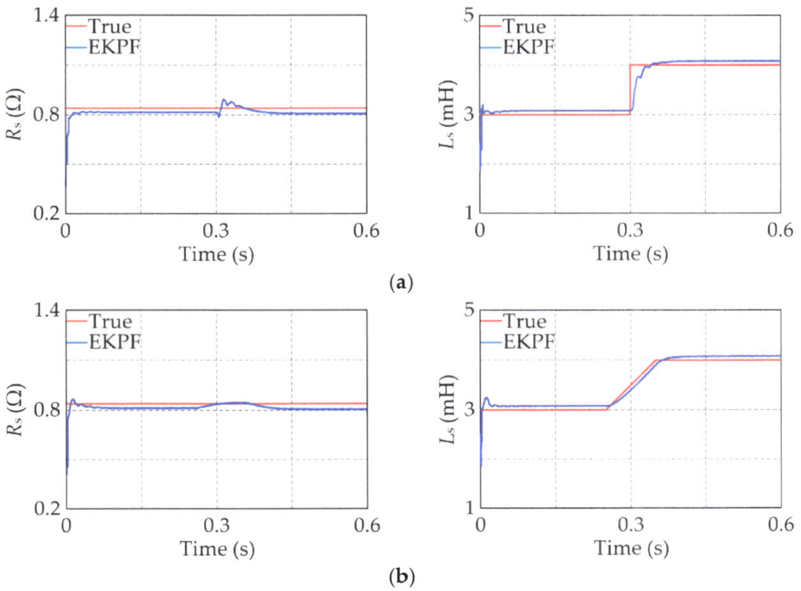

Figure 13. Parameter identification of Rs and Ls under stator inductance variation: (**a**) sudden inductance change; (**b**) gradual increase in inductance.

6. Conclusions

PMSMs are widely used in various fields, such as vessel autopilots. To achieve efficient PMSM control, the accurate identification of motor parameters is particularly important. This paper combines the PF algorithm with the EKF algorithm to form the EKPF algorithm for the online identification of motor resistance and inductance. The theoretical analysis and simulation results show that, compared to the PF algorithm, the EKPF algorithm can identify results more quickly and accurately, and its robustness is significantly improved. When the parameters to be identified change, whether suddenly or gradually, the EKPF algorithm maintains good identification accuracy and fast convergence speed.

Author Contributions: Conceptualization, T.Y. and T.W.; methodology, T.W.; software, T.W.; validation, T.Y., formal analysis, J.B. and T.W.; investigation, J.B. and J.F.; writing—original draft preparation, T.W.; writing—review and editing, J.B. and J.F.; visualization, T.W.; supervision, T.Y.; funding acquisition, T.Y. All authors have read and agreed to the published version of the manuscript.

Funding: This research was funded by Scientific Research Project of Education Department of Jilin Province, grant number (No. JJKH20230121KJ).

Institutional Review Board Statement: Not applicable.

Informed Consent Statement: Not applicable.

Data Availability Statement: Data are contained within the article.

Conflicts of Interest: Authors Jing Bai and Jingwen Fan were employed by the company Yuda Engineering (Jilin) Co., Ltd. The remaining authors declare that the research was conducted in the absence of any commercial or financial relationships that could be construed as a potential conflict of interest.

References

1. Liu, X.; Hu, W.; Ding, W.; Xu, H.; Zhang, Y. Research on multi-parameter identification method of permanent magnet synchronous motor. *Trans. China Electrotech. Soc.* **2020**, *35*, 1198–1207.
2. Zhang, Y.; Zhang, B.; Yang, H.; Norambuena, M.; Rodriguez, J. Generalized sequential model predictive control of IM drives with field-weakening ability. *IEEE Trans. Power Electron.* **2018**, *34*, 8944–8955. [CrossRef]
3. Zhang, Z.; Liu, X.; Yu, J.; Yu, H. Time-varying disturbance observer based improved sliding mode single-loop control of PMSM drives with a hybrid reaching law. *IEEE Trans. Energy Convers.* **2023**, *38*, 2539–2549. [CrossRef]
4. Chen, Z.; Zhang, X.; Liu, C.; Zhang, H.; Luo, G. Research on Current decoupling and harmonic suppression strategy of permanent magnet synchronous motor by proportional resonance adaptive disturbance rejection control. *Proc. CSEE* **2021**, *42*, 9062–9072.
5. Ullah, K.; Guzinski, J.; Mirza, A.F. Critical review on robust speed control techniques for permanent magnet synchronous motor (PMSM) speed regulation. *Energies* **2022**, *15*, 1235. [CrossRef]
6. Wang, M.; Sun, D.; Ke, W.; Nian, H. A universal lookup table-based direct torque control for OW-PMSM drives. *IEEE Trans. Power Electron.* **2020**, *36*, 6188–6191. [CrossRef]
7. Liu, X.P.; Hu, W.P.; Zou, Y.L.; Zhang, Y. Multi-parameter identification of permanent magnet synchronous motor based on improved particle swarm optimization. *Electr. Mach. Control* **2020**, *24*, 112–120.
8. Fang, G.H.; Wang, H.C.; Gao, X. Parameter identification algorithm of permanent magnet synchronous motor based on dynamic forgetting factor recursive least square method. *Comput. Appl. Softw.* **2021**, *38*, 280–283.
9. Shen, Y.; Jin, B. Permanent magnet synchronous motor fuzzy forgetting factor recursive least squares parameter identification. *J. Syst. Simul.* **2019**, *30*, 3404–3410.
10. Cao, X.Y.; Zhang, X.X. Research on multi-parameter identification metthod of parmanent magnet synchronous motor based on mutation particle swarm optimization. *Micromotors* **2021**, *54*, 83–88+96.
11. Li, W.; Du, Z. PMSM parameter identification based on improved grey wolf optimization algorithm. *Modul. Mach. Tool Autom. Manuf. Tech.* **2020**, *4*, 113–117.
12. Odhano, S.A.; Pescetto, P.; Awan, H.A.A.; Hinkkanen, M.; Pellegrino, G.; Bojoi, R. Parameter identification and self-commissioning in AC motor drives: A technology status review. *IEEE Trans. Power Electron.* **2018**, *34*, 3603–3614. [CrossRef]
13. Yin, H.; Wei, Y.; Zhang, Y.; Jing, P.; Cai, D.; Liu, X. Identification of control parameters of the permanent magnetic synchronous generator using least square method. *Energy Rep.* **2022**, *8*, 1538–1545. [CrossRef]
14. Holakooie, M.H.; Ojaghi, M.; Taheri, A. Modified DTC of a six-phase induction motor with a second-order sliding-mode MRAS-based speed estimator. *IEEE Trans. Power Electron.* **2018**, *34*, 600–611. [CrossRef]
15. Khan, Y.A.; Verma, V. Stator resistance estimation for MRAS-based speed sensorless vector-controlled switched reluctance motor drive. *Electr. Eng.* **2021**, *103*, 1949–1963. [CrossRef]
16. Yildiz, R.; Barut, M.; Zerdali, E. A comprehensive comparison of extended and unscented Kalman filters for speed-sensorless control applications of induction motors. *IEEE Trans. Ind. Inform.* **2020**, *16*, 6423–6432. [CrossRef]
17. Yin, Z.; Li, G.; Zhang, Y.; Liu, J. Symmetric-strong-tracking-extended-Kalman-filter-based sensorless control of induction motor drives for modeling error reduction. *IEEE Trans. Ind. Inform.* **2018**, *15*, 650–662. [CrossRef]
18. Liu, X.; Wang, X.P.; Wang, S.H. Inductance indentification of permanent magnet synchronous motor based on least square method. *Electr. Mach. Control Appl.* **2020**, *47*, 1–5+32.
19. Zhu, Z.-Q.; Liang, D.; Liu, K. Online parameter estimation for permanent magnet synchronous machines: An overview. *IEEE Access* **2021**, *9*, 59059–59084. [CrossRef]
20. Gao, D.X.; Zhou, L.; Chen, J.; Pan, C.Z. Design of derivative-free model-reference adaptive control for a class of uncertain systems based on disturbance compensation. *Control Theory Appl.* **2023**, *40*, 735–743.
21. Dang, D.Q.; Rafaq, M.S.; Choi, H.H.; Jung, J.-W. Online parameter estimation technique for adaptive control applications of interior PM synchronous motor drives. *IEEE Trans. Ind. Electron.* **2015**, *63*, 1438–1449. [CrossRef]
22. Liu, X. *Research on Torque Ripple Suppression of Brushless DC Motor Based on Buck-Boost Topology and Kalman Filter*; Jiangsu University: Zhenjiang China, 2018.
23. Hu, G.; Gao, B.; Zhong, Y.; Gu, C. Unscented kalman filter with process noise covariance estimation for vehicular ins/gps integration system. *Inf. Fusion* **2020**, *64*, 194–204. [CrossRef]
24. Wasim, M.; Ali, A.; Choudhry, M.A.; Saleem, F.; Shaikh, I.U.H.; Iqbal, J. Unscented Kalman filter for airship model uncertainties and wind disturbance estimation. *PLoS ONE* **2021**, *16*, e0257849. [CrossRef]
25. Huang, Y.; Zhang, Y.; Wu, Z.; Li, N.; Chambers, J. A novel adaptive Kalman filter with inaccurate process and measurement noise covariance matrices. *IEEE Trans. Autom. Control* **2017**, *63*, 594–601. [CrossRef]
26. Xia, J.; Gao, S.; Zhong, Y.; Zhang, J.; Gu, C.; Liu, Y. A novel fitting H-infinity Kalman filter for nonlinear uncertain discrete-time systems based on fitting transformation. *IEEE Access* **2019**, *8*, 10554–10568. [CrossRef]
27. Xia, J.; Gao, S.; Gao, B.; Wei, W.; Tian, T. Fitting H-infinity filter for nonlinear discrete-time systems. In Proceedings of the 2019 Chinese Control and Decision Conference (CCDC), Nanchang, China, 3–5 June 2019; pp. 4022–4027.
28. Zhao, J.; Zhang, G.; Dong, Z.Y.; La Scala, M. Robust forecasting aided power system state estimation considering state correlations. *IEEE Trans. Smart Grid* **2016**, *9*, 2658–2666. [CrossRef]
29. Zhao, J.; Mili, L. A robust generalized-maximum likelihood unscented Kalman filter for power system dynamic state estimation. *IEEE J. Sel. Top. Signal Process.* **2018**, *12*, 578–592. [CrossRef]

30. Zhao, J.; Mili, L. A theoretical framework of robust H-infinity unscented Kalman filter and its application to power system dynamic state estimation. *IEEE Trans. Signal Process.* **2019**, *67*, 2734–2746. [CrossRef]
31. Bian, Z.T.; Chen, H. Design and implementation of multi-core parallel particle filter algorithm. *Comput. Telecommun.* **2024**, *1*, 63.
32. Wei, Z.S.; Hou, W.; Zhao, Y.; Zheng, S.S. State of charge estimation of lithium-iron batteries based on extended Kalman filter. *Acta Sci. Nat. Univ. Sunyatseni* **2023**, *62*, 92–100.

Journal of
Marine Science and Engineering

MDPI

Article

Current Harmonic Suppression in Maritime Vessel Rudder PMSM Drive System Based on Composite Fractional-Order PID Repetitive Controller

Tianqing Yuan [1,2], Tianli Wang [1,2], Jingwen Fan [3,*] and Jing Bai [3]

1 Key Laboratory of Modern Power System Simulation and Control & Renewable Energy Technology, Ministry of Education, Northeast Electric Power University, Jilin 132012, China
2 Department of Electrical Engineering, Northeast Electric Power University, Jilin 132012, China
3 Yuda Engineering (Jilin) Co., Ltd., Siping 136000, China
* Correspondence: tskfjw@163.com

Abstract: To address the control performance and harmonic suppression issues in maritime vessel rudder permanent magnet servo systems, a fractional-order PID controller was introduced into the existing improved repetitive control strategy. We used the Oustaloup approximation algorithm and particle swarm optimization for tuning the fractional-order PID controller. The optimized parameters substantially improved the control performance. By integrating the fractional-order PID controller with the improved repetitive controller, a composite fractional-order PID repetitive control strategy was formed. Finally, MATLAB/Simulink simulations were conducted to compare and verify the disturbance rejection and harmonic suppression capabilities of the improved control strategy. The results demonstrate its superior control performance, thereby increasing the practicality of the control system in dealing with various situations.

Keywords: permanent magnet servo system; harmonic suppression; repetitive control; fractional-order PID control

1. Overview

1.1. Introduction

With the development of modern production technology, permanent magnet servo systems have been widely applied in servo robots, high-precision industrial equipment, aerospace, and other fields. In recent years, the application of permanent magnet servo systems in the shipbuilding and marine industries has also gradually increased, particularly excelling in autopilot systems and precision control equipment. To maintain safe and stable operation under various complex application scenarios, the primary focus is on improving the control performance of permanent magnet servo systems. When navigating in the marine environment, vessels must cope with changing sea conditions and stringent operational requirements, which place higher demands on the control accuracy and reliability of permanent magnet servo systems. Current harmonics are key factors affecting the energy utilization and control performance of permanent magnet servo systems, making harmonic suppression a major research direction. For maritime vessel autopilot systems, current harmonics not only affect motor efficiency and temperature increase but also negatively impact the precise control of the steering gear, thereby affecting the ship's course stability and steering sensitivity. In this context, researching efficient current harmonic suppression methods becomes particularly important. With the advancement of industrial technology and control theory, harmonic suppression methods are primarily divided into two categories: one approach focuses on motor structure design, optimizing defects in the stator skew slot and skewed pole, and the shape of the rotor's permanent magnets to improve the distribution of the magnetic field of the permanent magnets, enhance the sinusoidal nature of the back electromotive force (EMF), and reduce the harmonic content of the back EMF. The other approach focuses on control strategies, where optimized control methods compensate for current harmonics, thereby improving the current waveform and enhancing the harmonic suppression capability of the servo system [1–3].

In order to suppress cogging torque, an innovative approach was proposed in reference [4], which involves introducing auxiliary slots in the stator teeth and using a strategy combining a response surface model with the particle swarm algorithm to optimize and adjust the structural parameters of the auxiliary slots. Variance analysis of the constructed multidimensional quadratic response model was conducted to examine the interaction effects among variables and determine the optimal combination of structural parameters. The experimental results demonstrated its positive effect on suppressing cogging torque. Research on permanent magnet servo control systems has shown that improving control strategies can effectively suppress current harmonics and torque fluctuations. The Proportional Integral (PI) controller, as an effective current control method, can achieve zero steady-state error tracking of the system and has good robustness. Therefore, in permanent magnet servo systems, PI controllers are generally used as closed-loop feedback controllers [5]. However, traditional PI controllers can only achieve zero steady-state error tracking for DC signals and struggle to eliminate AC harmonic components [6]. To mitigate AC harmonics and reduce their impact on permanent magnet servo systems, various control methods such as sliding mode control, fuzzy control, repetitive control, fractional-order Proportional Integral Derivative (PID) control, etc., have been applied in servo control systems. Sliding mode control can enhance system disturbance rejection and is widely used in permanent magnet servo systems. In order to reduce the influence of external disturbances, parameter variations, and other uncertainties on the system control performance, reference [7] combines adaptive control with sliding mode control, proposing an adaptive sliding mode control strategy. The strategy adjusts system parameters online using adaptive laws and introduces hyperbolic tangent functions to reduce chattering. Subsequently, the adaptive laws are corrected via a projection factor to limit the range of adaptive gains. The system's disturbance rejection capability and dynamic response speed are verified through simulation. Reference [8] presents a continuous fast nonsingular terminal sliding mode tracking control scheme based on an uncertainty observer and utilizes finite-time exact observers to simplify the setting of control gain parameters. It achieves rapid convergence and high tracking accuracy under parameter uncertainties, exhibiting excellent disturbance rejection performance. Sliding mode control has good robustness but involves complex controller design. From the perspective of application effectiveness, fuzzy control demonstrates robustness but struggles to meet the requirements of high-precision control scenarios and generally needs to be combined with other algorithms. To address the balancing issue between the dynamic response and steady-state error of traditional PI controllers, reference [9] combines fuzzy control with PI control, designing a PI–fuzzy hybrid controller. This controller balances a low settling time and high steady-state accuracy, eliminates overshoot, and improves the system's dynamic performance.

1.2. Objectives

In practice, improving the motor structure for higher machining accuracy poses significant challenges, making implementation difficult. Therefore, it is generally achieved through control strategies to suppress current harmonics [10–13]. However, the aforementioned control strategies cannot simultaneously improve controller effectiveness while maintaining low system complexity. Currently used harmonic suppression strategies include PI control, PID control, fractional-order Proportional Integral Derivative (FOPID) control, sliding mode control, fuzzy control, repetitive control (RC), etc. [14–19]. Among them, repetitive control demonstrates high control accuracy and strong disturbance rejection capability, making it widely used in harmonic suppression in control systems. Fractional-order PID control, on the basis of the simple structure and robustness of PID controllers, introduces fractional order theory, thus offering broader applicability [20–23].

Based on the background mentioned above, this paper focuses on a permanent magnet servo system utilizing a combination of repetitive control and fractional-order PID control. The aim is to reduce overshoot, enhance the practicality of the control system under various conditions, and improve the system's harmonic suppression capability.

1.3. Paper Organization

The main research and overall research approach included in the paper are as follows.

For addressing the control performance and current harmonic suppression issues of the permanent magnet servo system, first, a fractional-order PID controller is introduced into the existing improved repetitive control strategy, with the fractional calculus operator approximated using the Oustaloup approximation algorithm. Then, the parameters of the fractional-order PID controller are tuned using the particle swarm optimization algorithm, selecting the optimal control result as the controller parameters. Furthermore, the fractional-order PID controller is combined with the improved repetitive controller to form a composite fractional-order PID repetitive control strategy. Finally, the improved control strategy's disturbance rejection and harmonic suppression capabilities are tested through simulation.

2. Fractional-Order PID Controller Design

The fractional calculus theory serves as the theoretical foundation for the fractional-order PID controller. It extends traditional calculus by incorporating fractional-order differentiation and integration operators, denoted as s^{λ} and s^{μ}, respectively. By adjusting the orders of these operators, transformations between controllers can be achieved. However, in the fractional-order PID controller, the differentiation and integration operators are irrational functions. Therefore, to implement the functionality of these operators within the controller, rational approximation is necessary.

Common approximation methods include the power series approximation of the Euler operator, the continued fraction approximation of the Tustin operator, and indirect approximation methods such as Carlson's method and Oustaloup's method. Among these, the Oustaloup approximation method exhibits a superior approximation performance and is more desirable. Hence, in this study, the Oustaloup approximation method is employed to approximate the differentiation and integration operators as rational functions.

2.1. Principle of Integer-Order PID Controller

The integer-order PID controller is typically employed in situations where accurate modeling of the system can be achieved. Its closed-loop control principle is as follows: the error $E(s)$ between the target signal $i_{ref}(s)$ and the actual output signal $i_g(s)$ of the system is adjusted by the proportional gain K_p, integral gain $K_i s^{-1}$, and derivative gain $K_d s$ of the integer-order PID controller. The output control signal $Y(s)$ is then used as the input signal to the controlled object, resulting in the actual output signal. The control principle of the integer-order PID controller is illustrated in Figure 1.

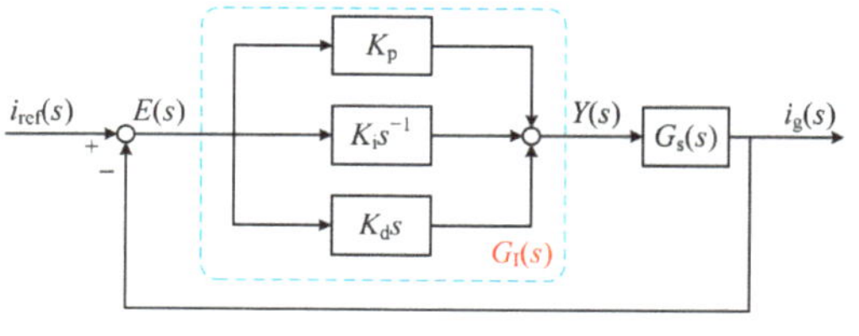

Figure 1. Schematic diagram of the integer-order PID controller control principle.

The transfer function of the integer-order PID controller can be expressed as follows:

$$G_I(s) = \frac{Y(s)}{E(s)} = K_p + \frac{K_i}{s} + K_d s \tag{1}$$

where $E(s)$ and $R(s)$ represent the input and output of the integer-order PID controller, respectively, while K_p, K_i, and K_d are the proportional, integral, and derivative coefficients, respectively. The time-domain expression of the integer-order PID controller is as follows:

$$Y(t) = K_p\left[e(t) + \frac{1}{N_i}\int_0^t e(t) + N_d\frac{de(t)}{dt}\right] = K_p + K_i\int_0^t e(t) + K_d e(t) \quad (2)$$

where N_i and N_d respectively denote the integral time constant and the derivative time constant, satisfying $N_i = K_p/K_i$ and $N_d = K_p/K_d$.

The proportional coefficient K_p allows adjustment of the error signal's magnitude to enhance the controller's response speed. The integral component integrates the error signal, and adjusting the integral coefficient K_i can improve the stability of the control system. The derivative component differentiates the error signal, influencing its rate of change, and adjusting the derivative coefficient K_d can shorten the adjustment time. By appropriately tuning these three coefficients, the control action of the integer-order PID controller can be realized.

2.2. Principle of Fractional-Order PID Controller

In practical applications, it is often challenging to accurately model the controlled system, making it difficult to achieve the ideal control performance of integer-order PID controllers. The emergence of fractional-order PID controllers addresses this issue. In principle, the fractional-order PID controller represents an innovative adaptation of fractional order theory to the traditional integer-order PID controller. Structurally, the fractional-order PID controller introduces the additional fractional orders λ for the integral component and μ for the derivative component. This not only enhances robustness but also provides greater flexibility and applicability.

Similar to the control principle of the integer-order PID controller, the fractional-order PID controller adjusts the error signal through the proportional gain K_p, integral gain K_i/s^λ, and derivative gain $K_d{\cdot}s^\mu$. The control principle diagram for the fractional-order PID controller is illustrated in Figure 2.

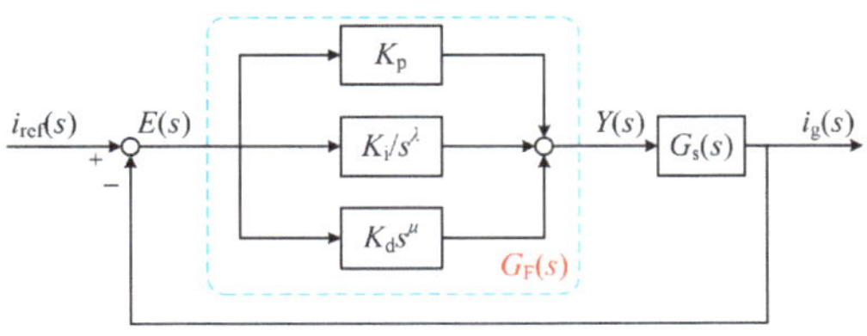

Figure 2. Schematic diagram of the fractional-order PID controller control principle.

From the figure, it can be observed that the transfer function $G_F(s)$ of the fractional-order PID controller can be represented as follows:

$$G_F(s) = \frac{Y(s)}{E(s)} = K_p + \frac{K_i}{s^\lambda} + K_d s^\mu \ \lambda > 0, \mu > 0 \quad (3)$$

where $E(s)$ and $R(s)$ represent the input and output of the fractional-order PID controller, respectively, while s^λ and s^μ represent the integral and derivative operators, respectively. According to Equation (3), the expression of the $PI^\lambda D^\mu$ controller in the time domain can be obtained as follows:

$$Y(t) = K_p + K_i D^{-\lambda} e(t) + K_d D^\mu e(t) \quad (4)$$

Figure 3 shows the range of values for the orders of the integer-order PID controller and the fractional-order PID controller. From the figure, it can be observed that the orders λ and μ of the integer-order PID controller can only be 0 or 1, corresponding to the four combinations on the graph: P, PI, PD, and PID controllers. In contrast, in the fractional-order

PID controller, λ and μ can take any value greater than 0, as indicated by any point within the shaded area on the graph, including the boundaries and endpoints. Therefore, the fractional-order PID controller exhibits greater flexibility, allowing for specific adjustment of the values of λ and μ according to the actual situation to achieve better control performance.

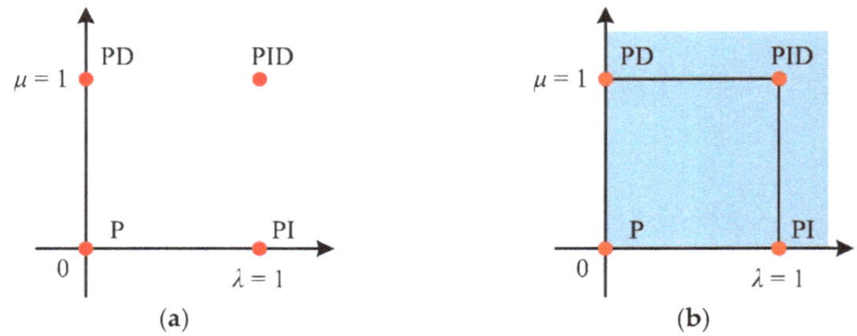

Figure 3. Controller calculus order value range. (**a**) Integer-order PID controller order value range. (**b**) Fractional-order PID controller value range.

In the fractional-order PID controller, when $\lambda = 1$ and $\mu = 1$, $G_F(s) = K_p + K_i/s + K_d s$, exhibiting the form of an integer-order PID controller; when $\lambda = 1$ and $\mu = 0$, $G_F(s) = K_p + K_i/s$, exhibiting the form of an integer-order PI controller; when $\lambda = 0$ and $\mu = 1$, $G_F(s) = K_p + K_d s$, exhibiting the form of an integer-order PD controller; when $\lambda = 0$ and $\mu = 0$, $G_F(s) = K_p$, exhibiting the form of a proportional controller; when $\lambda > 0$ and $\mu = 0$, $G_F(s) = K_p + K_i/s^{\lambda}$, exhibiting the form of a fractional-order PI^{λ} controller; and when $\lambda = 0$ and $\mu > 0$, $G_F(s) = K_p + K_d s^{\mu}$, exhibiting the form of a fractional-order PD^{μ} controller.

2.3. Oustaloup Approximation for Fractional-Order Calculus Operators

The approximation methods for fractional-order calculus operators can be categorized into direct approximation and indirect approximation methods. The difference between indirect and direct methods lies in the approach: indirect methods approximate first and then transform, meaning that the transfer function is first approximated into an integer-order form in the complex frequency domain s, followed by a transformation from the s domain to the z domain. In indirect approximation methods, the Oustaloup approximation method exhibits superior approximation performance. Therefore, this paper adopts the Oustaloup approximation method to approximate the fractional-order calculus operators s^{λ} and s^{μ} in the fractional-order PID controller.

The Oustaloup filter transfer function can be expressed as follows:

$$s^{\alpha} = H(s) = K\prod_{k=1}^{N}\frac{s + \omega_k'}{s + \omega_k} \tag{5}$$

$$K = \omega_h^{\alpha},\ \omega_k' = \omega_b\left(\frac{\omega_h}{\omega_b}\right)^{\frac{2k-1-\alpha}{2N}},\ \omega_k = \omega_h\left(\frac{\omega_h}{\omega_b}\right)^{\frac{2k-1+\alpha}{2N}} \tag{6}$$

where α represents the fractional order, N denotes the approximation order, and ω_h and ω_b respectively denote the upper and lower limits of the approximation frequency band.

2.4. The Orders λ and μ of the Fractional-Order PID Controller's Calculus Operators Affect the Control System

In order to gain a visual understanding of the effects of the orders λ and μ of the fractional-order PID controller's integral and derivative operators on the control system, this paper employs the fractional-order system from reference [24] as the controlled object. Through simulation and comparative analysis, the effects of increasing or decreasing parameters on the system are studied. The transfer function of the fractional-order system in integer-order form can be represented as follows:

$$G_{Fs}(s) = \frac{1}{0.8s^{2.2} + 0.5s^{0.9} + 1} \quad (7)$$

Utilizing the method of least squares, the integer-order form of the fractional-order system can be approximately obtained as follows:

$$G_{Is}(s) = \frac{1}{0.7414s^{2} + 0.2313s + 1} \quad (8)$$

The corresponding transfer function of the fractional-order PID controller can be obtained as follows:

$$G_F(s) = 138.18 + 2.89s^{-0.2} + 12.38s^{1.1} \quad (9)$$

The step response waveforms of the controlled fractional-order system and the controlled fractional-order system using the fractional-order PID controller are depicted in Figure 4. From the graph, it can be observed that in the step response curve of the fractional-order system with the application of the fractional-order PID controller, the overshoot is significantly reduced, the amplitude of the curve fluctuation is decreased, and the settling time is notably shortened, enabling a faster attainment of stability.

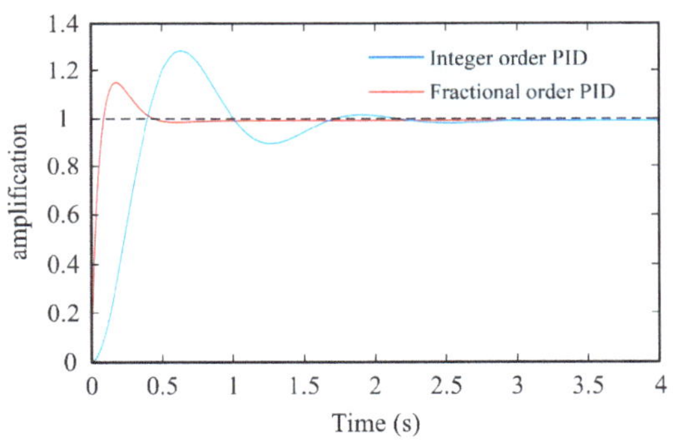

Figure 4. Comparison of system step response.

From references [25,26], it can be inferred that selecting the appropriate orders λ and μ for the integral and derivative operators can positively influence the system stability. However, if the values of λ and μ are excessively large, it may instead have a negative impact on the control performance of the fractional-order PID controller, reducing the system's stability. Therefore, typically, λ and μ are chosen to be within the range of (0, 2) to maintain good control performance.

3. The Improved Fractional-Order PID Repetitive Controller

Based on the analysis above, it is evident that reasonable selection of the parameters for the fractional-order PID controller can have a positive impact on the system's stability. In order to enhance the system stability while maintaining harmonic suppression performance and better suppressing periodic disturbances, this section applies the fractional-order PID controller to improve the repetitive controller, proposing a compound repetitive controller.

To achieve the desired control performance of the fractional-order PID controller, it is necessary to tune its parameters. Classical tuning methods for fractional-order controllers include time-domain and frequency-domain tuning methods, such as the dominant pole method and the gain and phase margin method. However, due to the complexity of traditional fractional-order PID controller parameter tuning, the introduction of various constraint formulas increases the computational complexity, resulting in a slow tuning speed. Moreover, the parameters obtained after tuning are often not accurate and may deviate from the ideal values.

Intelligent algorithms, such as particle swarm optimization, have been increasingly used for tuning fractional-order PID controllers. By setting objective functions, optimization solutions can be quickly found, and the tuned parameters are often more optimal, meeting the control performance requirements of the controller. Common intelligent algorithms include ant colony optimization, grey wolf optimization, particle swarm optimization, etc. In this section, the particle swarm optimization algorithm is introduced to tune the five parameters of the fractional-order PID controller.

3.1. Particle Swarm Optimization Algorithm

In 1995, American scholars Kennedy and Eberhart proposed the particle swarm optimization (PSO) algorithm based on modeling and simulation of the foraging behavior observed in bird flocks. Inspired by the decision-making process in human behavior, the PSO algorithm simulates the cooperation and information sharing among particles in a swarm to find the optimal solution to a given problem. With its versatility and effectiveness in optimization, the PSO algorithm has been widely applied in solving optimization problems associated with nonlinear and high-order systems. The iterative process of particles is illustrated in Figure 5.

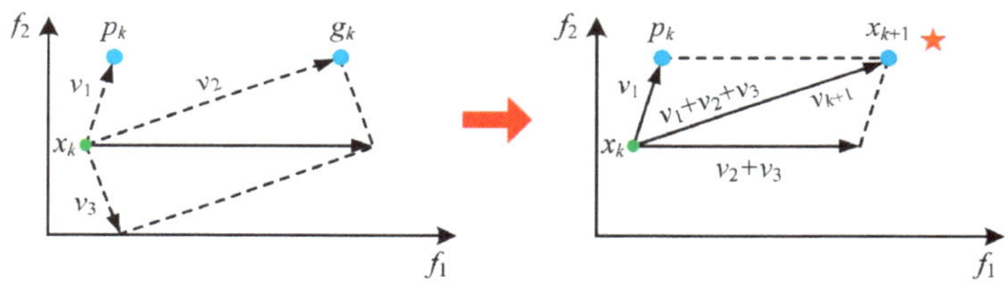

Figure 5. Diagram of the iterative process of particles.

From the above diagram, it can be intuitively understood how the positions and velocities change from the i-th optimization iteration to the $(i + 1)$-th optimization iteration. The red star denotes the global optimum; v_1 represents the velocity influenced by the particle's own movement history, directing it towards its personal best position; v_2 represents the velocity influenced by the collective movement experience of the entire population, directing the particle towards the global best position; and v_3 represents the particle's original velocity in the i-th iteration.

In a population optimization space set to N-dimensional space, comprising M particles, the updated formula for the particle swarm algorithm can be expressed as follows:

$$v_k(i+1) = wv_k(i) + c_1r_1(p_k(i) - x_k(i)) + c_2r_2(g_k(i) - x_k(i)) \tag{10}$$

$$x_k(i+1) = x_k(i) + v_k(i+1) \tag{11}$$

where w represents the inertia weight, indicating the degree to which the velocity of particles in the next iteration inherits from the current iteration velocity; $p_k(i)$ denotes the individual best position at the i-th iteration; $g_k(i)$ signifies the global best position at the i-th iteration; r_1 and r_2 are random numbers within the interval (0, 1); and c_1 and c_2 are learning factors, representing the degree to which particles learn from the individual best position and the global best position, respectively [27].

The basic process of PSO for optimization is illustrated in Figure 6.

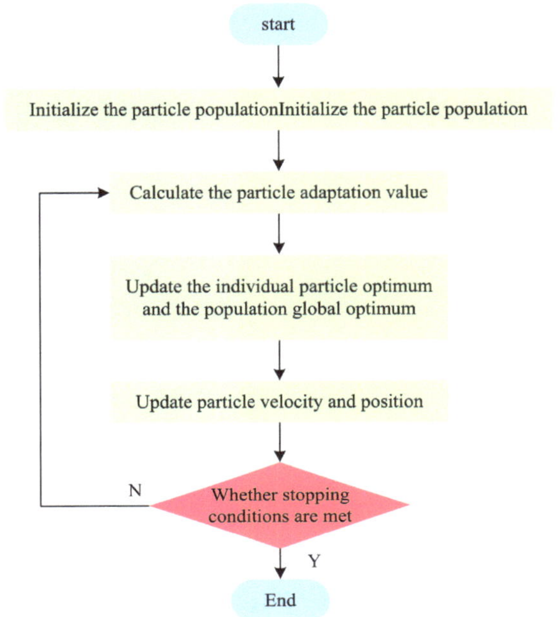

Figure 6. Basic flowchart of particle swarm algorithm.

3.2. Fractional-Order PID Controller Parameterization

In this section, the five parameters of the fractional-order PID controller are tuned using the particle swarm optimization algorithm to achieve the desired control performance. The schematic diagram of tuning the controller parameters using the particle swarm optimization algorithm is shown in Figure 7.

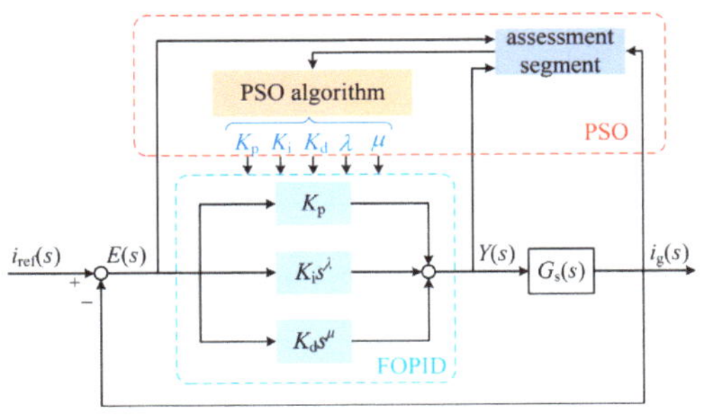

Figure 7. Particle swarm algorithm-rectified fractional-order PID controller structure.

The evaluation stage in the figure assesses the quality of errors during the particle swarm optimization tuning process. By setting error metrics, the problem of finding the optimal parameters for the fractional-order PID controller is transformed into minimizing the error performance index problem. To achieve better selectivity, performance indices such as ISE, ITSE, IAE, ITAE, etc., are commonly used in research as evaluation criteria. In order to enhance the disturbance rejection capability of the control system, this paper selects

ITAE along with overshoot as performance criteria for tuning the controller parameters. The ITAE criterion can be expressed as follows:

$$J = \int_0^{\infty} t \cdot |e(t)| dt \tag{12}$$

Evaluation indicators can be expressed as follows:

$$J_e = J + J_o = \int_0^{\infty} t \cdot |e(t)| dt + 0.001 \cdot \max(e_o) \tag{13}$$

where J_o denotes the criterion for evaluating the criterion as an overshooting quantity and e_o denotes the overshooting quantity.

The parameter tuning process can be implemented by combining simulation models with MATLAB files. The specific implementation process is as follows: First, a certain scale of particle swarm is generated through the particle swarm optimization algorithm and transmitted to the Simulink model via MATLAB files to initialize the parameters. Then, the simulation model is executed, and the fitness values are updated based on error performance metrics. Finally, the fitness values are returned to the particle swarm algorithm for iteration through the program. This process continues iteratively until the algorithm termination conditions are met, thereby concluding the tuning process.

In the process of parameter tuning, the inertial weight w is set to 0.7, and the learning factors c_1 and c_2 are set to 2. These parameters are used to adjust the fractional-order PID controller applied to the specified controlled object. After multiple iterations of parameter tuning, the following set of parameters are selected as the controller parameters: K_p = 2.1574, K_i = 200.0125, K_d = 0.7, λ = 1.1, and μ = 0.2.

3.3. Improved Fractional-Order PID Repeat Controller

The tuned fractional-order PID controller is integrated into the control loop by replacing the current PI regulator, forming a parallel combination with the improved discrete Fourier transform repetitive controller [28]. This integration results in the formation of an improved fractional-order PID repetitive controller. The structure of the improved fractional-order PID repetitive controller is illustrated in Figure 8.

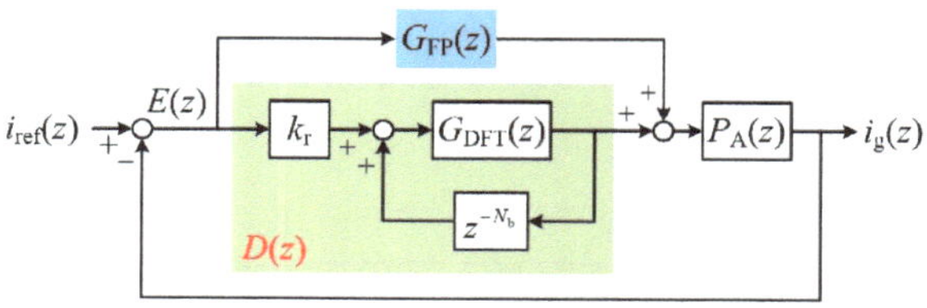

Figure 8. Structure diagram of fractional-order PID repetitive controller.

In the figure, $G_{FP}(z)$ denotes the transfer function of the fractional-order PID controller.

From the diagram, we can derive the relationship between the current control error $E(z)$ and the input signal $i_{ref}(z)$ as follows:

$$\frac{E(z)}{i_{ref}(z)} = \frac{1}{1 + [D(z) + G_{FP}(z)]P_A(z)} = \frac{1}{1 + \left[\frac{k_r G_{DFT}(z)}{1 - z^{-N_b} G_{DFT}(z)} + G_{FP}(z)\right] P_A(z)} \tag{14}$$

From the above equation, we can obtain the characteristic equation of the permanent magnet synchronous motor (PMSM) drive system using the improved fractional-order PID repetitive controller as follows:

$$\begin{aligned} 1 + [D(z) + G_{FP}(z)]P_A(z) &= [1 + G_{FP}(z)P_A(z)]\left[1 + D(z)\frac{P_A(z)}{1 + G_{FP}(z)P_A(z)}\right] \\ &= [1 + G_{FP}(z)P_A(z)][1 + D(z)G_M(z)] \end{aligned} \quad (15)$$

where $G_M(z)$ represents the equivalent controlled object of the improved fractional-order PID repetitive controller.

Similarly, when all the characteristic roots of the characteristic equation lie within the unit circle of the Z-plane, the control system is stable. Therefore, the improved fractional-order PID repetitive controller should satisfy the following stability conditions:

(1) All roots of $1 + G_{FP}(z)P_A(z) = 0$ lie within the unit circle.
(2) $|1 + D(z)G_M(z)| \neq 0$.

Based on stability condition (1), by analyzing the pole distribution of $P_A(z)$, the position of the roots for condition (1) can be determined. The pole distribution of $P_A(z)$ is illustrated in the figure on the left side of Figure 9.

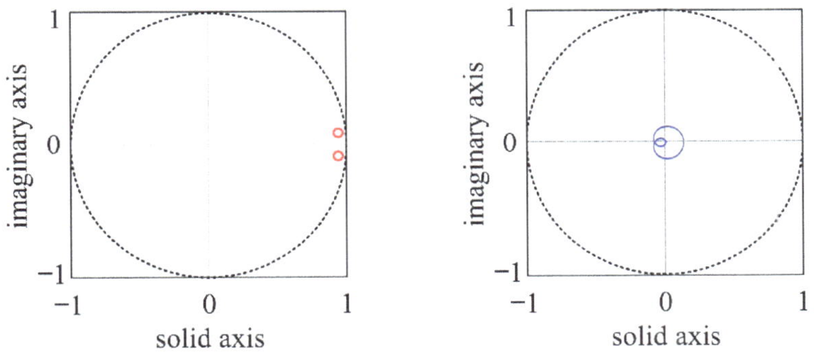

Figure 9. Pole-zero plot of $P_A(z)$ and Nyquist plot of $N(z)$.

From the pole-zero plot of $P_A(z)$, it can be observed that the poles of $P_A(z)$ are distributed within the unit circle, thus satisfying stability condition (1). According to the minimum gain theorem analysis, the establishment condition of stability condition (2) can be expressed as follows:

$$N(z) = \left|G_{DFT}(z)[z^{-N_b} - k_r G_M(z)]\right| < 1 \quad (16)$$

The Nyquist plot $N(z)$ is depicted in the figure on the right side of Figure 9, showing that the trajectory of $N(z)$ closely approximates that of $F(z)$, with only a slight expansion by 0.5 along the real and imaginary axes, remaining within the unit circle. Additionally, the root trajectory maintains a certain margin from the unit circle, indicating that the proposed improved fractional-order PID repetitive controller possesses a certain degree of stability margin.

4. Experimental Analysis and Comparison

In this section, the construction of a simulation model of a PMSM on the Simulink platform is described, with the traditional PI current regulator replaced by a fractional-order PID controller. This model was utilized to evaluate the control performance of the permanent magnet synchronous motor equipped with the improved fractional-order PID repetitive controller. The motor drive schematic is shown in Figure 10. The permanent magnet servo systems defined using the PI controller, the improved DFT repetitive controller, the fractional-order PID controller, and the improved fractional-order PID repetitive controller are denoted as $PMSS_1$, $PMSS_2$, $PMSS_3$, and $PMSS_4$, respectively.

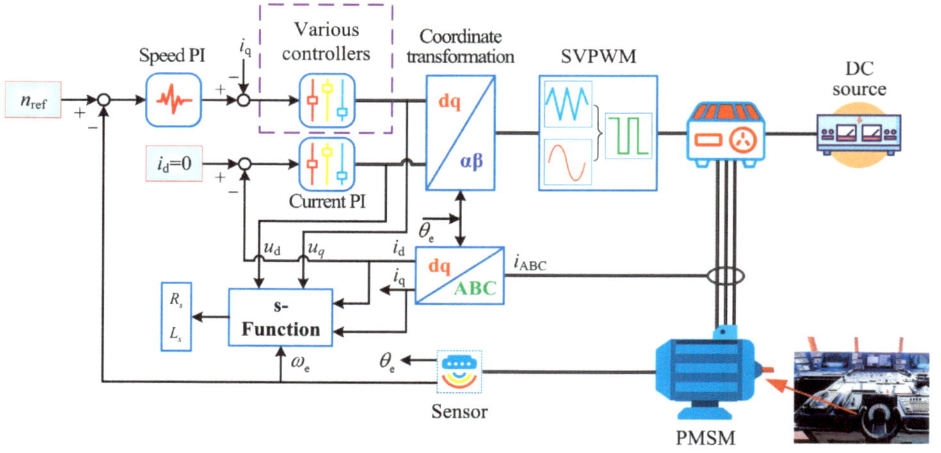

Figure 10. Motor drive schematic diagram.

4.1. Steady-State Performance Analysis

In this subsection, a comparative analysis of the steady-state performance of $PMSS_1$, $PMSS_2$, $PMSS_3$, and $PMSS_4$ under the same operating conditions is conducted. The reference speed was set to 1500 rpm, and the reference torque was set to 1.0 N·m. The torque waveform, phase A current waveform, and current spectrum analysis plots of $PMSS_1$, $PMSS_2$, $PMSS_3$, and $PMSS_4$ are shown in Figures 11 and 12.

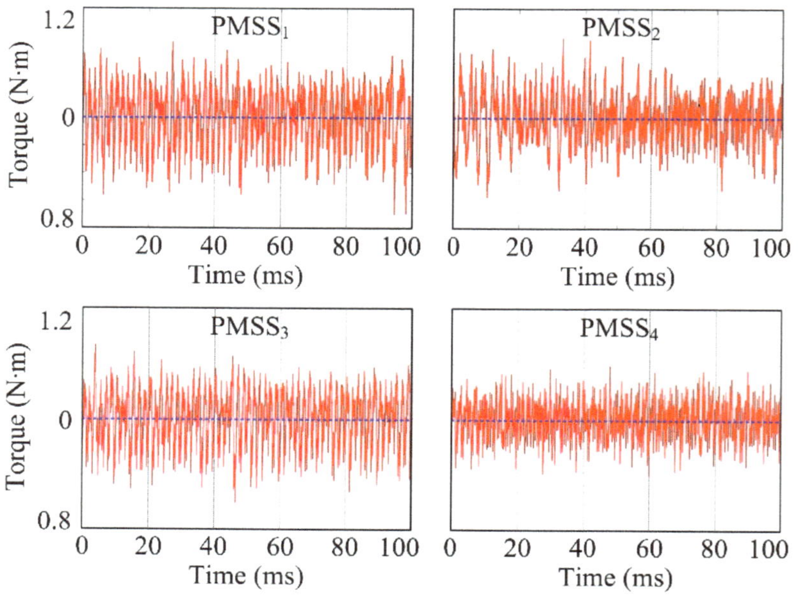

Figure 11. Comparative analysis of $PMSS_1$, $PMSS_2$, $PMSS_3$, and $PMSS_4$ torques.

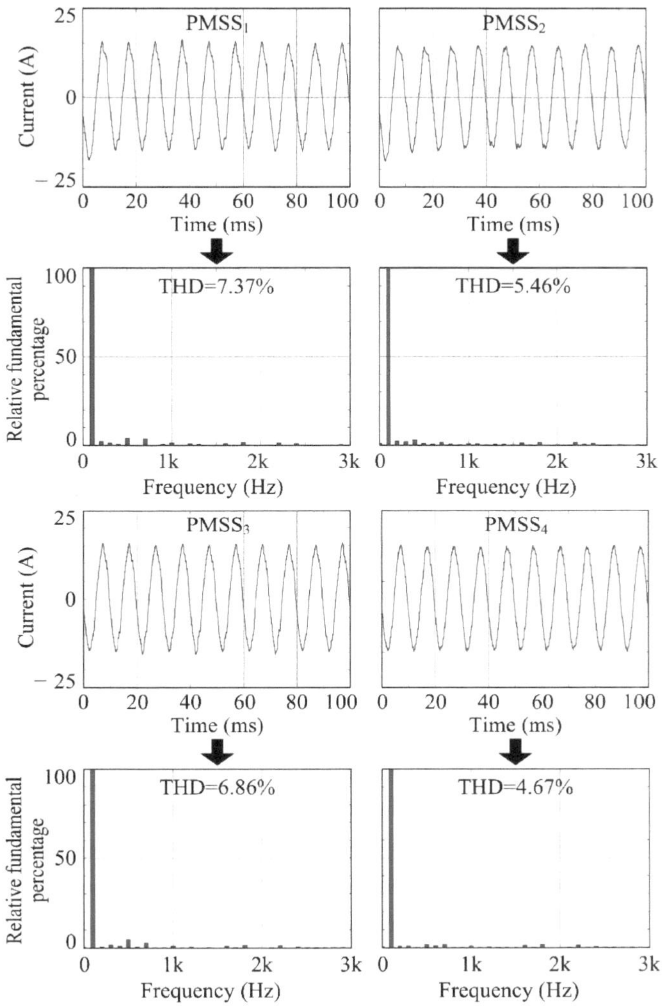

Figure 12. Graph of comparative analysis of current total harmonic distortion (THD) for $PMSS_1$, $PMSS_2$, $PMSS_3$, and $PMSS_4$.

The steady-state performance parameters of the permanent magnet servo systems using different controllers are presented in Table 1.

Table 1. Steady-state performance parameters of $PMSS_1$, $PMSS_2$, $PMSS_3$, and $PMSS_4$.

Stability Performance Parameters	$PMSS_1$	$PMSS_2$	$PMSS_3$	$PMSS_4$
Torque ripple (N·m)	0.21	0.16	0.18	0.12
Average torque (N·m)	0.98	1.01	0.99	1.00
Current THD	7.37%	5.46%	6.86%	4.67%

The simulations show that $PMSS_3$'s torque ripple decreased by 9% and the harmonic distortion by 6.91% compared to $PMSS_1$. Similarly, $PMSS_4$ exhibited a 25% reduction in torque ripple and a 14.47% reduction in current harmonic distortion compared to $PMSS_2$. Therefore, the fractional-order PID controller and the improved fractional-order PID repetitive controller outperformed the PI controller and the PI-DFT-RC controller, respectively, demonstrating superior torque and harmonic suppression performance and exhibiting excellent control capabilities.

4.2. Dynamic Performance Analysis

To verify the dynamic performance of the improved fractional-order PID controller, simulations of acceleration and deceleration were conducted for $PMSS_1$, $PMSS_2$, $PMSS_3$, and $PMSS_4$. A torque of 0.8 N·m was set, and a step change from 1200 rpm to 1500 rpm was applied to the reference speed, followed by a decrease from 1500 rpm to 900 rpm. The simulation results are depicted in Figure 13.

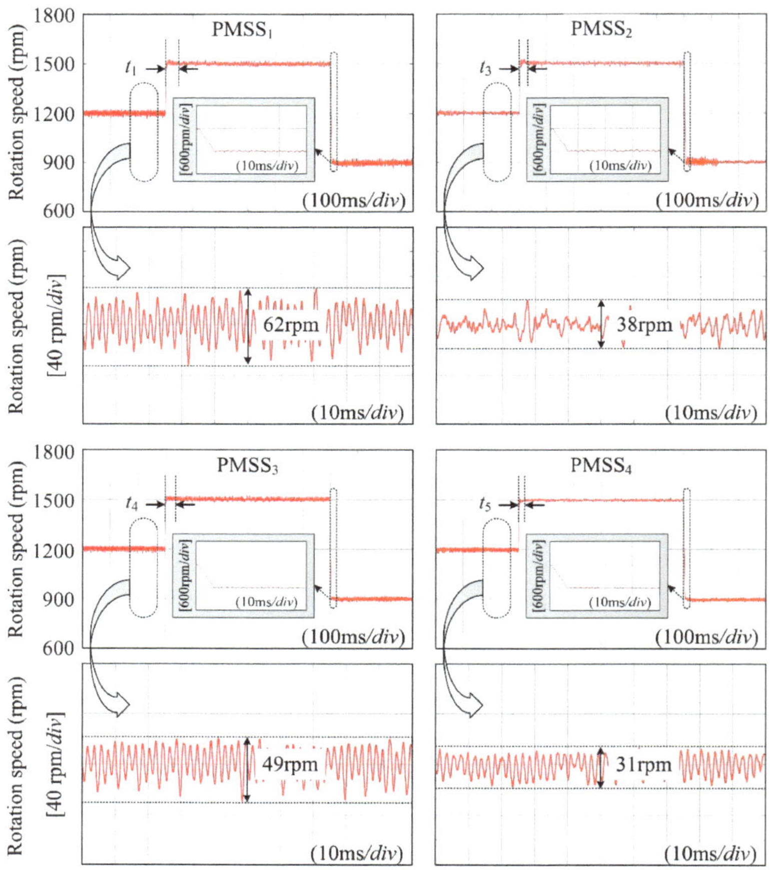

Figure 13. Comparative analysis of dynamic performance of $PMSS_1$, $PMSS_2$, $PMSS_3$, and $PMSS_4$.

Based on the simulated waveform data, a summary calculation was performed to derive the experimental results, as presented in Table 2.

Table 2. Comparative table of transient performance of $PMSS_1$, $PMSS_2$, $PMSS_3$, and $PMSS_4$.

Performance Parameters	$PMSS_1$	$PMSS_2$	$PMSS_3$	$PMSS_4$
Rise time of speed/s	0.0051	0.0035	0.0043	0.0029
Speed fluctuation/rpm	62	38	49	31
Overshoot in speed/%	4.13	2.53	3.27	2.07
Steady-state time in speed/s	0.048	0.03	0.04	0.025

Combining Figure 13 and Table 2, the speed settling times for $PMSS_1$, $PMSS_2$, $PMSS_3$, and $PMSS_4$ were 0.048 s, 0.03 s, 0.04 s, and 0.025 s, respectively, with speed fluctuations of 62 rpm, 38 rpm, 49 rpm, and 31 rpm, respectively. $PMSS_3$ and $PMSS_4$ also exhibited smaller speed overshoot compared to $PMSS_1$ and $PMSS_2$. Through comparative analysis, it can be concluded that the fractional-order PID controller and the improved fractional-order PID repetitive controller have stronger disturbance rejection capabilities and superior dynamic stability performance compared to the PI controller and PI-DFT-RC controller.

5. Conclusions

Early studies demonstrated that an improved control strategy can significantly reduce current harmonics and torque ripple, although most studies have not considered the application of the composite fractional-order PID repetitive control method. Therefore, this paper proposed a composite repetitive control strategy. Through simulation verification, it was found that the proposed composite fractional-order PID repetitive controller exhibits a superior harmonic suppression capability and dynamic stability performance compared to the conventional fractional-order PID repetitive controller, with stronger signal tracking performance. The results indicate that the proposed composite controller possesses excellent disturbance rejection, harmonic suppression, and dynamic-static stability capabilities, thereby improving the overall control performance of the system to a certain extent.

The controller proposed in this paper has good harmonic suppression capabilities and improved the system's control performance to some extent, but there are still some shortcomings that need to be addressed. For example, the Oustaloup approximation algorithm used performs poorly at both ends of the approximation frequency band. Therefore, further research can be conducted on better approximation methods to improve the accuracy of the approximation and obtain more ideal controller parameters.

Author Contributions: Study conception and design, T.Y. and J.B.; data collection, J.F.; analysis and interpretation of results, J.F.; draft manuscript preparation, T.W.; writing—review and editing, T.Y. and J.F. All authors have read and agreed to the published version of the manuscript.

Funding: This research was funded by the Scientific Research Project of the Education Department of Jilin Province (grant number JJKH20230121KJ).

Data Availability Statement: Data are contained within the article.

Conflicts of Interest: Author Jingwen Fan and Jing Bai were employed by the company Yuda Engineering (Jilin) Co., Ltd.. The remaining authors declare that the research was conducted in the absence of any commercial or financial relationships that could be construed as a potential conflict of interest.

References

1. Flieh, H.M. Permanent Magnet Servo Motors Design for Dynamic Loss Minimization and Self-Sensing Control. Ph.D. Thesis, The University of Wisconsin-Madison, Madison, WI, USA, 2019.
2. Fang, S.; Wang, Y.; Wang, W.; Chen, Y.; Chen, Y. Design of permanent magnet synchronous motor servo system based on improved particle swarm optimization. *IEEE Trans. Power Electron.* **2021**, *37*, 5833–5846. [CrossRef]
3. Bu, F.; Yang, Z.; Gao, Y.; Pan, Z.; Pu, T.; Degano, M.; Gerada, C. Speed ripple reduction of direct-drive PMSM servo system at low-speed operation using virtual cogging torque control method. *IEEE Trans. Ind. Electron.* **2020**, *68*, 160–174. [CrossRef]
4. Du, X.B.; Huang, K.S.; Tan, G.R.; Huang, X. Reducing Cogging Torque of Outer Rotor Permanent Magnet Motor Based on Response Surface Method. *Small Spec. Electr. Mach.* **2018**, *56*, 16–19.

5. Li, Z.; Wang, J.; Zhou, L.; Liu, X.; Jiang, F. Enhanced generalized vector control strategy for torque ripple mitigation of IPM-type brushless DC motors. *IEEE Trans. Power Electron.* **2019**, *34*, 12038–12049. [CrossRef]
6. Wu, Y.L.; Li, H.; Song, X.D.; Chen, B.D. Suppression of Harmonic Current in Permanent Magnet Synchronous Motors Using Improved Repetitive Controller. *Trans. China Electrotech. Soc.* **2019**, *34*, 2277–2286. [CrossRef]
7. Zheng, R.X. *Research on Adaptive Sliding Mode Control for Permanent Magnet Linear Synchronous Motor*; Shenyang University of Technology: Shenyang, China, 2022.
8. Fang, G.H.; Wang, H.C.; Gao, X. Parameter identification algorithm of permanent magnet synchronous motor based on dynamic forgetting factor recursive least square method. *Comput. Appl. Softw.* **2021**, *38*, 280–283.
9. Wang, J.; Fan, P.G. Research on PI-fuzzy Hybrid Controller for Position Loop of PMSM Servo System. *Micromotors* **2020**, *53*, 33–36
10. Li, S.; Lu, W.; Yan, S.; Zhao, Z. Improving dynamic performance of boost PFC converter using current-harmonic feedforward compensation in synchronous reference frame. *IEEE Trans. Ind. Electron.* **2019**, *67*, 4857–4866. [CrossRef]
11. Babu, N.; Padhy, N.P. An Approach to Improve Harmonic Attenuation and Stability Performance in Multi-Parallel Inverter System. *IEEE Trans. Power Deliv.* **2023**, *38*, 3634–3646. [CrossRef]
12. Liu, X. Research on Repetitive Sliding Mode Variable Structure Control of Permanent Magnet Synchronous Motor. *Microcomput. Appl.* **2023**, *39*, 41–44.
13. Fang, S.L.; Fan, J.D. A Composite Control Strategy of Permanent Magnet Synchronous Motor Based on Active Disturbance Rejection. *Electr. Mach. Control Appl.* **2023**, *50*, 62–69.
14. Khanh, P.Q.; Anh, H.P.H. Advanced PMSM speed control using fuzzy PI method for hybrid power control technique. *Ain Shams Eng. J.* **2023**, *14*, 102222. [CrossRef]
15. Wang, H.; Xu, S.; Hu, H. PID controller for PMSM speed control based on improved quantum genetic algorithm optimization *IEEE Access* **2023**, *11*, 61091–61102. [CrossRef]
16. Chen, P.; Luo, Y. Analytical fractional-order pid controller design with bode's ideal cutoff filter for pmsm speed servo system *IEEE Trans. Ind. Electron.* **2022**, *70*, 1783–1793. [CrossRef]
17. Yu, L.; Chen, Q. PMSM Vector Control Based on Fuzzy Sliding Mode Control. *Autom. Instrum.* **2024**, *39*, 65–70.
18. Xu, B.; Xia, C. Research on Variable Angle Transmission Ratio of Vehicle Steer-by-wire Based on Fuzzy Conti. *Automob. Appl. Technol.* **2024**, *49*, 15–23.
19. Li, Z.C.; Peng, J.L. Repetitive Sliding Mode Control Method of Three-phase Active Power Filter. *Autom. Instrum.* **2024**, *39*, 39–44
20. Tian, S.; Liu, K.Z.; Zhang, M.; Lu, C.; Wu, M.; She, J. Harmonic Disturbance Suppression for High-Performance Nonlinear Repetitive-Control Systems. *IFAC-PapersOnLine* **2023**, *56*, 4545–4550. [CrossRef]
21. Tian, M.; Wang, B.; Yu, Y.; Dong, Q.; Xu, D. Discrete-time repetitive control-based ADRC for current loop disturbances suppression of PMSM drives. *IEEE Trans. Ind. Inform.* **2021**, *18*, 3138–3149. [CrossRef]
22. Lendek, A.; Tan, L. Mitigation of derivative kick using time-varying fractional-order PID control. *IEEE Access* **2021**, *9*, 55974–55987 [CrossRef]
23. Yao, G.; Li, H.R.; Zhou, L.D.; Cao, Z.J.; Wang, J.; Huang, S.H. Research on Fractional Order PID Control of Six Phase Permanent Magnet Wind Turbine Driven by MP-MMC. *Electr. Mach. Control* **2022**, *26*, 93–107.
24. Xiao, M.; Tao, B.; Zheng, W.X.; Jiang, G. Fractional-order PID controller synthesis for bifurcation of fractional-order small-world networks. *IEEE Trans. Syst. Man Cybern. Syst.* **2019**, *51*, 4334–4346. [CrossRef]
25. Li, X. *Research on Frequency Domain Design Method of Fractional Order PID Controller*; University of Science and Technology of China: Hefei, China, 2022.
26. Hu, H.T. *Research on AC Servo System with Adaptive Fractional Order PID Controller*; Xi'an Technological University: Shaanxi, China, 2018.
27. Li, C. *Improvement Strategies for Particle Swarm Optimization Algorithms with Applications*; Jiangnan University: Wuxi, China, 2021
28. Yuan, T.Q.; Yuan, L.; Zhang, Y.P. Current harmonic suppression for PMSM driven system utilizing PI-DFT-RC controller. *Energy Rep.* **2023**, *9*, 335–342. [CrossRef]

Journal of
Marine Science and Engineering

MDPI

Article

Nonlinear Model Predictive Control Energy Management Strategy for Hybrid Power Ships Based on Working Condition Identification

Yucheng Yan [1], Zhichao Chen [2] and Diju Gao [1,*]

[1] Key Laboratory of Transport Industry of Marine Technology and Control Engineering, Shanghai Maritime University, Shanghai 201306, China; onlyones@foxmail.com
[2] Marine Design and Research Institute of China, Shanghai 200011, China; m15851346533@163.com
* Correspondence: djgao@shmtu.edu.cn

Abstract: Hybrid power technology for ships is an effective way to promote the green and low-carbon development of the maritime industry. The development of pattern recognition technology provides new research ideas for the rational allocation and utilization of energy in hybrid power ships. To reduce fuel consumption, a nonlinear model predictive control energy management strategy based on working condition identification is proposed for optimal energy management to solve the problem of real-time optimal adjustment of generators and batteries. The core of the strategy is to identify the ship's working conditions and the nonlinear model predictive control algorithm. Firstly, to achieve the working condition identification task, a ship working condition dataset based on a hybrid supply power ship data is constructed. The labeled dataset is trained using deep learning techniques. Secondly, based on the identification results, a nonlinear model predictive control algorithm is designed to adjust the generator speed and the battery current to achieve energy optimization control under constraints. Finally, the effectiveness of the proposed strategy in optimizing energy control and reducing fuel consumption is verified through simulation. The proposed strategy can reduce the generator fuel consumption by 5.5% under no noise disturbance when compared with conventional predictive control. Under 10% noise disturbance, it is still able to reduce the fuel consumption by 2.6%.

Keywords: hybrid power ships; energy management; working condition identification; nonlinear model predictive control; deep learning technology

1. Introduction

In 2018, the International Maritime Organization (IMO) adopted a greenhouse gas reduction strategy, marking the shipping industry's entry into carbon reduction efforts. By 2023, IMO passed a series of amendments aimed at reducing air pollution and improving ship energy efficiency [1], highlighting the industry's increasing demand for energy-saving and emission-reduction technologies. Clearly, further development and improvement of energy control technologies to meet the requirement of reducing ship fuel consumption remains a key direction for the industry. Hybrid power technology is driving the shipping industry toward greener development [2–4]. The integration of operating condition identification technology and energy management strategies (EMS) for hybrid ships has provided new opportunities for reducing fuel consumption [5].

In the field of hybrid vehicles, many studies suggest that identifying vehicle working conditions provides valuable decision-making information for energy management strategies, enabling the reduction of fuel consumption. Experimental validations show that accurate identification of working conditions positively impacts the reduction of vehicle operating costs [6–9]. However, for hybrid ships, distinguishing working conditions is more challenging due to the complexity of the water environment and ship operations. This presents difficulties and challenges in applying operating condition identification technology to optimize fuel consumption in hybrid ships. On one hand, the lack of comprehensive operating condition datasets hinders the analysis of ship operation patterns and anomaly detection. On the other hand, the low accuracy of operating condition identification impairs the performance of energy management strategies, leading to poor optimization and increased fuel consumption, which in turn exacerbates carbon and nitrogen oxide emissions, resulting in significant environmental pollution.

Numerous scholars have conducted research on EMS and operating condition identification technologies, aiming to effectively coordinate multiple power sources in hybrid systems to reduce fuel consumption and lower emissions. EMS can be categorized into rule-based, global optimization-based, and instantaneous optimization strategies [5]. Rule-based EMS often relies on fuzzy logic rules, but such strategies lack flexibility and may not adapt well to complex environments and changing energy demands [10–12]. EMS based on global optimization is difficult to implement and is not conducive to practical engineering applications [13,14]. Instantaneous optimization strategies, such as model predictive control (MPC), equivalent fuel consumption minimization, and swarm intelligence algorithms, require real-time performance and high computational speed [15,16]. Operating condition identification methods can be divided into physics-based and data-driven approaches. Physics-based methods require a deep understanding of the equipment and system, involving significant prior knowledge and computation. For example, in [16], nonlinear model predictive control (NMPC) was used to reduce fuel consumption and optimize CO_2 emissions by considering the impact of random waves on the propeller load of hybrid ships. In contrast, data-driven methods rely on large datasets for training and validation and do not require an accurate pre-established system model. These methods can be further categorized into supervised and unsupervised learning. Both types rely on data as a foundation [17]. However, unsupervised learning methods are highly dependent on data quality and features, and the resulting models may be difficult to interpret or apply to all datasets [18,19]. Supervised learning has the advantage of allowing for model correction and optimization, but it requires manual data labeling [20]. For instance, in [20], Least Squares Support Vector Machine (LSSVM) was used to distinguish between fast-varying and slow-varying conditions, and RBF neural networks and Markov Chain models were applied to predict load demands, but without addressing energy management optimization. In [21], a hierarchical distributed control method for hybrid systems was proposed to reduce battery current fluctuations and stabilize bus voltage. Yuan et al. [22] applied Support Vector Machine (SVM) models for operating condition identification and used multi-step Markov Chain models to predict power demands under various conditions, though with relatively low accuracy.

Overall, previous research has rarely focused on utilizing operating condition identification technology to provide decision-making information for the Energy Management System (EMS) of hybrid ships. On one hand, much of the existing research on ship working conditions has concentrated on prediction and analysis. On the other hand, the integration of operating condition identification with EMS has not been sufficiently explored, and the low accuracy of identification hinders effective energy optimization under varying operating conditions. The model accuracy rates of Gao et al. [20] and Yuan et al. [22] are 41.5% and 90.88%, respectively. The accuracy of the models still needs to be further improved. Therefore, the focus of this study is on improving the accuracy of operating condition identification using deep learning techniques and integrating it with NMPC to optimize the energy control of hybrid ships, to reduce fuel consumption. The research begins by analyzing ship operational data and creating an operating condition dataset. To enhance identification accuracy, a deep learning approach utilizing Convolutional Neural Networks (CNN) is employed to train an operating condition recognition model, incorporating Efficient Channel Attention (ECA) [23–27]. In terms of energy management, NMPC is used as the core strategy to address real-time performance requirements and nonlinear system constraints.

The proposed method integrates operating condition identification with EMS, enabling the identification of operating conditions to provide valuable decision-making information for EMS. Furthermore, this approach has significant implications for reducing fuel consumption in hybrid ships and is expected to contribute to advancing future research on ship working conditions.

The main contributions of this paper are as follows:

1. A working condition dataset was constructed based on the historical data of a hybrid supply vessel. The dataset was classified using clustering methods, and the ship's status corresponding to each category was analyzed.
2. An offline working condition identification model was trained based on CNN, with an identification accuracy of up to 99.8%.
3. Taking a hybrid supply vessel as the research object, this study emphasizes practical engineering applications. The speed of the diesel generator and the output current of the energy storage battery are set as control objectives, aiming to optimize fuel consumption. The NMPC-based EMS achieves optimal control under different working conditions, reducing fuel consumption by 2.6–5.5%.

This study is structured as follows: Section 2 introduces the research object and scheme. Section 3 elaborates in detail on the case ship power system model. Section 4 presents the EMS based on the working condition identification, and Section 5 shows the simulation results and analysis. Section 6 presents the conclusions and offers future perspectives.

2. Research Object and Scheme

2.1. Research Object

In this study, a hybrid supply power ship is used as the research object, named "CNOOC 257". The ship mainly navigates between docks and offshore oil platforms. It is tasked with supplying and transporting workers to and from the platforms. Its propulsion system features dual propellers and engines in a parallel arrangement, composed of propellers and bow thrusters. The main diesel engines provide the propulsion power, while the thrusters, via an AC bus, offer lateral movement capabilities. The key parameters of the ship such as length, breadth, and draft are listed in Table 1.

Table 1. Ship parameters.

Description	Value
Ship length	79 m
Ship breadth	16 m
Draft	8 m
Mass	15,000~20,000 ton

The topological structure of the power system is shown in Figure 1; shaft generators, auxiliary generators, and the energy storage battery are the power supply equipment in the grid. The shaft generators are SG1 and SG2. The auxiliary generators are DG1, DG2, and DG3. There is a 1021 kWh energy storage battery equipped with this power management system, which can handle a maximum input and output power of 400 kW through a bidirectional DC/AC converter.

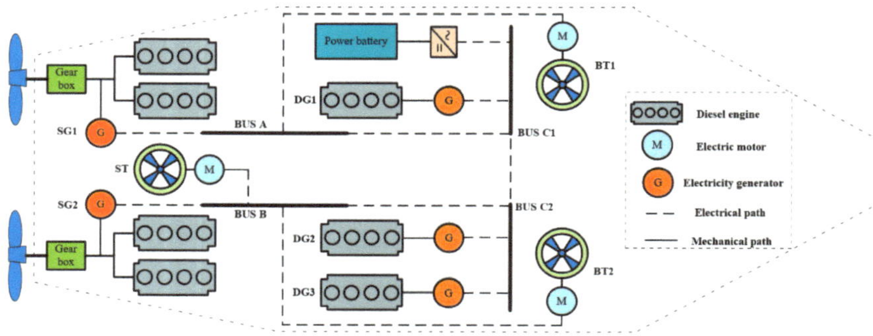

Figure 1. Hybrid power system electrical and dynamic system topology for CNOOC 257.

BUS A, BUS B, BUS C1, and BUS C2 serve as the ship's AC buses. Power for the energy storage battery is typically supplied from shore or a shaft generator SG1 or SG2. It can be provided flexibly and independently to shipboard electrical equipment via any AC bus. The power system structure includes two main thrust propellers and three thruster units with two bow thrusters. Thruster units BT1 and BT2 and one stern thruster, ST, are shown in Figure 1.

In Figure 1, the operations of the hybrid power ship are more complicated and are summarized as follows: (1) The propellers that provide forward power are driven by the diesel main engines. (2) The shaft generator, auxiliary generator, and energy storage battery can provide electricity to the thruster units. However, the shaft generators only work in power take-off operating mode and only run when the ship is being propelled, thereby increasing fuel consumption as power is supplied to electrical devices through the AC buses. (3) The power topology is complex, and the power difference between the shaft generator and the auxiliary generator is huge and does not allow parallel operation for a long time. These working conditions pose challenges in reducing fuel consumption and maintaining ship stability, necessitating an effective EMS.

2.2. Scheme

Simplified operation: Setting up the shaft generator does not participate in the power supply of the side-thrust device but only charges the energy storage battery.

The proposed scheme in this paper is shown in Figure 2. Reducing diesel fuel consumption is the research goal and motivation of this paper. It achieves this by identifying the working conditions to maintain the stable operation of the hybrid power ship. The methodology involves modeling the electrical system, utilizing operating condition identification methods, and implementing an NMPC energy management strategy for optimal control. Finally, the contribution is demonstrated by validating the effectiveness of the proposed strategy through energy consumption analysis and real-time performance indicators.

Working condition identification is achieved through deep learning technologies. To begin with, a ship working condition dataset will be established, and then a CNN-based model will be trained for the identification task. Then, the proposed strategy can adjust in real time based on the identification results to complete the task of energy allocation under different conditions. At the same time, the optimal control of the ship is also realized.

Finally, the effectiveness of the working condition identification on the EMS will be verified by Simulink simulations. Fuel consumption, battery state of charge (SOC), and real-time performance metrics will be analyzed to evaluate the strategy.

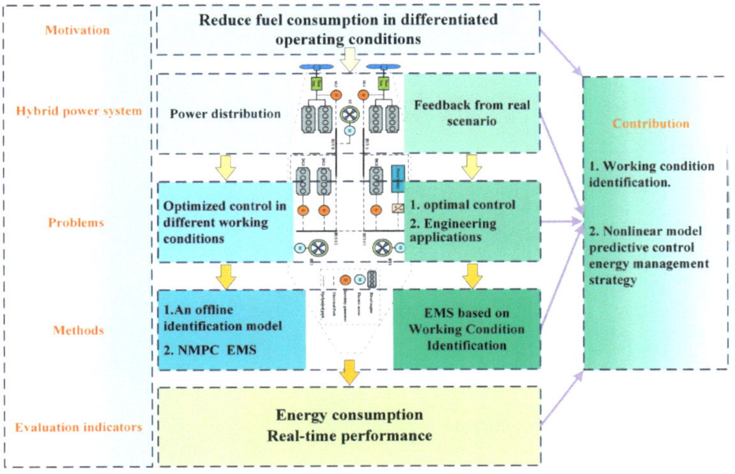

Figure 2. Energy management scheme based on working condition identification.

3. Model Establishment

The ship power system model will be established from the point of view that the hybrid ship power system conforms to the conservation of energy. The hybrid power system topology for CNOOC 257 is shown in Figure 1. At the same time, the model of the energy storage battery and auxiliary generator will be constructed by the data fitting method.

3.1. Hybrid Power Ship Power System Model

The ship's thruster is powered by the auxiliary generator, an energy storage battery. Based on the law of conservation of energy, and ignoring mechanical losses and some electrical losses, the model description of its power can be expressed by

$$\begin{cases} P_{load} = P_{dg} + P_{bat}, P_{bat} \geq 0 \\ |P_{bat}| = P_{sg}, P_{bat} < 0 \end{cases} \tag{1}$$

wherein P_{load} is the power demanded by the ship's thrusters and on-board electrical equipment. P_{dg} is the power provided by the auxiliary generator. The ship's energy storage device is a lithium iron phosphate power battery. P_{bat} is the power provided by the battery. When P_{bat} is greater than zero, the battery is discharged and outputs power. Otherwise, the battery is charged and inputs power. P_{sg} is the power provided by the shaft generator. This is used here to illustrate the energy source of the energy storage battery.

3.2. Energy Storage Battery

The internal resistance equivalent model is adopted as shown in Figure 3. Its current is calculated as follows [13]:

$$I_{bat} = \frac{V_{oc} - \sqrt{V_{oc}^2 - 4P_{bat}R_0}}{2R_0} \tag{2}$$

wherein V_{oc} is the open circuit voltage of the internal resistance model. P_{bat} is the output power of the storage battery. R_0 is the internal resistance of the storage battery.

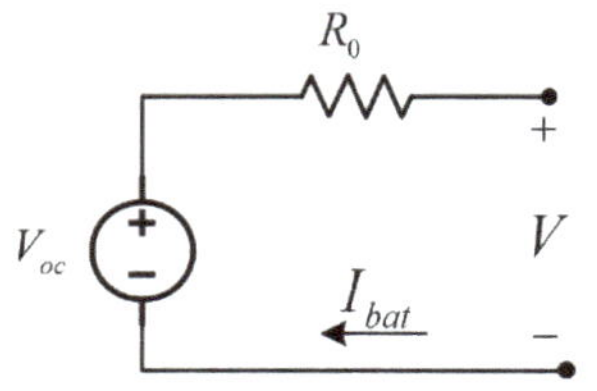

Figure 3. Equivalent circuit of the battery.

The SOC of the battery is calculated as follows:

$$SOC = SOC_0 - \frac{\int_{t_0}^{t} I_{bat}dt}{3600Q} \tag{3}$$

wherein SOC_0 is the initial state of charge. Q is the battery capacity. I_{bat} is the energy storage battery current.

The powers of the battery are constrained as follows:

$$P_{bat,min} \leq P_{bat} \leq P_{bat,max} \tag{4}$$

wherein $P_{bat,min}$ is the minimum charging power. $P_{bat,max}$ are the maximum charging power.

The limits for battery SOC and current are constrained as follows:

$$\begin{aligned} SOC_{min} &\leq SOC \leq SOC_{max} \\ I_{bat,min} &\leq I_{bat} \leq I_{bat,max} \end{aligned} \tag{5}$$

The ship's energy storage battery has four operating modes, as shown in Figure 4: (a) Mode 1 denotes the operating mode where the shaft generator charges the energy storage battery through BUS A; (b) Mode 2 denotes the operating mode where the shaft generator charges the energy storage battery through BUS B; (c) Mode 3 denotes the operating mode where the energy storage battery provides power through BUS C1; and (d) Mode 4 denotes the operating mode where the energy storage battery provides power through BUS C2 to provide power.

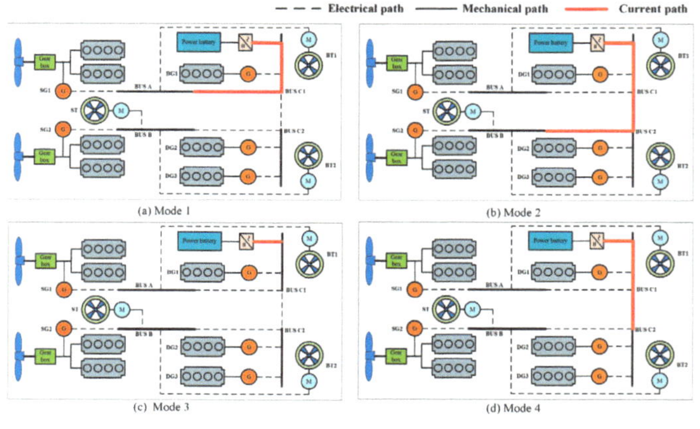

Figure 4. Energy storage battery's operating modes.

3.3. Auxiliary Generator Model

The auxiliary generator primarily consists of a diesel main engine and generator. The output power of the diesel generator is related to the engine speed and exhibits a nonlinear relationship. The rotational speed and output power of the auxiliary generator are fitted by the scheme of modeling the auxiliary generator in reference [13]. The formula for calculating the output power of the auxiliary generator is denoted as follows:

$$P_{dg} = \sum_{i=0}^{3} a_i \omega_{dg}^i \tag{6}$$

where ω_{dg} is times the rotational speed of the generator, and a_i and i are the coefficients and exponents of the approximate polynomial, respectively.

The specific fuel oil consumption (SFOC) of the diesel engine and the auxiliary generator output power are calculated by the following formula [13]:

$$m_f = \sum_{j=0}^{3} b_j P_{dg}^j \tag{7}$$

where P_{dg}^j is the output power of the auxiliary generator, and b_j and j are the coefficients and exponents of the approximate polynomial, respectively.

The constraints of the diesel generator are shown by:

$$P_{dg,min} \leq P_{dg} \leq P_{dg,max} \tag{8}$$

wherein $P_{dg,min}$ is the minimum output power. $P_{dg,max}$ is the maximum output power.

4. EMS Based on Working Condition Identification

The EMS is regarded as the core of energy distribution schemes for hybrid power ships. Working condition identification can obtain the current condition of the ship and provide decision-making information for the EMS, but it has rarely been considered in previous literature. To address this challenge, this paper combs and classifies the data of the supplying ship to train the identification model. To cope with the energy optimization problem under different conditions, this paper targets the relevant parameters from diesel generators and batteries and uses NMPC for rolling optimization.

4.1. Working Condition Identification

Due to the complexity of ship operation conditions, the IMO has not established a unified standard of working conditions for ships, which is very unfavorable for the research on the working conditions of ships. However, identifying the ship's condition only from the status of the shipboard equipment is not able to incorporate the subjective initiative of human beings. Therefore, the study from the perspective of data is an effective method.

4.1.1. Working Condition Dataset for Hybrid Power Ship

The operational data of the vessel "CNOOC 257" was collected and analyzed through hifleet.com (accessed on 27 April 2023). On average, the vessel spends 2.24 h per day in port, 13.6 h offshore, and 8.16 h sailing. A ship operating condition dataset was created by collecting the vessel's operational data over a week. The sampling frequency was set to 1 Hz, and key feature parameters included ship speed, load power, load voltage, load current, propulsion power, propulsion voltage, propulsion current, and propulsion torque. This dataset comprises 98,143 data entries.

The K-means clustering algorithm and clustering evaluation methods were applied to determine data categories. The elbow method was used as the clustering evaluation algorithm. This method evaluates clustering performance by calculating the sum of squared errors (SSE), which measures the squared distance between each data point and its respective cluster center, across different numbers of clusters. The "elbow" is identified as the point where the SSE curve shows a significant reduction, indicating the optimal number of clusters.

Figure 5 illustrates the clustering results using the elbow method. As shown in this figure, a distinct elbow point occurs when the number of clusters is four. Thus, four clusters were chosen as the basis for classification.

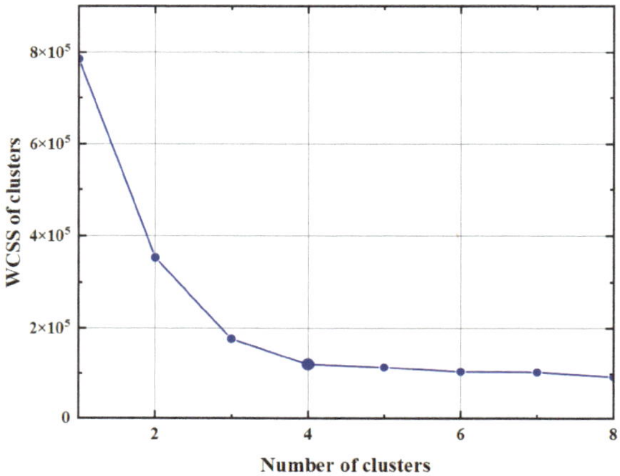

Figure 5. Elbow method.

As shown in Figure 6, the ship speed and load power data for the four categories are presented. Ship speed and load power are two critical factors for determining the operational state of the vessel. Statistical analyses were performed for the four categories of data. The power range for Type 1 is approximately 0–300 kW, and the ship speed is around 0–7 Kn. The power range for Type 2 is between 280–800 kW, and the ship speed is 0–7 Kn. For Type 3, the power range is approximately 0–300 kW, and the ship speed is around 6–14 Kn. Type 4 has a power range of approximately 150–450 kW, and the ship speed is around 7–14 Kn.

Figure 7 illustrates the median, mean, and proportion of ship speed and demand power within the dataset for the four operating conditions. These metrics provide a comprehensive overview of the characteristics of each operational category. As shown in Figure 7, the proportions of operating conditions of Type 1 and Type 4 are relatively large, while those of Type 2 and Type 3 are relatively small. By comparing the mean and median values, the differences for operating conditions of Type 1 and Type 3 are relatively small, while the differences for Type 2 and Type 4 are more significant. This indicates that the power and ship speed data of operating conditions of Type 2 and Type 4 show a skewed distribution. Additionally, the average demand power for operating conditions of Type 2 and Type 4 is around 400 kW, while the average power demand for operating conditions of Type 1 and Type 3 is 100 kW or less.

Through cluster analysis and detailed examination of each data category, the ship's status was linked to each data type based on two critical indicators: ship speed and load power. The results are presented in Table 2.

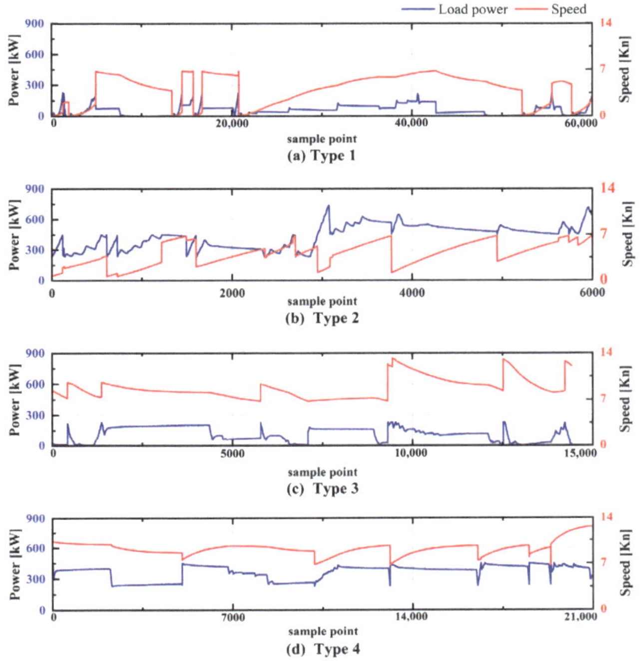

Figure 6. The ship's working condition dataset.

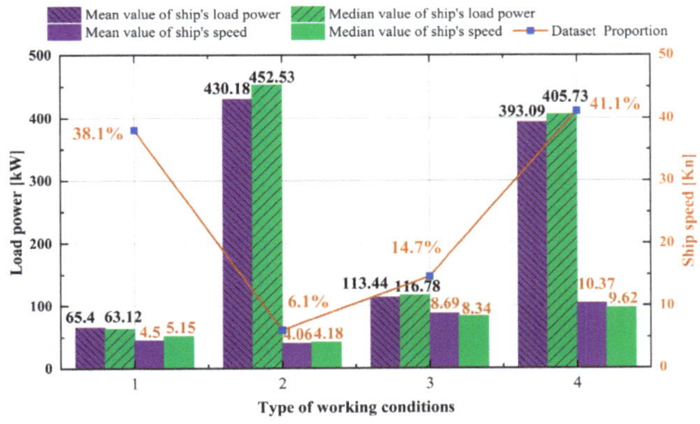

Figure 7. Data statistics of the ship's working condition dataset.

Table 2. Working condition information and number.

Conditions	Description	Ship Status
Type 1	The shipload power is low, and the speed is low.	Mooring operations or low-speed navigation
Type 2	The shipload power is high, and the speed is low.	Accelerating navigation and the operating equipment running
Type 3	The shipload power is low, and the speed is high.	Ship's high-speed navigation status
Type 4	The shipload power is high, and the speed is high.	High-speed navigation operational equipment running

4.1.2. Working Condition Identification Model

The core of the operating condition identification model is a CNN model enhanced with an ECA attention mechanism. CNN, as a representative algorithm in deep learning, is known for its feature learning capabilities. It can perform translation-invariant classification of input data based on its hierarchical structure, making it suitable for learning from the dataset established in this study. Moreover, numerous studies have demonstrated that attention mechanisms can help models focus on important features and improve generalization performance. Therefore, this study employs ECA-net, whose principle is based on a dimensionality-preserving local cross-channel interaction strategy and an adaptive method for selecting the size of one-dimensional convolution kernels, thereby enhancing model performance [24–26].

The following steps describe the training and prediction process of the proposed operating condition identification model using the ship operating condition dataset:

1. Split the dataset into a training set and a validation set with an 8:2 ratio.
2. Normalize the data, and train the classification model.
3. Normalize the test dataset, and use the trained model to make predictions, evaluate the model, and obtain the results.
4. Normalize the collected ship operating condition data or test dataset, use the trained classification model to make predictions, and evaluate the model.

The CNN model used in this study is based on the models proposed by Wang et al. [26], Yang et al. [27], and Zhang et al. [28]. Through multiple experiments, the structure of the CNN is determined. The criteria for the structure are twofold: firstly, the training time should be as short as possible, and secondly, the identification should be accurate. Therefore, the identification model consists of one input layer, three convolutional layers, two ECA-net modules, three pooling layers, and two fully connected layers. The details of dataset preprocessing, the structure of the CNN model, and its parameter settings during training are shown in Figure 8.

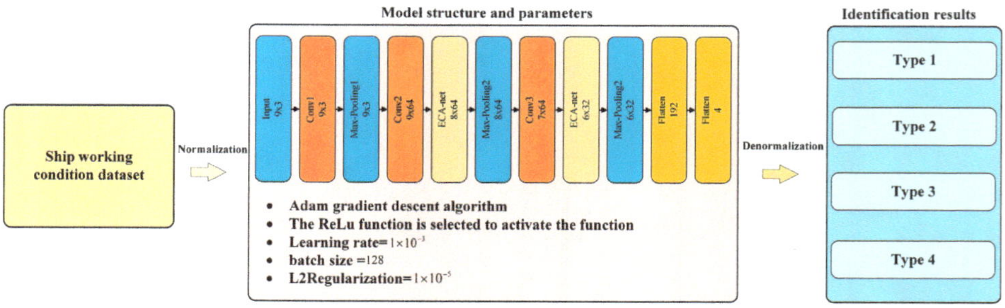

Figure 8. Structure, parameters, and training process of the proposed identification model.

4.2. NMPC-Based EMS Based on Working Condition Identification

In this study, NMPC is selected as the optimization algorithm for energy management strategies, which is the main representative algorithm for transient optimization strategies. The core of NMPC is its rolling optimization capability, and its cost function can be given by a non-quadratic programming form, unlike the MPC. However, the design of the NMPC cost function affects the optimization results. At each moment, NMPC solves the optimal control sequence of the optimization problem takes the first set of values in the solved control sequence as control inputs and then repeats the process at the next moment. Therefore, in this paper, the SOC of the battery and the instantaneous fuel consumption of the auxiliary generator are taken as the main optimization objectives, and the diesel generator and the battery are used as the control objects, the state variables $x(k) = [SOC, m_f]^T$, control variables $u(k) = [I_{bat}, \omega_{dg}]^T$, output variables $y(k) = [SOC, m_f]^T$. The nonlinear model is given by the following:

$$x(k+1) = f(x(k), u(k)) = \begin{bmatrix} x_1(k) - \frac{\int u_1(k)dt}{3600Q} \\ \sum\limits_{j=1}^{3} b_j \left(\sum\limits_{i=1}^{3} a_i (u_2(k))^i \right)^j \end{bmatrix} \tag{9}$$

The cost function is designed in the following form:

$$\begin{aligned} \min J = {} & \beta_{1,w} \sum_{N-1}^{i=1} (SOC_i - SOC_{ref})^2 + \beta_{2,w} \sum_{N-1}^{i=1} m_{f,i}^2 + \mu_{1,w} \sum_{N-1}^{i=1} P_{bat,i}^2 \\ & + \mu_{2,w} \sum_{N-1}^{i=1} P_{dg,i}^2 + \varphi_{1,w} \left(SOC_N - SOC_{ref,N} \right)^2 + \varphi_{2,w} m_{f,N}^2 \\ & s.t. \begin{cases} u_{m,min} \leqslant u_m \leqslant u_{m,max} \\ \dot{u}_{m,min} \leqslant \dot{u}_m \leqslant \dot{u}_{m,max} \\ SOC_{min} \leqslant SOC \leqslant SOC_{max} \\ P_{load} = P_{bat,i} + P_{dg,i} \end{cases} \end{aligned} \tag{10}$$

wherein $\left(SOC_i - SOC_{ref}\right)^2$ is a penalty for battery SOC deviation, which ensures avoiding excessive battery discharge variations in energy demand throughout the entire operating cycle. SOC_{ref} is the reference input for the battery SOC. $m_{fc,i}^2$ is the penalty term for minimizing the fuel consumption. $P_{bat,i}^2$ and $P_{dg,i}^2$ are minimizing system inputs. $\left(SOC_N - SOC_{ref,N}\right)^2$ and $m_{f,N}^2$ are terminal constraints. $\beta_{1,w}, \beta_{2,w}, \mu_{1,w}, \mu_{2,w}, \varphi_{1,w}, \varphi_{2,w}$ are the penalty coefficients. Their values are determined by the working condition type w. u_m is the control variable, which should satisfy the constraints of (4) and (8). Meanwhile, to keep the control variables from changing too drastically and to keep the rate of change within a certain range, the energy storage battery SOC is constrained by (5). Finally, the energy conservation of the ship is incorporated into the control strategy in the form of equation constraints.

Based on the results of operating condition identification, the adjustment of the penalty coefficient is determined by the different motion states of the vessel under various operating conditions. These differences affect the proportion of power output from the diesel generator and the battery. The selection of penalty coefficients is performed manually. Through the Monte Carlo experimental method, 1000 sets of penalty coefficients are randomly generated. For each set of penalty coefficients, the objective function values are calculated based on the mean load power under different operating conditions and the fuel economy range of the auxiliary generator. Statistical analysis is then performed on these objective function values. Ultimately, a set of penalty coefficients is determined for each type of operating condition. In the Type 1 operating condition, for example, the battery output should be increased, while the diesel generator's power output should be reduced accordingly.

The identification must match the computational speed of the EMS, and the time series of the proposed strategy is shown in Figure 9, with the time units in seconds. The time required for one calculation of the experimental platform is as follows:

$$\begin{aligned} & \Delta t = \Delta t_1 + \Delta t_2 \\ & s.t. \Delta t \leq 1 \end{aligned} \tag{11}$$

wherein Δt_1 represents the working condition identification time, and Δt_2 represents the solution time for the NMPC controller.

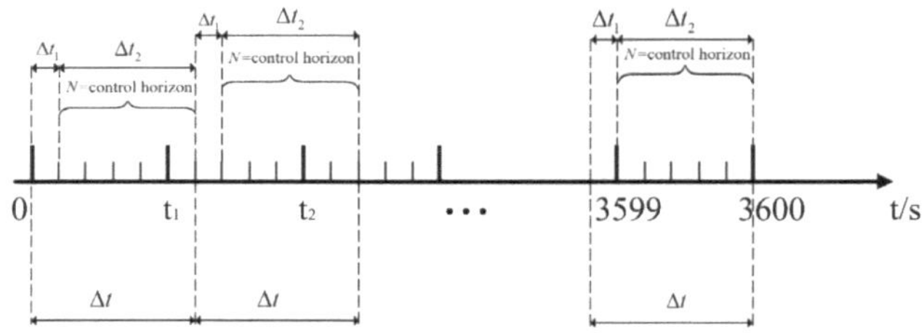

Figure 9. Time series of the proposed strategy.

The block diagram of the proposed EMS based on working condition identification is shown in Figure 10. It is mainly divided into the working condition identification module, the NMPC controller, and the controlled system consisting of the diesel generator and the energy storage battery.

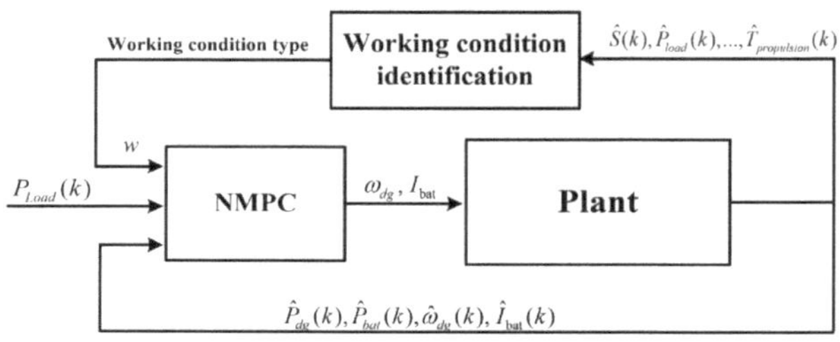

Figure 10. Block diagram of NMPC control based on working condition identification.

The inputs to the condition identification module are the eight data parameters mentioned in Section 4, i.e., ship speed $\hat{S}(k)$, load power $\hat{P}_{load}(k)$, load voltage $\hat{V}_{load}(k)$, load current $\hat{I}_{load}(k)$, propulsion power $\hat{P}_{propulsion}(k)$, propulsion voltage $\hat{V}_{propulsion}(k)$, propulsion current $\hat{I}_{propulsion}(k)$, and propulsion torque $\hat{T}_{propulsion}(k)$, which are fed back from the controlled system. Its output is the result of the four types of conditions described, called w. The results of the identification are used to adjust the constraints and the penalty coefficients of the cost function.

The NMPC controller is used for rolling optimal control of the energy control of the auxiliary generator, $\hat{P}_{dg}(k)$ and storage battery, $\hat{P}_{bat}(k)$. The controlled system feeds back the measured output power of the auxiliary generator and the measured output power of the storage battery to the NMPC controller to complete the closed-loop control.

5. Experimentation and Analysis

The experimental platform is equipped with a GPU, NVIDIA GeForce RTX 4060, and a CPU, Intel Core i7-13700H. The construction and training of the identification model are both carried out in MATLAB software (R2022b).

5.1. Accuracy of the Working Condition Identification Model

The performance of the model based on deep learning technology on the ship condition dataset can be evaluated through indicators such as the confusion matrix, accuracy, and loss. The confusion matrix is an important indicator for evaluating the classification model. As shown in Figure 11, it includes the number of true positives (TP), true negatives (TN), false positives (FP), and false negatives (FN). Therefore, when analyzing and evaluating the working condition identification model using the confusion matrix metrics, certain conclusions can be drawn from the proportion of the diagonal elements TP and TN in Figure 11.

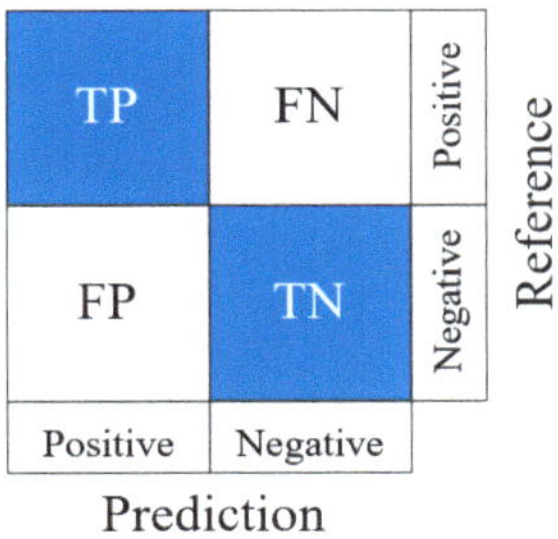

Figure 11. Confusion matrix.

The definition and calculation formula for the model's accuracy are as follows:

$$Accuracy = \frac{TP + TN}{TP + TN + FP + FN} \tag{12}$$

wherein TP, TN, FP, and FN are consistent with the definitions in Figure 11, respectively.

One of the important indicators for measuring the training process of deep learning is the loss, calculated as follows:

$$Loss = -\sum_{i}^{n} y_i \log(p_i) \tag{13}$$

wherein y_i is the data sample label. When the data sample belongs to the class i, y_i is 1; otherwise, it is 0. p_i is the probability that the data sample belongs to the class i. As the name suggests, the smaller the loss of the model, the better the model performance on the dataset, and the better the model fits the data.

The loss and accuracy during the training of the proposed identification model are shown in Figure 12. As the number of training increases, the accuracy of the model increases, and the loss degree decreases. The loss of the model can decrease to 0.035; the accuracy can increase to 98.6%.

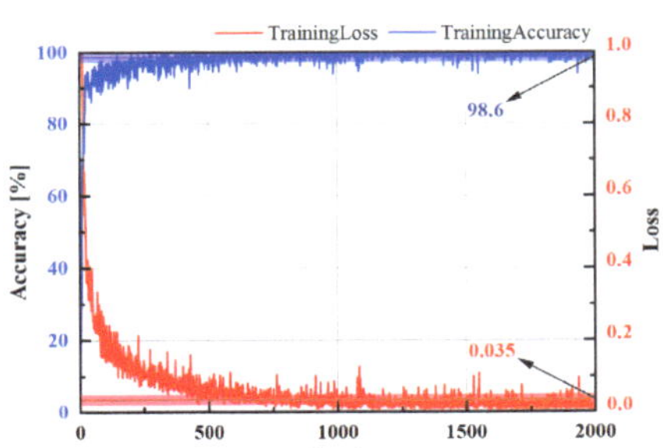

Figure 12. Degree of loss and accuracy of the training process.

The comparison of the training set and test set prediction results is shown in Figure 13. The training set contains a total of 78,514 data entries, and the test set includes a total of

19,629 data entries. The proposed identification model achieves an accuracy of 99.8268% for both datasets, predicting a result different from the actual situation for only a small number of data entries.

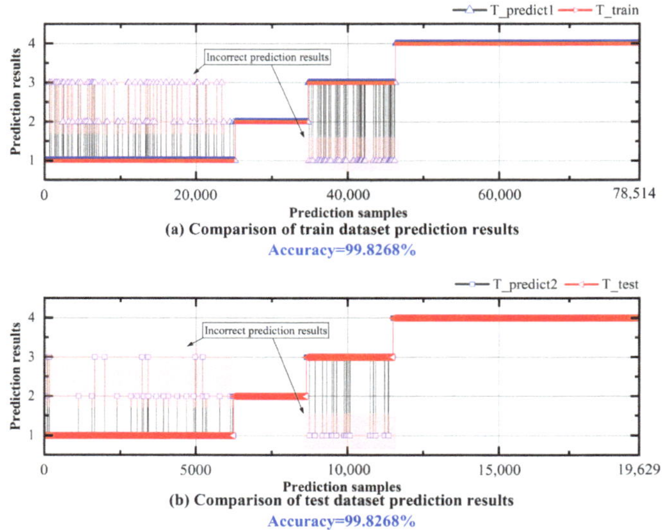

Figure 13. The proposed identification model training set and test set prediction results.

For further analysis, the confusion matrices of the training set data and the test set data are plotted in Figure 14. In Figure 14a, the training set contains 25,036 data entries for condition Type 1, of which 24,996 are correctly identified, 10 are incorrectly identified as Type 2, and 30 are incorrectly identified as Type 3. There are 4808 data entries, of which 4799 are correctly identified and 9 are incorrectly identified as Type 1; for condition Type 3. There are 16,329 data entries, of which 16,095 are correctly identified and 234 are incorrectly identified as Type 1. For condition Type 4, there are 32,341 data entries, all of which are correctly identified.

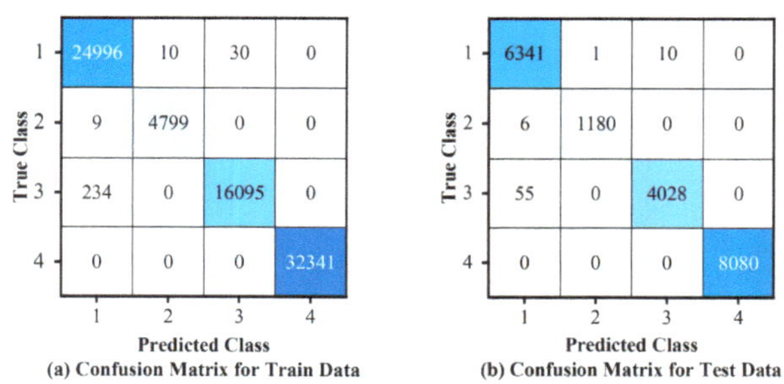

Figure 14. Confusion matrix based on the proposed model.

Figure 14b indicates that in the test set, there are 6352 data entries for condition Type 1, of which 6341 are correctly identified, one is incorrectly identified as Type 2, and 10 are incorrectly identified as Type 3. For condition Type 2, there are 1186 data entries, with 1180 correctly identified and 6 incorrectly identified as Type 1. For condition Type 3, there are

4083 data entries, with 4028 correctly identified and 55 incorrectly identified as Type 1. For condition Type 4, there are 8080 data entries, all of which are correctly identified.

In summary, the identification accuracy of working conditions Type 1, Type 2, and Type 4 is higher than 99%; Type 3 has 98% identification results. Type 1 has 1% of the data classified as Type 3. Type 3 has 1.4% of the data classified as Type 1. Due to the low number of incorrectly identified data, it can be recognized as a small probability event. From the analysis, the proposed identification model can accomplish the working condition identification task.

5.2. Computational Time of Working Condition Identification

The proposed identification model must match the computational speed of the EMS. For this reason, the identification speed of the working condition identification model is evaluated through comparative experiments. They are divided into two parts: (1) testing the time required to train the model and (2) the time required to use the model to complete the working condition identification task. This study designs two sets of controlled experiments on the same experimental platform, comparing the models based on the SVM identification model and the proposed model used in this paper.

- Experiment 1: This sets the parameters of the identification model proposed as shown in Figure 8. The identification model is based on SVM, and the RBF function is chosen as the kernel function. Both models are trained 100 times. The training time of each time is counted as a comparison term.
- Experiment 2: Among the two identification models trained in Experiment 1, the two models with relatively good accuracy are selected. Select 1000 pieces of data from the ship's working condition dataset to form the test data of Experiment 2, with condition types 1, 2, 3, and 4 each accounting for 25% of the test data. Record the time required for each identification model to complete the identification task.

The statistical results of the experiments are shown in Figure 15. The training time of the proposed model in this paper ranges from 30 to 90 s in Figure 15a, with an average training time of 58.76 s. The training time of the SVM-based model for the identification ranges from 1 to 7 s, with an average training time of 2.48 s. In Figure 15b, the time required for the model used to identify the test data of Experiment 2 ranges from 0.01 to 0.4 s, with an average identification time of 0.08 s. The identification time of the SVM-based model ranges from 0.05 to 3.5 s, with an average identification time of 1.31 s.

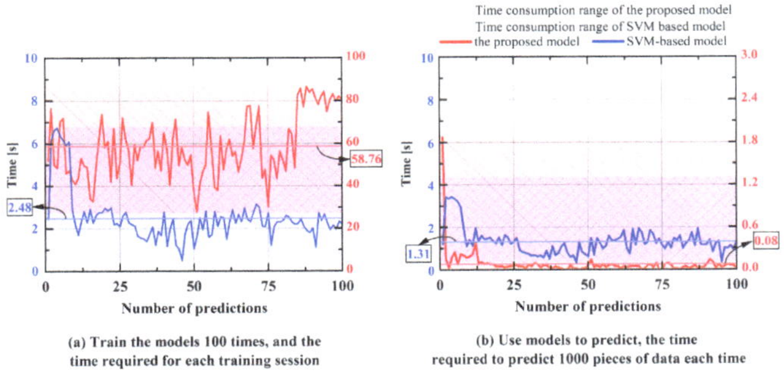

Figure 15. Comparison of model training and identification time.

Scheme 1 and 2 are shown in Table 3. Considering that working condition identification only needs to be performed offline, the training time is used as a reference for evaluating the models. The accuracy and real-time performance are the focus of the comparison experiments. In summary, it can be seen that the experimental results of the model used in this paper are all better than those of the SVM-based identification model. Therefore, the proposed identification model in this paper can improve the accuracy and real-time performance of working condition identification and better complete the task of identification.

Table 3. Results of the identification experiments.

Model	Real-Time Performance
SVM	0.05–3.5 (1000 times/s)
The proposed method	0.01–0.4 (1000 times/s)

5.3. Experimentation and Analysis of EMS Based on Working Condition Identification

To validate the optimization effect and fuel economy of the proposed strategy, simulation experiments were conducted using MATLAB software. The simulation model parameters and control parameters are shown in Tables 4 and 5.

Table 4. Simulation model parameters.

Description	Parameter	Symbol	Value
Battery	Capacity	Q	1100 kWh
	Internal resistance	R_0	0.0065 Ω
	Maximum voltage	V_{oc}	750 V
Diesel generators	Speed range	ω_{dg}	0–1800 RPM
	Efficiency	η_{dg}	0.98
	Fitted coefficient	$a_i, i = 0,1,2,3$	$[-409.3, 1.599 \times 10^{-1}, 6.458 \times 10^{-4}, -2.608 \times 10^{-7}]$
		$b_j, j = 0,1,2,3$	$[302.3, 4.78 \times 10^{-1}, 1.36 \times 10^{-3}, -1.213 \times 10^{-6}]$

Table 5. Results of the experiments.

Description	Symbol	Value
Battery initial SOC	SOC_0	0.9
Battery maximum SOC	SOC_{max}	0.99
Battery minimum SOC	SOC_{min}	0.20
Battery maximum current	$I_{bat,max}$	600 A
Battery minimum current	$I_{bat,min}$	−600 A
Coulombic efficiency	η_{bat}	0.98
DG maximum output power	$P_{dg,max}$	450 kW
DG minimum output power	$P_{dg,min}$	0 kW
DG maximum speed	$\omega_{dg,max}$	1800 RPM
DG minimum speed	$\omega_{dg,min}$	0 RPM

The load demand of the simulated ship is shown in Figure 16, with a sampling frequency of 1 Hz. The load power covers the four types of working conditions, with each type accounting for approximately 25%. This setup was used to test the effectiveness and real-time performance of the working condition identification NMPC-based EMS. In the simulation experiment, at each sampling point, the eight feature data including load power were either test set data or data generated by interpolation from the test set data. The source of the ship speed data is based on the analysis of ship speed shown in Figures 6 and 7.

To test the robustness of the proposed strategy, the method from reference [21] is used to add 80 dB of Gaussian noise to the load demand, which accounts for 10% of the maximum demand power. Simulations are performed under this condition using the conventional NMPC strategy and the proposed strategy. The power tracking of the ship demand power is performed using the NMPC-based EMS and the strategy proposed, with the simulation results shown in Figure 17.

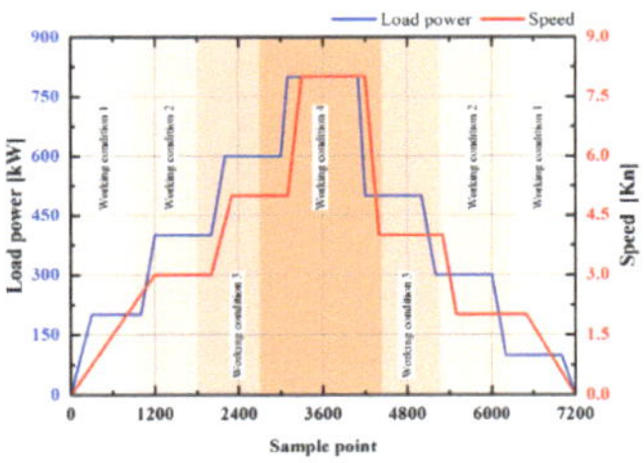

Figure 16. Ship demand power load and speed.

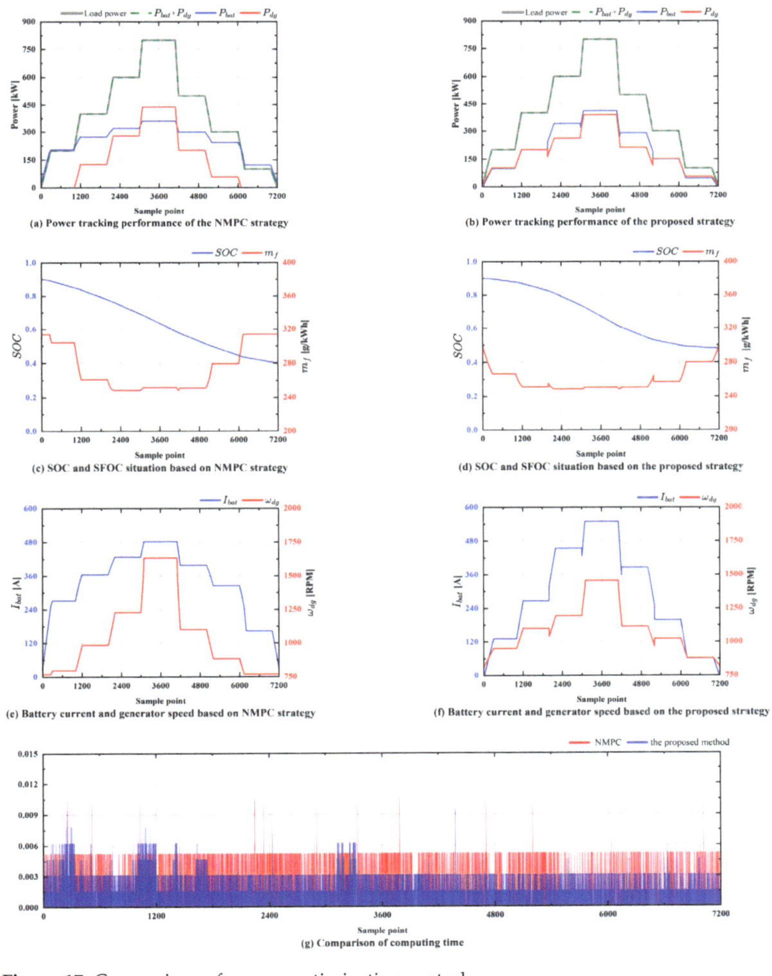

Figure 17. Comparison of energy optimization control.

As in Figure 17a,b, it shows the diesel generator output power and battery output power of the proposed strategy versus the conventional NMPC strategy when tracking the demand power. It is seen that the diesel generator and battery can meet the demand power at all times, with relatively stable output power. Figure 17c,d depict the SFOC and battery SOC changes; compared to each other, the proposed strategy can reduce SFOC, with the end SOC differing by about 10%. In Figure 17e,f, the output current of the battery and the speed of the diesel generator are shown. In Figure 17g, the computation time required for each sampling point is shown. It is seen that the average computation time of the proposed strategy in this paper is less.

Comparing the NMPC EMS and the strategy proposed in this study for power tracking the demand power of the ship with 10% noise, the simulation results are presented in Figure 18.

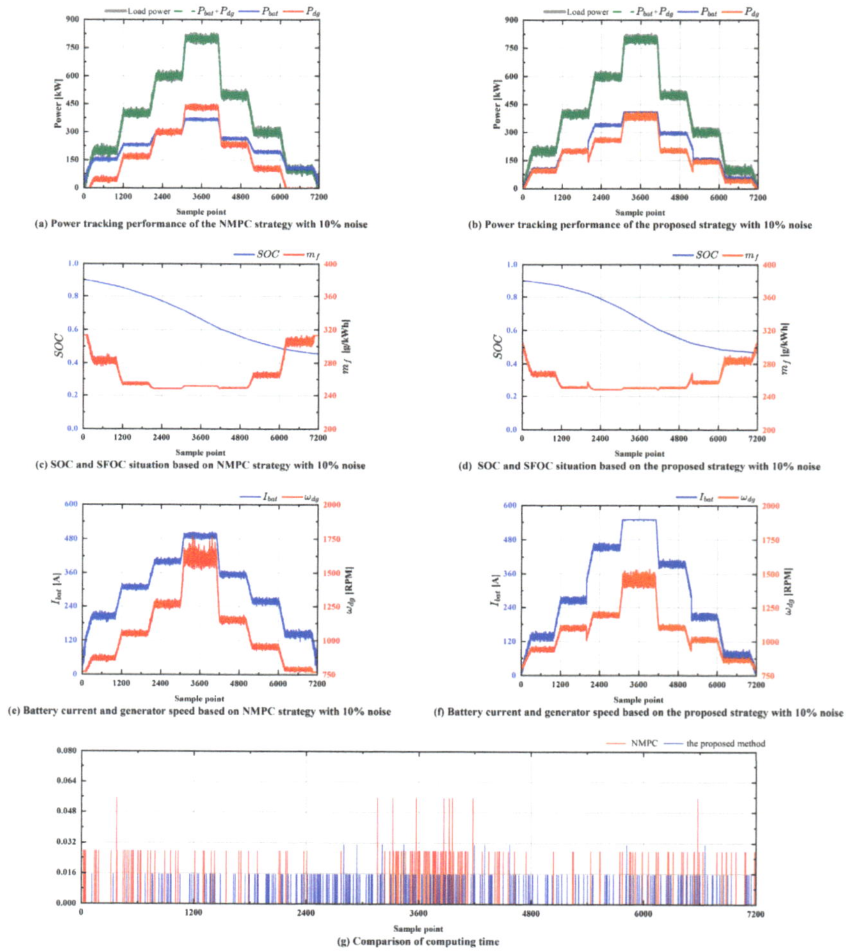

Figure 18. Comparison of energy optimization control with 10% noise.

In Figure 18a,b, the diesel generator output power and the battery output power using the NMPC strategy and the strategy proposed in this paper are shown. It is seen that the diesel generator and battery can meet the demand for power with 10% noise. In Figure 18c,d, the instantaneous fuel consumption and battery SOC variations are demonstrated. It is shown that the SOC values at the end of the simulation do not differ much, but the proposed strategy significantly reduces fuel consumption. In Figure 18e,f, the battery output current and diesel generator speed cases are displayed. The proposed strategy has a sudden decrease in generator speed and battery current at the 2000th and 5000th sampling points. But the sudden increase and decrease in this case is not significant. In Figure 18g, the computation time required at each sampling point is shown.

The rule-based control strategy proposed by Khan et al. [10] is applied to the ship in the case study. The rules are set as follows: (1) When the battery's SOC is above 50%, the diesel generator and the battery bear 40% and 60% of the load power, respectively. (2) When the SOC is lower than 50% but higher than 20%, the diesel generator and the battery bear 60% and 40% of the load power, respectively. (3) When the SOC is lower than 20%, the battery stops working, and the load power is borne by multiple diesel generators. The simulation results are summarized in Table 6.

Table 6. Comparison results of different strategies.

Method	Noise	Fuel Consumption	Final SOC	Real-Time Performance
Rule-based	no noise	2056.809 kg	56.17%	1.7×10^{-5} s
	10% noise	2057.435 kg	56.18%	1.9×10^{-5} s
NMPC	no noise	1973.789 kg	40.30%	2.6×10^{-3} s
	10% noise	1923.495 kg	45.70%	4.0×10^{-3} s
The proposed	no noise	1865.518 kg	48.10%	1.16×10^{-3} s
	10% noise	1874.001 kg	47.15%	6.0×10^{-4} s

Based on the simulation results of Figures 17 and 18, the cumulative fuel consumption data, battery SOC, and real-time performance for different scenarios are summarized in Table 6.

In scenarios where demand power is free from noise interference, the proposed EMS achieves a fuel consumption reduction of 209.366 kg compared to the rule-based strategy, corresponding to a decrease of 10.2%. In scenarios where demand power is affected by noise interference, the proposed EMS reduces fuel consumption by 183.434 kg compared to the rule-based strategy, resulting in a decrease of approximately 8.9%. However, regarding the final SOC value and real-time performance, the proposed method does not perform as well as the rule-based strategy. If the SOC difference between the rule-based control and the proposed method is higher, at 8.07% and 9.03%, respectively, after converting the energy and translating it into fuel consumption, the fuel consumption is still higher than that of the method proposed in this study. The proposed method inherently requires a certain amount of computation. Therefore, compared to the rule-based strategy, the proposed method can reduce fuel consumption, but it performs poorly in terms of final SOC and computation time.

In the scenario without noise interference in demand power, the proposed EMS reduces fuel consumption by 108.271 kg compared to the conventional NMPC strategy, a decrease of 5.5%. In the scenario with noise interference in demand power, the proposed EMS reduces fuel consumption by 49.494 kg compared to the conventional NMPC strategy, a decrease of approximately 2.6%. In terms of battery SOC changes, the final SOC value of the proposed strategy is higher than that of the conventional NMPC strategy. In terms of real-time performance, the working condition identification helps to reduce the solution time of the NMPC controller and meets the constraints. In the scenario without and with noise, the identification time accounts for 34.4% and 13.3% of the entire solution time, respectively. Overall, the proposed strategy outperforms the conventional NMPC strategy.

Experimental simulations were conducted in this study. The effectiveness of the hybrid power ship in real-time identification of working conditions that can provide decision-making information to the EMS is verified. In addition, the effectiveness of the proposed strategy is also demonstrated from the perspective of cumulative fuel consumption. The identification model proposed in this paper shows better accuracy, which may be due to two main reasons: Firstly, the self-constructed dataset used in this paper contains more data features. Secondly, the experimental platform used in this paper is superior, which makes the model performance and real-time computational capability better.

6. Conclusions

Aiming at the problem of energy optimization control of hybrid power ships, a nonlinear model predictive control EMS based on working condition identification is proposed. Data acquisition, processing, and analysis were carried out based on the ship operation process. Data labeling work was performed to construct a ship working condition dataset. A condition identification model based on CNN was established, and the model was trained and tested using the dataset for offline identification of ship conditions, providing decision-making information for the EMS. Finally, by constructing the NMPC-based EMS, the energy optimization control of shipboard generators and the battery under different working conditions was realized, effectively reducing fuel consumption. The main conclusions are as follows:

(1) The study takes the CNOOC 257 supply ship as its object, collects data about the ship operation process, creates a working condition identification dataset, and uses the proposed identification model to complete the task of identifying the working conditions of the ship, with an accuracy rate of over 99%.

(2) Simulation results show that the EMS based on working condition identification using NMPC can reduce fuel consumption by 5.5% compared to the conventional NMPC strategy. Under the condition of adding 10% noise to the demanded power, it can further reduce fuel consumption by 2.6%. Additionally, the proposed strategy is able to meet the real-time requirements.

This study compares fuel consumption, battery SOC variation, and real-time performance. There is also a possibility that the identification may be disturbed by the environment in real applications. In the future, more factors will be considered in our work. For example, resistance to environmental interference, simulation scenario realism, and multi-model fusion to improve the accuracy of working condition identification.

Author Contributions: Conceptualization, Y.Y. and D.G.; methodology, Y.Y. and D.G.; software, Z.C.; validation, Y.Y., Z.C. and D.G.; formal analysis, D.G.; writing—original draft preparation, Y.Y.; writing—review and editing, D.G.; project administration, D.G. All authors have read and agreed to the published version of the manuscript.

Funding: The study was founded by the National Key R&D Program of China (Grant 2022YFB4301403).

Institutional Review Board Statement: Not applicable.

Informed Consent Statement: Not applicable.

Data Availability Statement: The paper codes and data will be uploaded to https://github.com/inspireyc (accessed on 3 May 2024).

Acknowledgments: The authors thank the anonymous reviewers for suggesting valuable improvements to the paper.

Conflicts of Interest: The authors declare no conflicts of interest.

Abbreviations

The following abbreviations are used in this manuscript:

IMO	International Maritime Organization
EMS	Energy Management Strategy
MPC	Model Predictive Control
NMPC	Nonlinear Model Predictive Control
SVM	Support Vector Machine
LSSVM	Least Squares Support Vector Machine
CNN	Convolutional Neural Networks
ECA	Efficient Channel Attention Networks
AC	alternating current
DC	direct current
SOC	battery state of charge
SFOC	specific fuel oil consumption

References

1. Cret, L.; Baudry, M.; Lantz, F. How to Implement the 2023 IMO GHG Strategy? Insights on the Importance of Combining Policy Instruments and on the Role of Uncertainty. *Marine Policy* **2024**, *169*, 106332. [CrossRef]
2. Zhou, C.; Huang, H.; Liu, Z.; Ding, Y.; Xiao, J.; Shu, Y. Identification and Analysis of Ship Carbon Emission Hotspots Based on Data Field Theory: A Case Study in Wuhan Port. *Ocean Coast. Manag.* **2023**, *235*, 106479. [CrossRef]
3. Zhang, W.; Xi, B. The Effect of Carbon Emission Trading on Enterprises' Sustainable Development Performance: A Quasi-Natural Experiment Based on Carbon Emission Trading Pilot in China. *Energy Policy* **2024**, *185*, 113960. [CrossRef]
4. Bai, M.; Li, C. Research on the Allocation Scheme of Carbon Emission Allowances for China's Provincial Power Grids. *Energy* **2024**, *299*, 131551. [CrossRef]
5. Inal, O.B.; Charpentier, J.-F.; Deniz, C. Hybrid Power and Propulsion Systems for Ships: Current Status and Future Challenges. *Renew. Sustain. Energy Rev.* **2022**, *156*, 111965. [CrossRef]
6. Zheng, Y.; He, F.; Shen, X.; Jiang, X. Energy Control Strategy of Fuel Cell Hybrid Electric Vehicle Based on Working Conditions Identification by Least Square Support Vector Machine. *Energies* **2020**, *13*, 426. [CrossRef]
7. Gharibeh, H.F.; Yazdankhah, A.S.; Azizian, M.R. Energy Management of Fuel Cell Electric Vehicles Based on Working Condition Identification of Energy Storage Systems, Vehicle Driving Performance, and Dynamic Power Factor. *J. Energy Storage* **2020**, *31*, 101760. [CrossRef]
8. Chang, C.; Zhao, W.; Wang, C.; Song, Y. A Novel Energy Management Strategy Integrating Deep Reinforcement Learning and Rule Based on Condition Identification. *IEEE Trans. Veh. Technol.* **2023**, *72*, 1674–1688. [CrossRef]
9. Nivolianiti, E.; Karnavas, Y.L.; Charpentier, J.-F. Fuzzy Logic-Based Energy Management Strategy for Hybrid Fuel Cell Electric Ship Power and Propulsion System. *J. Mar. Sci. Eng. JMSE* **2024**, *12*, 1813. [CrossRef]
10. Khan, M.M.S.; Faruque, M.O.; Newaz, A. Fuzzy Logic Based Energy Storage Management System for MVDC Power System of All Electric Ship. *IEEE Trans. Energy Convers.* **2017**, *32*, 798–809. [CrossRef]
11. Hu, W.; Shang, Q.; Bian, X.; Zhu, R. Energy Management Strategy of Hybrid Energy Storage System Based on Fuzzy Control for Ships. *Int. J. Low-Carbon Technol.* **2022**, *17*, 169–175. [CrossRef]
12. Tang, R.; Li, X.; Lai, J. A Novel Optimal Energy-Management Strategy for a Maritime Hybrid Energy System Based on Large-Scale Global Optimization. *Appl. Energy* **2018**, *228*, 254–264. [CrossRef]
13. Gao, D.; Jiang, H.; Shi, W.; Wang, T.; Wang, Y. Adaptive Equivalent Consumption Minimization Strategy for Hybrid Electric Ship. *Energy Sci. Eng.* **2022**, *10*, 840–852. [CrossRef]
14. Zhang, Y.; Xue, Q.; Gao, D.; Shi, W.; Yu, W. Two-Level Model Predictive Control Energy Management Strategy for Hybrid Power Ships with Hybrid Energy Storage System. *J. Energy Storage* **2022**, *52*, 104763. [CrossRef]

15. Chen, L.; Gao, D.; Xue, Q. Energy Management Strategy for Hybrid Power Ships Based on Nonlinear Model Predictive Control. *Int. J. Electr. Power Energy Syst.* **2023**, *153*, 109319. [CrossRef]
16. García-Pérez, D.; Saeed, M.; Díaz, I.; Enguita, J.M.; Guerrero, J.M.; Briz, F. Machine Learning for Inverter-Fed Motors Monitoring and Fault Detection: An Overview. *IEEE Access* **2024**, *12*, 27167–27179. [CrossRef]
17. Zhou, X.; Zhang, R.; Wang, X.; Huang, T.; Yang, C. Kernel Intuitionistic Fuzzy C-Means and State Transition Algorithm for Clustering Problem. *Soft Comput* **2020**, *24*, 15507–15518. [CrossRef]
18. Cerdà-Alabern, L.; Iuhasz, G.; Gemmi, G. Anomaly Detection for Fault Detection in Wireless Community Networks Using Machine Learning. *Comput. Commun.* **2023**, *202*, 191–203. [CrossRef]
19. Fayed, H.A.; Atiya, A.F. Decision Boundary Clustering for Efficient Local SVM. *Appl. Soft Comput.* **2021**, *110*, 107628. [CrossRef]
20. Gao, D.; Jiang, Y.; Zhao, N. A Novel Load Prediction Method for Hybrid Electric Ship Based on Working Condition Classification. *Trans. Inst. Meas. Control* **2022**, *44*, 5–14. [CrossRef]
21. Liu, H.; Fan, A.; Li, Y.; Bucknall, R.; Chen, L. Hierarchical Distributed MPC Method for Hybrid Energy Management: A Case Study of Ship with Variable Operating Conditions. *Renew. Sustain. Energy Rev.* **2024**, *189*, 113894. [CrossRef]
22. Yuan, Y.; Ye, T.; Tong, L.; Yuan, C.; Teng, L. Optimizing Energy Management Strategies for Hybrid Electric Ships Based on Condition Identification and Model Predictive Control. *Int. J. Green Energy* **2023**, *20*, 1763–1775. [CrossRef]
23. Tervo, J.; Junttila, J.; Lämsä, V.; Savolainen, M.; Ronkainen, H. Hybrid Methodology Development for Lubrication Regimes Identification Based on Measurements, Simulation, and Data Clustering. *Tribol. Int.* **2024**, *195*, 109631. [CrossRef]
24. Yang, Q.; Wei, M.; Zhu, R.; Zhou, B. Facial Expression Recognition Algorithm Based on Efficient Channel Attention. *J. Electron. Imaging* **2022**, *31*, 053021. [CrossRef]
25. Guang, J.; Xi, Z. ECAENet: EfficientNet with Efficient Channel Attention for Plant Species Recognition. *J. Intell. Fuzzy Syst.* **2022**. [CrossRef]
26. Wang, J.; Wu, X. A Deep Learning Refinement Strategy Based on Efficient Channel Attention for Atrial Fibrillation and Atrial Flutter Signals Identification. *Appl. Soft Comput.* **2022**. [CrossRef]
27. Yang, Y.; Ding, K.; Chen, Z. Ship Classification Based on Convolutional Neural Networks. *Ships Offshore Struct.* **2022**, *17*, 2715–2721. [CrossRef]
28. Zhang, Y.; Wang, S.; Zhao, H.; Guo, Z.; Sun, D. CT Image Classification Based on Convolutional Neural Network. *Neural Comput. Appl.* **2021**, *33*, 8191–8200. [CrossRef]

Journal of
Marine Science and Engineering

MDPI

Article

Enhancing Efficiency in Hybrid Marine Vessels through a Multi-Layer Optimization Energy Management System

Hoai Vu Anh Truong [1,*], Tri Cuong Do [2] and Tri Dung Dang [2,*]

[1] Department of Mechanical Engineering, Pohang University of Science and Technology, Pohang 37673, Republic of Korea
[2] College of Technology and Design, University of Economics Ho Chi Minh City, Ho Chi Minh City 70000, Vietnam; cuongdt298@ueh.edu.vn
* Correspondence: truonganh@postech.ac.kr or truonganh241292@gmail.com (H.V.A.T.); dungdt@ueh.edu.vn (T.D.D.)

Abstract: Configuring green power transmissions for heavy-industry marines is treated as a crucial request in an era of global energy and pollution crises. Following up on this hotspot trend, this paper examines the effectiveness of a modified optimization-based energy management strategy (OpEMS) for a dual proton exchange membrane fuel cells (dPEMFCs)-battery-ultra-capacitors (UCs)-driven hybrid electric vessels (HEVs). At first, the summed power of the dual PEMFCs is defined by using the equivalent consumption minimum strategy (ECMS). Accordingly, a map search engine (MSE) is proposed to appropriately split power for each FC stack and maximize its total efficiency The remaining power is then distributed to each battery and UC using an adaptive co-state, timely determined based on the state of charge (SOC) of each device. Due to the strict constraint of the energy storage devices' (ESDs) SOC, one fine-corrected layer is suggested to enhance the SOC regulations. With the comparative simulations with a specific rule-based EMS and other approaches for splitting power to each PEMFC unit, the effectiveness of the proposed topology is eventually verified with the highest efficiency, approximately about 0.505, and well-regulated ESDs' SOCs are obtained.

Keywords: energy management strategy; optimization; equivalent consumption minimum strategy (ECMS); hybrid electric vessels; multi-stack PEMFC

1. Introduction

Recently, the marine industry and vessel traffic services have facilitated economic growth, global trading markets, ecosystem stabilization, and so on. They have been recognized internationally as an aid to the safety of life at sea [1] and fueled an efficient movement for the cross-border trade of merchandise on an unprecedented scale. However, the transportation sectors, in general, and vessel services in particular, are facing the challenges of a full-blown global energy crisis and environmental pollution from the heavy use of fossil fuel and oil scarcity [2]. Although an energy storage device was integrated for hybrid vessels [3,4], the mentioned prototypes still employed traditional diesel engines. As a result, identifying an alternative power source for the industry's shift away from fossil fuels is crucial sustainable development.

Proton exchange membrane fuel cells (PEMFCs) have been acknowledged as a preferable alternative to traditional fuels and realistically put into small-scaled and large-scaled installations [5], due to their properties of high efficiency, high power density, low operating temperature, low noise, and zero emission. In terms of the shipping sector, the PEMFCs are viewed as the most promising clean resource and as friendly to the marine ecosystem [6], and their use sees increasing growth [7] on ships that currently make coastal voyages. Nevertheless, the complex operations of naval vessels against the severe fluctuations of propulsion load and environmental impacts pull the standalone PEMFC lifetime down due to its slow dynamic response. Additionally, the required quick start-up time and regenerating excessive power also represent challenges for the standalone PEMFCs, requiring significant effort.

To overcome these inherent demerits, the hybridization of the PEMFCs with one or more energy storage devices (ESDs) has been switching into an upward trend for the research community in terms of designing feasible hybrid configurations and developing effective energy management strategies (EMSs) for practical realization. Various EMSs with enhancements have been proposed such as [8,9], fuzzy logic-based state machine EMS [10,11], equivalent consumption minimum strategy (ECMS) [11–13], model predictive control [14], and a learning-based algorithm with the balance-of-plant [15] for the hybrid electric ships and vessels powered by an integrated PEMFC battery power source. Extended to a hybrid topology of PEMFC, battery, and ultra-capacitors (UCs), Peng et al. [16] constructed a particle swarm optimization (PSO)-based EMS with a wavelet transform technique to obtain an optimal power reference for each device. Other achievements can be referred to in the literature [17–19]. Moreover, it has been established that a single-stack PEMFC hardly satisfies the heavy marine load request regarding the recently gained technology. It is estimated that the required propulsion power for vessels usually varies from 18 MW to around 30 MW and may be up to higher than 80 MW [20], which exceeds the workability of the latest PEMFCs [21]. Consequently, the single-stack PEMFC has been turned into multi-stack PEMFC for further compatible realization and developments. Yet, appropriately splitting power for each PEMFC stack and retaining high efficiency has been an increasingly active focus in the related fields.

The research on designing EMSs for multi-stack PEMFC hybrid power sources has been ongoing for several hybrid electric transportations. Generally, there are three methodologies for power-sharing on a multi-stack PEMFC: equal distribution (EqD), daisy chain, and instantaneous optimization-based method [22]. The first approach offers simple calculation with equally distributed power for each stack; however, the overall efficiency was recorded as low. The daisy chain method manipulates all stacks in sequential operations. The overall efficiency was improved only at low-power operation compared to the former. At the high-power range, the former outperformed the daisy chain [23]. Based on the demerits of each method, adaptive control was proposed. In [24], the authors developed an adaptive state machine to allocate power for multi-stack PEMFC. In [25], the authors introduced a degree of performance degradation with virtual resistance to assist the power-sharing process. Other techniques of using a fuzzy logic-based state machine and hierarchical EMS approaches for a multi-stack PEMFC battery–UCs power system, realized on hybrid tramways, can be referred to [26,27]. Although the overall efficiency was improved, the mentioned studies could not achieve the optimal power reference and maximize the overall efficiency.

On the contrary, the optimization-based category is able to achieve the optimal solution for power coordination, enhance efficiency, and prolong the durability and lifetime of all PEMFC stacks [28]. This approach can be classified into offline control, well known as global optimization, and real-time control, known as local optimization. As the existing demerits of offline control are burdensome computation, being time-consuming, and it being impossible to instantly change updates in real time [29], real-time control is preferable with various accomplishments recorded. In [30], the authors first built the efficiency map and established an online control to define the optimal power reference for each stack through equivalent dual-power for hybrid locomotives powered by multi-stack PEMFC interconnected with supercapacitors. Subsequently, on the basis of the map-search engine (MSE), an Extremum-seeking-based EMS was exploited to optimize the PEMFC power by seeking out the highest efficiency for each PEMFC primary source [31]. Similarly, in [32], Do et al. developed an ES-based EMS with an enhanced map-search method to split power for dual PEMFCs at which each PEMFC stack could operate in their highest efficiency regions. The effectiveness of these contributions was realized on hybrid electric locomotives and tramways. In the field of marine vessels, the authors in [33] initiated an optimization-based multi-layer EMS for passenger ships driven by a hybridization of multi-stack PEMFC and battery. Unfortunately, the proposed method followed the daisy chain method, which returns low total efficiency and causes the possibility of increased stack degradation. As observed, the emergence of dual PEMFCs into the optimization-based EMS, with the effective use of the power distribution, has not garnered much attention from scholars and has been limited to hybrid vessels.

Comprehensively motivated by the existing research gaps, this paper aims at conducting a multi-layer optimization-based EMS (MOEMS) for vessels driven by dual-PEMFC (dPEMFC) battery–UCs hybrid power sources. In this manner, the dPEMFC functions as a primary source while the battery and UCs are supplements and all devices are interconnected, in parallel, to a DC bus via DC/DC converters. The MOEMS is established based on the hybrid off–online algorithm to exert the highest efficiency and prolong the dPEMFC lifetime. For this purpose, an ECMS framework is first utilized as an online control in the upper layer to optimize the equivalent power of the dual stacks. Subsequently, an MSE, obtained through offline data acquisition, is integrated into the mediate layer to appropriately share power for each primary unit. An adaptive co-state is introduced to distribute the remaining load to supplements, whose dynamics are derived based on the instant battery and UCs' state of charge (SOC). The determined reference power is then designated to a lower layer for the pulse-width modulation (PWM). The effectiveness of the proposed control scheme is validated via numerical simulations on a specific vessel model with another rule-based benchmark.

The remainder of this paper is organized as follows: Section 2 describes the modeling of the system, including DC/DC converters dynamics. Then, Section 3 dedicates the proposed EMS for appropriate power allocation. The feasibility of the proposed method is certified by comparative simulation in Section 4. Finally, Section 5 summarizes key points and suggestions for further development.

2. System Configuration and Modeling

2.1. Overview of the Hybrid Vessel Power System

The proposed hybrid vessel power system integrates multiple energy sources to meet the varied power demands of maritime operations while prioritizing efficiency and environmental sustainability. The configuration of the dual FC/BAT/UC vessel is depicted in Figure 1, with the Alsterwasser passenger vessel, the world's first hydrogen fuel cell passenger vessel involved in the examined objective, whose dynamics and characteristics are detailed in [34], for the proposed EMS implementation. The core of this system are two PEMFC stacks, which serve as the primary power source while a battery pack and a UC bank are complemented to provide auxiliary power and energy storage capabilities.

The dual PEMFC configuration allows for greater power output and operational flexibility compared to single-stack systems. Each PEMFC stack is connected to the DC bus through a unidirectional DC/DC converter, which steps up the fuel cell output voltage to match the bus voltage. This arrangement enables efficient power delivery from the fuel cells to the vessel's electrical systems.

The battery pack and UC bank are both integrated into the system via bidirectional DC/DC converters. This configuration allows these components to either supply power to the DC bus or absorb excess energy for storage, depending on the vessel's instantaneous power requirements and the state of the overall system. The DC bus serves as the central power distribution point, connecting to an inverter that converts the DC power to AC for use by the vessel's electric propulsion motor and other onboard systems. This architecture provides a flexible platform for implementing sophisticated energy management strategies to optimize power flow and system efficiency.

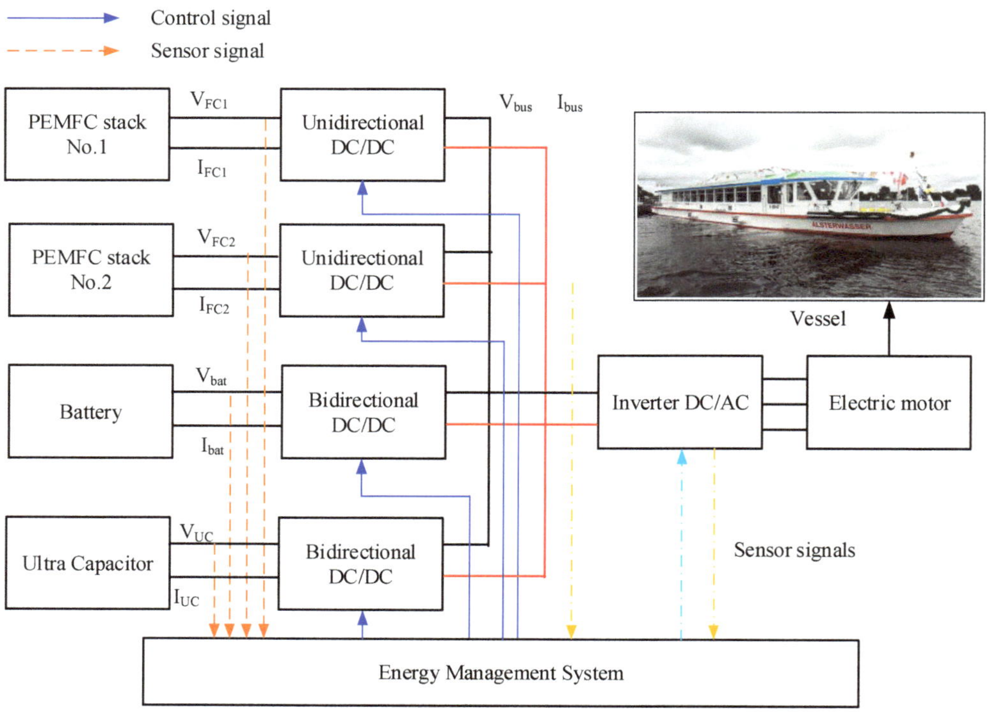

Figure 1. Configuration of dual PEMFC/battery/UCs hybrid vessels.

An energy management strategy is then designed and validated based on this model to meet the power needs of the hybrid vessel, as shown in Figure 2.

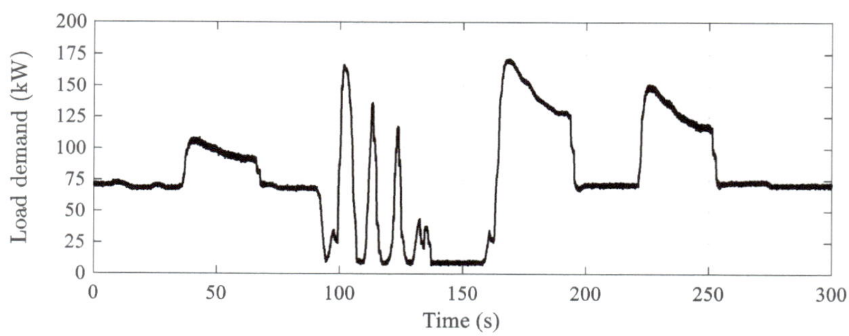

Figure 2. Part of the power requirement of hybrid electric vessels.

2.2. Component Modeling

To effectively analyze and optimize the hybrid vessel power system, it is crucial to have accurate mathematical representations of each component. This section outlines the models used for the key elements of the system: fuel cells, batteries, UCs, and DC/DC converters.

2.2.1. Fuel Cell Model

The fuel cell model captures the complex electrochemical processes occurring within the PEMFC. A simplified electrical equivalent circuit represents the fuel cell, as illustrated in Figure 3. The output voltage of the fuel cell is determined by several factors and can be inheritably expressed based upon [35–38] as

$$V_{cell} = E_{Nernst} - V_{act} - V_{ohm} - V_{conc} \quad (1)$$

In this equation, E_{Nernst} represents the thermodynamic potential, while V_{act}, V_{ohm}, and V_{conc} account for activation, ohmic, and concentration voltage losses, respectively.

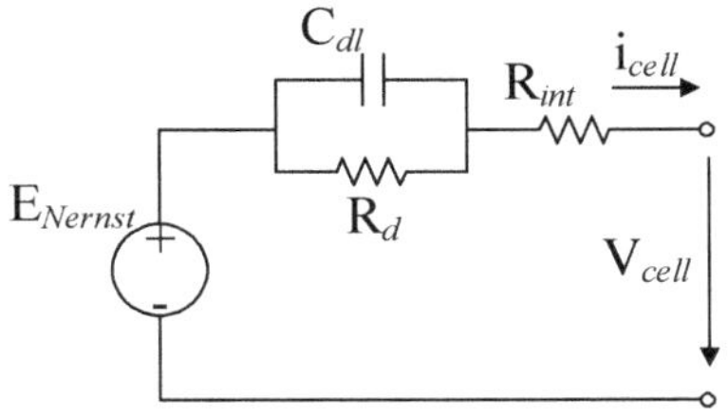

Figure 3. A simplified model of the FC.

Each of these terms is further defined by equations that incorporate factors such as temperature, pressure, and current density. The model also includes dynamics related to the double-layer capacitance effect and reactant flow rates.

The Nernst equation provides the reversible thermodynamic potential:

$$E_{Nernst} = 1.229 - 8.5 \times 10^{-4}(T_{FC} - 298.15) + \frac{RT_{FC}}{2F} \ln\left[p'_{H_2}(p'_{O_2})^{0.5}\right] \quad (2)$$

where p'_{H_2} and p'_{O_2} are partial pressures of hydrogen and oxygen, T_{FC} is the cell temperature, R is the universal gas constant, and F is the Faraday constant.

Ohmic voltage loss is estimated by

$$V_{ohm} = iR_{ohm} \quad (3)$$

where i is the cell current and R_{ohm} is the internal resistance of the electrolyte membrane.

Activation voltage loss is characterized as

$$V_{act} = \xi_1 + \xi_2 T_{FC} + \xi_3 T_{FC} \ln(c'_{O2}) + \xi_4 T_{FC} \ln(i) \quad (4)$$

where c'_{O2} is oxygen concentration at the cathode/membrane interface, and $\xi_1, \xi_2, \xi_3, \xi_4$ are parametric coefficients.

Concentration voltage loss is approximated by

$$V_{conc} = \frac{RT_{FC}}{nF} \ln\left(\frac{(i/A)_L}{(i/A)_L - (i/A)}\right) \quad (5)$$

The voltage drop, V_{drop}, due to the double capacitor layers' effect. is calculated as [37,38]

$$C\frac{dV_{drop}}{dt} = i - \frac{V_{drop}}{R_a} \quad (6)$$

where R_a is the sum of activation and concentration resistances.

$$R_a = \frac{V_{act} + V_{conc}}{i} \quad (7)$$

Then, the voltage output of a single cell can be calculated by

$$V_{cell} = E_{Nernst} - V_{drop} - V_{ohm} \quad (8)$$

The total voltage output for a stack of N cells is given by

$$V_{stack} = NV_{cell} \tag{9}$$

The dynamics of reactant flow within the fuel cell are crucial for accurate modeling. For the anode, one has

$$V_a \frac{dp'_{H_2}}{dt} = (\dot{m}_{H_2,in} - \dot{m}_{H_2,out} - \frac{Ni}{2F})RT_{FC} \tag{10}$$

where V_a is the anode volume, $\dot{m}_{H_2,in}$ and $\dot{m}_{H_2,out}$ are hydrogen inlet hydrogen outlet flow rates through the PEMFC stack, and F = 96,485 (C mol^{-1}) is the Faraday constant.

The hydrogen outlet flow rate is given by

$$\dot{m}_{H_2,out} = k_a\left(p'_{H_2} - p_{amb}\right) \tag{11}$$

where k_a is a flow constant for the anode and P_{amb} is the ambient pressure.

Similarly, for the cathode:

$$V_c \frac{dp'_{O_2}}{dt} = (\dot{m}_{O_2,in} - \dot{m}_{O_2,out} - \frac{Ni}{4F})RT_{FC} \tag{12}$$

where V_c is the cathode volume, while $\dot{m}_{O_2,in}$ and $\dot{m}_{O_2,out}$ are oxygen inlet and outlet flow rates through the FC stack.

$$\dot{m}_{O_2,out} = k_c\left(p'_{O_2} - p_{amb}\right) \tag{13}$$

The total power input to the system is proportional to the hydrogen consumed:

$$P_{tot} = \dot{m}_{H_2,used}\Delta H = \frac{Ni}{2F}\Delta H \tag{14}$$

where ΔH is the enthalpy of combustion for hydrogen.

For the dual fuel cell system, the individual and total power output are

$$P_{FC,i} = V_{stack,i}i_i \tag{15}$$

$$P_{dFC} = P_{FC,1} + P_{FC,2} \tag{16}$$

where P_{FC1} and P_{FC2} are the output power of PEMFC-1 and PEMFC-2.

The average efficiency of the dPEMFC system is calculated as (only when at least one PEMFC runs)

$$\eta_{FC\_total} = \frac{P_{FC,1} + P_{FC,2}}{\frac{P_{FC,1}}{\eta_{FC1}} + \frac{P_{FC,2}}{\eta_{FC2}}} \tag{17}$$

where η_{FC1} and η_{FC2} are the efficiencies of PEMFC-1 and PEMFC-2, respectively (obtained from calculating the ratio between the output net power and total power that includes power for auxiliaries such as pump, fan, compressor, and so on, i.e., $\eta_{FCi} = \frac{P_{FC,i}}{P_{tot,i}}$).

The hydrogen consumption rate can be derived from the power output and efficiency of each fuel cell [39]:

$$\dot{m}_{H_2\_total} = \frac{P_{FC,1}}{\eta_{FC1}LHV} + \frac{P_{FC,2}}{\eta_{FC2}LHV} \tag{18}$$

where LHV is the lower heating value of hydrogen.

The efficiency map of each PEMFC system is displayed in Figure 4.

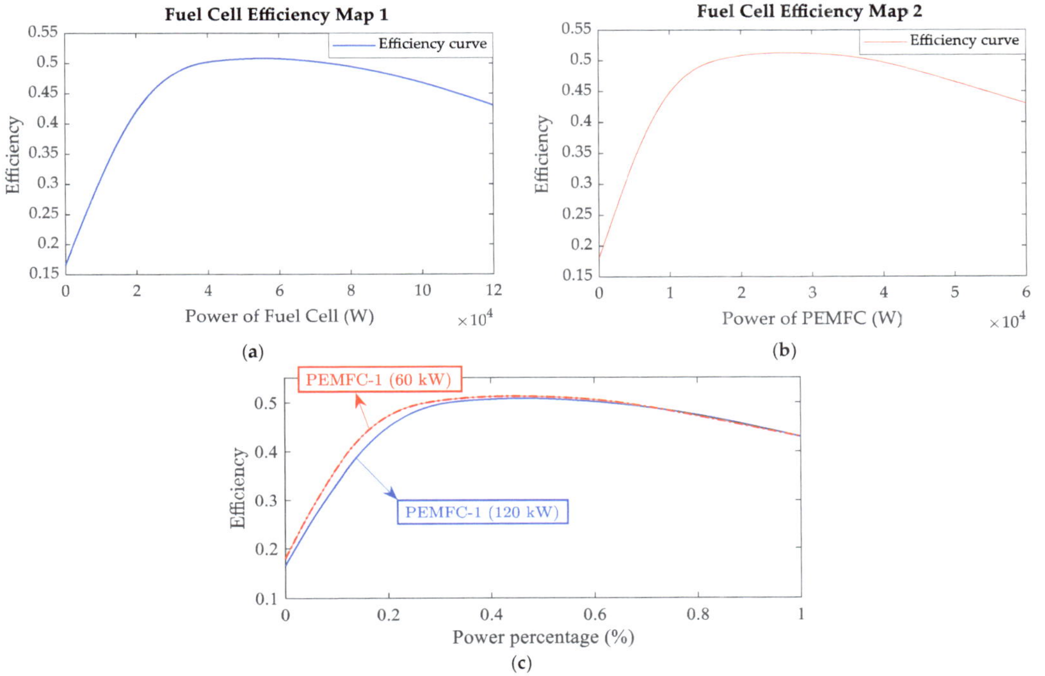

Figure 4. Efficiency maps of FC systems: (**a**) PEMFC-1 (120 kW), (**b**) PEMFC-2 (60 kW), (**c**) comparison of their efficiency maps.

2.2.2. Battery Model

In hybrid power systems, the battery serves as a critical energy buffer, complementing the fuel cell during periods of high power demand or when rapid load changes occur. To effectively design an energy management system (EMS), it is essential to develop an accurate battery model that captures key parameters and operational states. This model provides the foundation for implementing efficient control strategies.

The battery can be represented as a controlled voltage source, with its behavior described by several key equations [40]:

$$E = E_0 - K\frac{Q_{\max}}{Q} + Ae^{B(Q-Q_{\max})} \tag{19}$$

where Q and $Q_{\max}$ represent, in turn, instant and maximum battery capacity, E_0 is the free-load open-circuit voltage, A is a constant, B is a constant of battery exponential capacity, and K signifies the polarized factor.

The battery voltage (V_{bat}) is then obtained as a function of its open-circuit voltage, internal resistance, and current flow:

$$V_{bat} = E - Ri \tag{20}$$

with R (Ω) being the battery internal resistor and i (Amp) is the instant current flow.

To relate the battery voltage to its state of charge (SOC_{bat}), we can rewrite the equation as [41]

$$E = E_0 - K\frac{1}{SOC_{bat}} + Ae^{BQ_{\max}(SOC_{Bat}-1)} \tag{21}$$

The energy released from the battery during discharge can be calculated using:

$$E_{release} = E_0 - K\frac{1}{SOC_{bat}} \times it - Ri + Ae^{BQ_{\max}(SOC_{bat}-1)} - K\frac{1}{SOC_{bat}} \times i^* \quad (22)$$

where i^* denotes the low-frequency filtered current, and t is the time parameter.

The battery's output power is determined by

$$P_{bat} = V_{bat} i \quad (23)$$

Then, the state of charge SOC_{bat} is a crucial parameter that indicates the battery's remaining capacity. It can be derived from the charging current and the battery's maximum charge:

$$SOC_{bat} = \frac{Q_{\max} - it}{Q_{\max}} \quad (24)$$

To account for battery degradation over time, a capacity loss model is employed:

$$Q_{loss}(\sigma, Ah) = \sigma(I_c, \theta, SOC_{bat})Ah^z \quad (25)$$

where Ah is the accumulated charge throughput, z is the power law exponent representing Ah throughput dependence, σ is a nonlinear function of severity factors, I_c is the charging current, θ is the test temperature, and σ can be expressed as

$$\sigma = (A_{bat}SOC_{bat} + B_{bat})\exp\left(\frac{-E_a + \eta I_c}{R_g(273.15 + \theta)}\right) \quad (26)$$

where A_{bat}, B_{bat}, and η are constants determined through curve fitting, $\eta = -63.54$, $A_{bat} = -74.99$, $B_{bat} = 12895.92$, R_g is the universal gas constant and equals 8.314 J/mol/K, and E_a is the activation energy that equals 31,700 J/mol [42].

The accumulated charge throughput (Ah), which represents the battery capacity loss, is calculated as

$$Ah = \int_0^t \sigma|I_c(t)|dt \quad (27)$$

Finally, the State of Health (SOH) of the battery can be determined using

$$SOH(t) = \frac{Q_{nom} - Q_{loss}(t)}{Q_{nom}} \quad (28)$$

where Q_{nom} is the nominal capacity of the battery.

2.2.3. Ultra-Capacitor Model

Ultra-capacitors (UCs) are incorporated into the system as secondary power units due to their exceptional characteristics, including high power density, rapid charge and discharge capabilities, and impressive power release. The integration of UCs alleviates the burden on fuel cells and batteries during high peak power demands, thereby enhancing overall system performance, extending component lifespans, and potentially reducing system size and costs [43]. The UC model employed in this study is based on an equivalent circuit approach [44]. Each UC unit cell consists of two parallel RC branches, as illustrated in Figure 5 [45,46]. This configuration effectively captures the device's electrical behavior under various operating conditions.

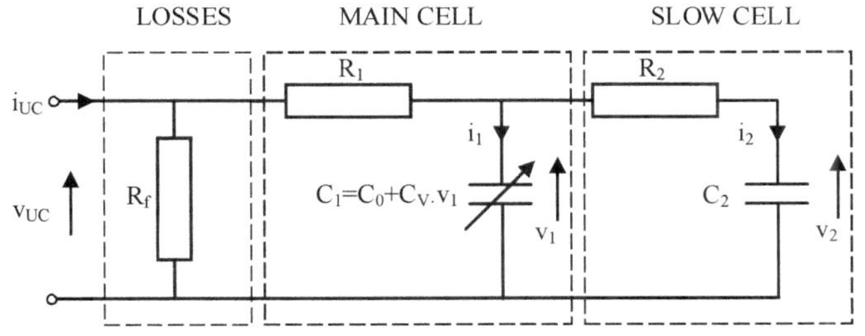

Figure 5. A simplified model of one UC.

The immediate branch, represented by R_1C_1, accounts for the rapid response during short-duration charge or discharge events. The delayed branch, denoted by R_2C_2, models the charge redistribution phenomenon that occurs after the initial charge/discharge process. A leakage resistor R_f is included to represent self-discharge behavior, although its impact is often negligible due to the typically low leakage current in high-capacity UCs.

The energy stored in a UCs bank at voltage U_{uc} is given by

$$E_{uc} = \frac{1}{2}C_{eq}U_{uc}^2 = \frac{1}{2}\frac{N_{p\_uc}}{N_{s\_uc}}C_{uc}U_{uc}^2 \tag{29}$$

where C_{eq} is an equivalent capacity of the UCs, N_{p_uc} and N_{s_uc} represent the parallel branches and the serial connection of the UCs, respectively, and C_{uc} is the capacitance of a single UC unit.

The voltage across the UC bank can be determined by considering the characteristics of a single UC pack:

$$U_{uc} = N_{s\_uc}\left(v_1 + R_1\frac{I_{uc}}{N_{p\_uc}}\right) \tag{30}$$

where U_{uc} and I_{uc} are the voltage and current of the UC bank, while v_{uc} and i_{uc} represent the voltage and current of an individual UC unit.

The voltage across the secondary capacitor C_2 is described by a non-linear function of its capacitance and resistance R_2:

$$v_2 = \frac{1}{C_2}\int \frac{1}{R_2}(v_1 - v_2)dt \tag{31}$$

The rate of change in the instantaneous charge of C_2 is proportional to the current i_2:

$$\frac{d}{dt}Q_2 = i_2(t) \tag{32}$$

The current through the main capacitor C_1 can be expressed as a function of its charge Q_1:

$$i_1 = \frac{dQ_1}{dt} = C_1\frac{dv_1}{dt} = (C_0 + C_v v_1)\frac{dv_1}{dt} \tag{33}$$

where the charge Q_1 is calculated using the equivalent capacitance C_1 and the voltage across it:

$$Q_1 = C_0 v_1 + \frac{1}{2}C_v v_1^2 \tag{34}$$

From this, we can derive the voltage v_1 across C_1:

$$v_1 = \frac{-C_0 + \sqrt{C_0^2 + 2C_v Q_1}}{C_v} \tag{35}$$

A critical parameter for UC operation is its state of charge (SOC_{UC}), which is defined as the ratio of its current capacity to its maximum capacity:

$$SOC_{UC} = \frac{1}{Q_{UC\max}} \int_{t_0}^{t} I_{UC}(\tau) d\tau \tag{36}$$

where I_{UC} is the charging current and $Q_{UC\max}$ is the maximum capacity of the UC.

This SOC_{UC} value serves as a crucial indicator for evaluating the state of the UC bank and plays a significant role in energy management strategies.

2.2.4. DC/DC Converter Model

In the hybrid power system under consideration, DC/DC converters play a crucial role in managing power flow between various components. Specifically, two DC/DC boost converters are employed to interface the PEM fuel cell system with the high-voltage DC bus, facilitating the necessary voltage step-up. Additionally, a bidirectional DC/DC converter is positioned between the battery and the DC bus, enabling both power distribution and regenerative energy capture [37,40]. When modeling these converters for energy management purposes, it is important to consider the different time scales at which various system components operate. The power management layer typically functions at a lower frequency compared to the local control loops of individual converters. This separation of time scales allows us to make certain simplifying assumptions in our model. Given that the switching frequency and modulation rate of modern DC/DC converters are significantly higher than the time constants of other system components (such as the inductor), we can employ an averaged model approach. This method effectively captures the converter's behavior from the perspective of the energy management system without the need to model high-frequency switching dynamics. Furthermore, assuming well-designed inner control loops, we can expect the converter to respond rapidly to reference changes. This allows us to further simplify our model by reducing the fast dynamics of the DC/DC converter to an equivalent static model.

The resulting simplified model for the DC/DC converters can be expressed using the following set of equations [47]:

$$V_I = V_h + L\frac{di_L}{dt} + i_L R_L \tag{37}$$

$$V_O = \frac{V_h}{\kappa} \tag{38}$$

$$\eta^{\beta} = \frac{i_O}{\kappa i_L}, \quad \begin{cases} \beta = 1, & \text{for boost converter} \\ & \text{or for bidirectional converter with } i_O V_O \geq 0 \\ \beta = -1, & \text{for bidirectional converter with } i_O V_O < 0 \end{cases} \tag{39}$$

where V_I and V_O represent the input and output voltages of the converter, respectively. L and R_L are the inductance and resistor of the inductor, κ denotes the conversion ratio of the converter, i_L, and i_O are the input (inductor) current and output current, respectively, and η represents the converter's efficiency.

This model, while simplified, captures the essential behavior of the DC/DC converters from an energy management perspective. It accounts for the voltage conversion ratio, current transformation, and power transfer efficiency, which are the key parameters of interest for system-level energy management strategies. By using this static equivalent model, we can effectively represent the DC/DC converters in the overall system simulation without the computational burden of modeling high-frequency switching dynamics. This approach strikes a balance between model fidelity and computational efficiency, making it well-suited for energy management system design and optimization.

It is worth noting that, while this model is adequate for many energy management studies, more detailed models may be necessary for analyzing specific phenomena such as transient responses, or for designing the converters' internal control loops. However, for the purposes of system-level energy management, this simplified model provides a solid foundation for strategy development and evaluation.

3. Enhanced Equivalence Consumption Minimum Strategy

In this section, the multi-layer EMS for vessels driven by the hybrid dPEMFC battery–UCs is dedicatedly discussed. The comprehensive control architecture is illustrated in Figure 6.

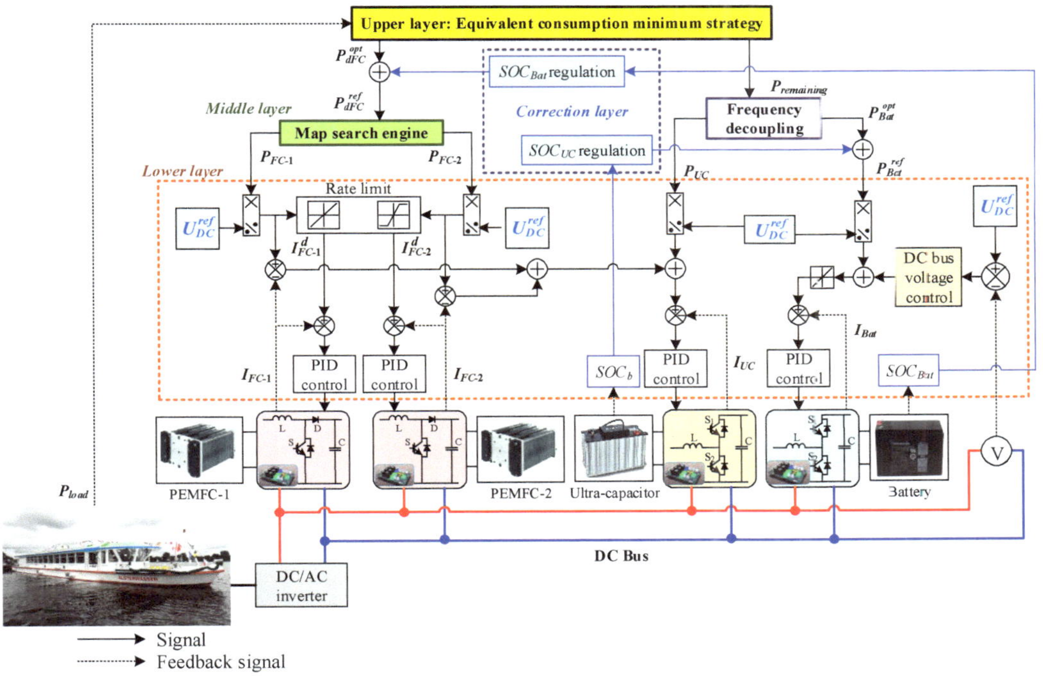

Figure 6. Comprehensive multi-layer EMS for vessels.

3.1. Upper Layer: ECMS-Based EMS for Optimal Power Observation

3.1.1. Optimal Power for Dual PEMFCs

In this manner, the ECMS is employed for seeking out the optimal working point of the dPEMFC (optimal power) while concerning the supplements SOCs. This methodology has been broadly applied and verified on various systems. In this manner, the total consumption is first calculated by converting the electric consumption of the battery and UCs into equivalent hydrogen consumption. Thus, the objective function can be defined as

$$\min\left(m_{H_2} + K_{Bat}\gamma_{Bat}P_{Bat}\right), \tag{40}$$

subject to

$$\begin{cases} P_{FC,1}^{\min} \le P_{FC,1} \le P_{FC,1}^{\max} \\ P_{FC,2}^{\min} \le P_{FC,2} \le P_{FC,2}^{\max} \\ SOC_{Bat}^{\min} \le SOC_{Bat} \le SOC_{Bat}^{\max} \end{cases}, \tag{41}$$

where m_{Bat} denotes the battery equivalent consumption, γ_{Bat} is the battery state-dependent coefficient, and K_{Bat} represents the penalty coefficient of battery. Regarding [48], the penalty coefficients K_{Bat} should be constrained between the initial $SOC_{Bat}(t_0)$ and current $SOC_{Bat}(t)$. The coefficient γ_{Bat} is performed as

$$\gamma_{Bat} = 1 - \sigma \frac{2SOC_{Bat} - \left(SOC_{Bat}^{\max} + SOC_{Bat}^{\min}\right)}{\left(SOC_{Bat}^{\max} + SOC_{Bat}^{\min}\right)}, \tag{42}$$

where σ, conventionally designed by 0.6 [29], denotes an adjustable balance coefficient.

The battery's equivalent consumption can be defined as [48]

$$m_{Bat} = \beta_{Bat} P_{Bat} \frac{\overline{m}_{dFC}}{\overline{P}_{dFC}} (\text{kg}), \tag{43}$$

where $\overline{m}_{dFC}$ (kg) is the average hydrogen consumption, $\overline{P}_{dFC}$ (kW) is the average power of the dPEMFC, and β_{Bat} is the battery equivalent conversion coefficient, which is expressed by [29]

$$\beta_{Bat} = \begin{cases} \frac{1}{\overline{\eta}_{chg} \eta_{dischg}} P_{Bat} \geq 0 \\ \eta_{chg} \overline{\eta}_{dischg} \; P_{Bat} < 0 \end{cases}, \tag{44}$$

with η_{chg} and $\overline{\eta}_{chg}$ being, in turn, the charged and its average coefficients; η_{chg} and $\overline{\eta}_{chg}$ being, in turn, the discharged and its average coefficients of the battery, whose dynamics are specified by [29]

$$\eta_{chg/dischg} = \begin{cases} 0.5\left(1 + \sqrt{1 - \frac{4R_{chg} P_{aux}}{V_{OC}^2}}\right) & P_{aux} \geq 0 \\ 2/\left(1 + \sqrt{1 - \frac{4R_{dischg} P_{aux}}{V_{OC}^2}}\right) & P_{aux} < 0 \end{cases}, \tag{45}$$

where R_{chg} and R_{dischg} are, in turn, internal charged and discharged resistors (Ω), V_{OC} is an open circuit voltage (V), and P_{aux} is the auxiliary power (kW).

The optimal solution for the battery can be computed by the following [49]:

$$P_{Bat}^{opt} = \begin{cases} U_{DC,\min} \frac{(E_{Bat} - U_{DC,\min})}{R_{dischg}}, K_1 \leq \alpha \chi_{\min} \\ E_{Bat}^2 \frac{(1 - K_1/\alpha)}{4R_{dischg}}, \alpha \chi_{\min} \leq K_1 \leq \alpha \\ 0, \alpha \leq K_1 \leq \frac{\alpha}{\overline{\eta}_{chg} \overline{\eta}_{dischg}} \\ E_{Bat}^2 \frac{\left\{1 - (K_1 \overline{\eta}_{chg} \overline{\eta}_{dischg}/\alpha)^2\right\}}{4R_{dischg}}, \frac{\alpha}{\overline{\eta}_{chg} \overline{\eta}_{dischg}} \leq K_1 \leq \frac{\alpha \chi_{\max}}{\overline{\eta}_{chg} \overline{\eta}_{dischg}} \\ -U_{DC,\max} \frac{(U_{DC,\max} - E_{Bat})}{R_{chg}}, K_1 \geq \frac{\alpha \chi_{\max}}{\overline{\eta}_{chg} \overline{\eta}_{dischg}} \end{cases}, \tag{46}$$

where K_1 is determined based on the battery charged and discharged status as follows:

- Discharged ($P_{Bat} \geq 0$):

$$\begin{cases} K_1 = \gamma_{Bat} \overline{m}_{H_2} \frac{1}{\overline{\eta}_{chg} P_{dFC}} \\ \chi_{\min} = \sqrt{1 + \frac{4U_{DC,\min}}{E_{Bat}^2} (U_{DC,\min} - E_{Bat})} \end{cases}, \tag{47}$$

- Charged ($P_{Bat} < 0$):

$$\begin{cases} K_1 = \gamma_{Bat} \overline{m}_{H_2} \frac{\overline{\eta}_{dischg}}{\overline{P}_{dFC}} \\ \chi_{\min} = \sqrt{1 + \frac{4U_{DC,\max}}{E_{Bat}^2}(U_{DC,\max} - E_{Bat})} \end{cases}, \quad (48)$$

Subsequently, the dPEMFC optimal power can be obtained by

$$P_{dFC}^{opt} = \max\left\{\min\left(P_{load} + P_{aux} - P_{Bat}^{opt}, P_{dFC}^{\max}\right), P_{dFC}^{\min}\right\}, \quad (49)$$

It is worth noting that, following the optimal control above, the battery power is optimized in such a way that the dPEMFC operates in the highest-efficiency region and the battery SOC, SOC_{Bat}, is strictly constrained within the pre-set interval, constrained by $SOC_{Bat}^{\min} \leq SOC_{Bat} \leq SOC_{Bat}^{\max}$. As a result, the optimal solution P_{Bat}^{opt} varies with positive and negative values to stabilize the SOC_{Bat} and P_{dFC}^{opt}.

3.1.2. Reference Power for Battery and UCs

Despite the optimal power for dPEMFC and battery defined regarding the above calculations, the battery may not fulfill the abrupt or sudden load change as its slow dynamic. Hence, to tackle this problem, the computed P_{Bat}^{opt} is considered the pre-optimal power, or the remaining load, $P_{\text{remaining}}$, in other words. Hence, a frequency decoupling mechanism is applied to decouple high frequency and low frequency. Thereby, the filtered remaining load is handled by the battery while the high-frequency load is tackled by the UCs. The frequency decoupling mechanism is initiated as shown in Figure 7:

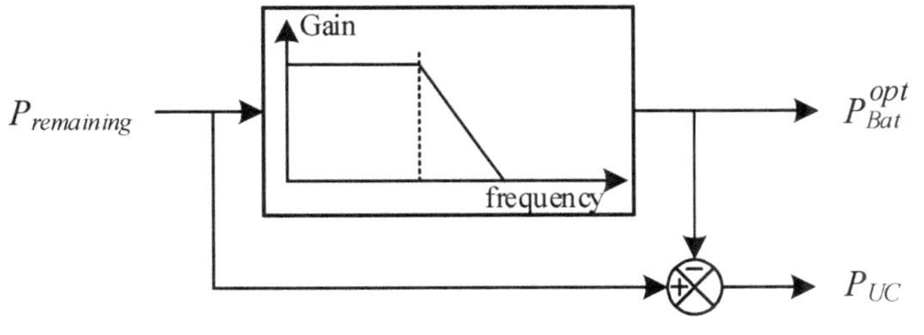

Figure 7. Frequency decoupling mechanism.

The UCs aim at not only addressing the high-frequency load but also compensating for the deviation between the computed optimal power references and their rated limits for safety operation. These errors are summed into the input reference of the UC and tackled in the lower layer.

3.2. Correction Reference Power for SOC Regulation

Despite the optimal power for each source defined, the SOCs of the ESDs may not be strictly constrained within the desired intervals. Hence, instead of directly designating these optimal parameters to the middle and lower layers, the optimal reference power of the dPEMFC and battery are fine-corrected to regulate the SOCs in such a way that the final SOCs at the end of a driving cycle should be equal to the initial ones, as displayed in the "Corrected layer" in Figure 1.

The structures of the fine-corrected power are illustrated in Figure 8. In this manner, the fine-corrected power of the optimal dPEMFC and battery power, namely P_{dFC}^{ref} and P_{Bat}^{ref}, are computed as follows:

$$P_{dFC}^{ref} = P_{dFC}^{opt} + \gamma_1^b \Delta_1(t) + \gamma_2^b \int_0^t \Delta_1(\tau)d(\tau) + \gamma_3^b \frac{d\Delta_1(t)}{dt}, \tag{50}$$

$$P_{Bat}^{ref} = P_{Bat}^{opt} + \gamma_1^{uc} \Delta_2(t) + \gamma_2^{uc} \int_0^t \Delta_2(\tau)d(\tau) + \gamma_3^{uc} \frac{d\Delta_2(t)}{dt} \tag{51}$$

where $\Delta_1(t) \triangleq SOC_{Bat}^{ref} - SOC_{Bat}(t)$ with SOC_{Bat}^{ref} and $SOC_{Bat}(t)$ being the reference and currently time-varying battery's SOC; γ_1^b, γ_2^b, and γ_3^b are positive constants for the battery's SOC regulation. $\Delta_2(t) \triangleq SOC_{UC}^{ref} - SOC_{UC}(t)$ with SOC_{UC}^{ref} and $SOC_{UC}(t)$ being the reference and currently time-varying battery's UC; γ_1^{uc}, γ_2^{uc}, and γ_3^{uc} are positive constants for the battery's SOC regulation.

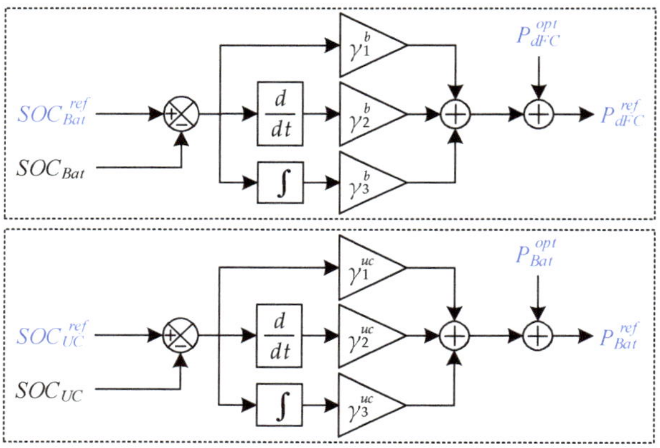

Figure 8. Correction of the power reference for each source.

3.3. Middle Layer: Map Search Engine

To meet the calculated power requirements for the dual fuel cell system, it is essential to implement an efficient energy allocation strategy. The MSM is used to enhance the performance of the dual PEMFC system and reduce hydrogen consumption. As presented in Equation (16), there are numerous power distribution combinations between the two fuel cells that still meet the required power of the dual FC system. Enhancing the efficiency of each FC not only improves the overall efficiency of the dual FC system but also reduces energy consumption (as shown in Equations (17) and (18)). In this study, the output power of the dual FC system ranges from 0 to 150 kW, divided between two FCs of different capacities: 60 kW and 110 kW. For each power requirement, the MSM seeks the optimal distribution between FC 1 and FC 2, ensuring that the combined output meets the required power while achieving maximum efficiency.

Consequently, power allocation curves for PEMFC-1 and PEMFC-2 are established based on the power demand, and the average efficiency of the dual FC system is determined (Figures 9 and 10).

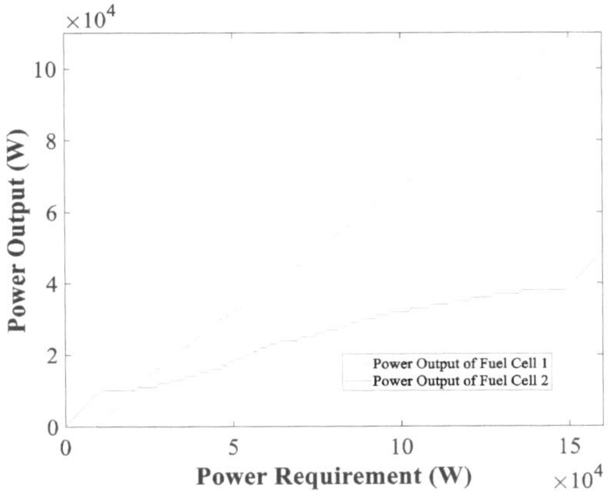

Figure 9. Power output distribution for each PEMFC illustration using the MSM method.

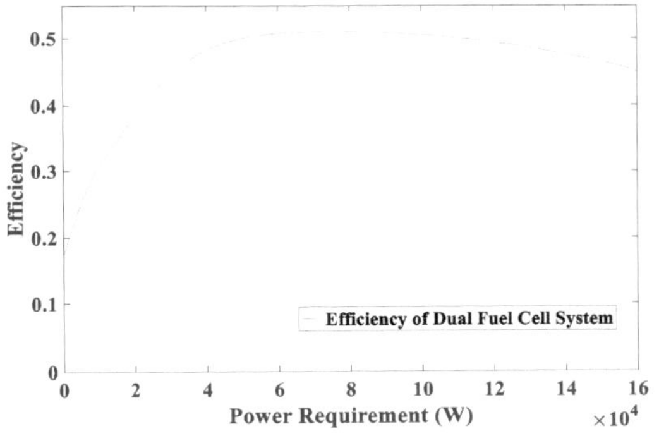

Figure 10. The average efficiency of the dual FC system.

3.4. Lower Layer: Pulse-Width Modulation

In this layer, the input duty cycles to DC/DC converters are generated through pulse-width modulation (PWM) generation to regulate the output power of each source to meet the determined reference one. Moreover, the voltage of the DC bus should be regulated to guarantee the control performance. In accordance with each device's characteristics, the battery is designated to maintain the DC bus voltage. In this scenario, the proportional–integral–derivative (PID) controls are employed due to their simplicity and robustness.

3.4.1. Duty Cycle for the Individual PEMFC

$D_{FC,i}$ is defined as the duty cycle input to the boost DC/DC converter of the i-th PEMFC. Thus, $D_{FC,i}$ is specified as

$$D_{FC,i} = f_{a-PWM}(Out_{FC,i}), \quad (52)$$

Herein, $Out_{FC,i}$ is the output of the PID control for the PEMFC current regulation as

$$PWM_{FC,i} = K_{FC,i}^{P}\left(I_{FC,i}^{d} - I_{FC,i}\right) + K_{FC,i}^{I}\int_{0}^{t}\left(I_{FC,i}^{d}(\tau) - I_{FC,i}(\tau)\right)d\tau + K_{FC,i}^{D}\frac{d\left(I_{FC,i}^{d} - I_{FC,i}\right)}{dt}, \quad (53)$$

where $K^{P}_{FC,i}$, $K^{I}_{FC,i}$, and $K^{D}_{FC,i}$ are positive constants; $I_{FC,i}$ is the measured current of the *i*-th PEMFC; and $I^{d}_{FC,i}$ is the desired current, which is obtained by

$$I^{d}_{FC,i} = \text{rate-limit}\left(I^{ref}_{FC,i}\right), \tag{54}$$

with $I^{ref}_{FC,i}$ being directly obtained from the individual *i*-th PEMFC reference power (fine-corrected optimal power) and reference DC bus voltage U^{ref}_{DC}.

Moreover, due to the use of the rate limit operator, there exists a deviation between the optimal current $I^{opt}_{FC,i}$ and desired current $I^{d}_{FC,i}$ as $\varepsilon_{FC,i} = I^{opt}_{FC,i} - I^{d}_{FC,i}$, which will be then addressed by the UC later. The control structure for the $D_{FC,i}$ is illustrated in Figure 11.

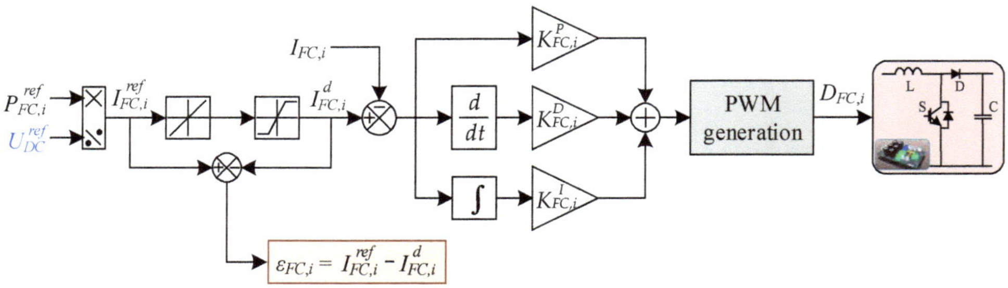

Figure 11. PWM generation for individual PEMFC.

3.4.2. PWM for the Battery and DC Bus Voltage Regulation

Define D_{Bat} is the duty cycle input to the bi-directional DC/DC converter of the battery. Thus, D_{Bat} is specified through an analog-PWM operator as

$$D_{Bat} = f_{a-PWM}(Out_{Bat}), \tag{55}$$

Out_{Bat} is the output of the PID control for the battery current regulation as

$$Out_{Bat} = K^{P}_{Bat}\left(I^{d}_{Bat} - I_{Bat}\right) + K^{I}_{Bat}\int_{0}^{t}\left(I^{d}_{Bat}(\tau) - I_{Bat}(\tau)\right)d\tau + K^{D}_{Bat}\frac{d\left(I^{d}_{Bat} - I_{Bat}\right)}{dt}, \tag{56}$$

where K^{P}_{Bat}, K^{I}_{Bat}, and K^{D}_{Bat} are positive constants; I_{Bat} is the measured current of the *battery*; and I^{d}_{Bat} is the desired current, which is obtained by

$$I^{d}_{Bat} = \text{rate-limit}\left(I^{ref}_{Bat}\right), \tag{57}$$

with I^{ref}_{Bat} being attained from the sum of the battery's optimal power and output of the DC bus voltage regulation I^{out}_{DC}, obtained by

$$I^{out}_{DC} = K^{P}_{Udc}\left(U^{ref}_{DC} - U_{DC}\right) + K^{I}_{Udc}\int_{0}^{t}\left(U^{ref}_{DC}(\tau) - U_{DC}(\tau)\right)d\tau + K^{D}_{Udc}\frac{d\left(U^{ref}_{DC} - U_{DC}\right)}{dt}, \tag{58}$$

with U_{DC} being the measured DC bus voltage, K^{P}_{Udc}, K^{I}_{Udc}, and K^{D}_{Udc} are the proportional, integral, and derivative gains, respectively.

Moreover, due to the use of the rate limit operator, there exists a deviation between the optimal current $I^{ref}_{FC,i}$ and desired current $I^{d}_{FC,i}$ as $\varepsilon_{Bat} = I^{ref}_{Bat} - I^{d}_{Bat}$, which will be then addressed by the UCs later. The control structure for the D_{Bat} is configured in Figure 12.

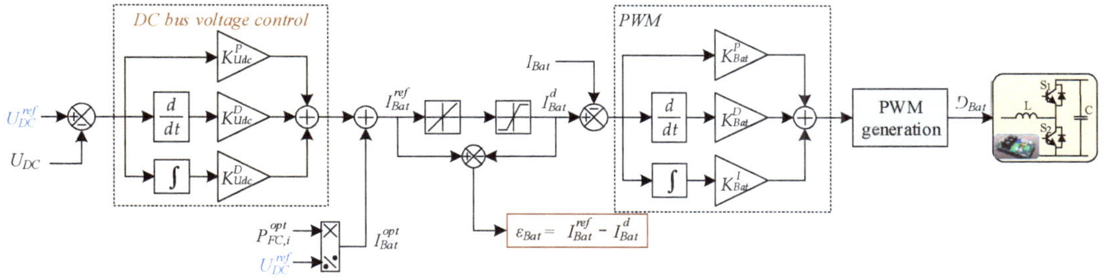

Figure 12. Control architecture for DC bus voltage regulation and PWM generation for the battery.

3.4.3. PWM for the UCs

Similarly, D_{UC} is the duty cycle input to the bi-directional DC/DC converter of the UCs, which is specified through an analog PWM operator, as illustrated in Figure 13. Thus, one has

$$D_{UC} = f_{a-PWM}(Out_{UC}). \tag{59}$$

Out_{UC} is the output of the PID control for the UC current regulation, expressed by

$$Out_{UC} = K_{UC}^{P}\left(I_{UC}^{d} - I_{UC}\right) + K_{UC}^{I}\int_{0}^{t}\left(I_{UC}^{d}(\tau) - I_{UC}(\tau)\right)d\tau + K_{UC}^{D}\frac{d\left(I_{UC}^{d} - I_{UC}\right)}{dt}, \tag{60}$$

where K_{UC}^{P}, K_{UC}^{I}, and K_{UC}^{D} are the proportional, integral, and derivative gains, respectively; I_{UC} is the measured current of the UC; and I_{UC}^{d} is the desired current, which is obtained by

$$I_{UC}^{d} = \frac{P_{UC}}{U_{DC}^{ref}} + \sum_{i=1}^{2}\varepsilon_{FC,i} + \varepsilon_{Bat}, \tag{61}$$

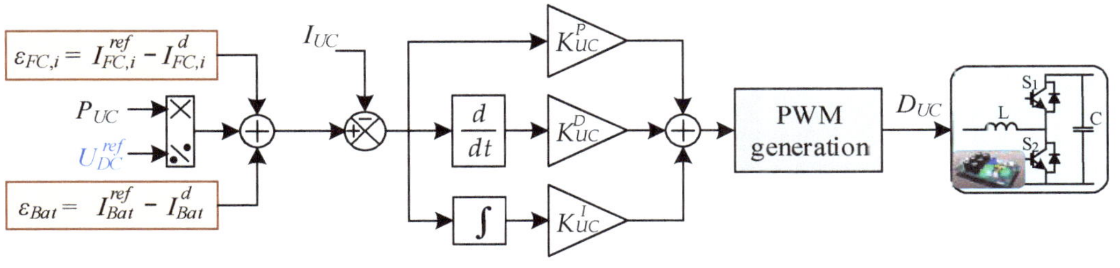

Figure 13. Control architecture of PWM generation for UCs.

4. Comparative Simulations and Discussions

Regarding the examined topology in Figure 1, in this simulation setup, two different characteristics of PEMFCs are employed: one with the maximum power of 120 kW (PEMFC-1) and another one with the maximum power of 60 kW (PEMFC-2). The reason is to demonstrate the MSE effectiveness compared to other conventional methods of EqD and daisy chain in splitting power. If two PEMFC stacks have the same specifications, then the MSE and EqD approaches have the same performance since the power for each stack is simply half that of the dPEMFC power. Therefore, two different PEMFCs should be considered for validating the MSE superiority in appropriately allocating power for each PEMFC power source.

The parameters of the hybrid power source are selected as shown in Tables 1–3 with the sampling time of t_s = 0.01 s, subject to the load profile shown in Figure 8.

Table 1. The 120 kW PEMFC system parameters [36,37].

Parameters		Symbol	Value	Unit
Cells number	PEMFC-1		30	
	PEMFC-2		18	
Rated power	PEMFC-1		110	kW
	PEMFC-2		60	kW
Membrane thickness			178	μm
Area		S	232	cm^2
Coefficients		ξ_1	−0.948	-
		ξ_2	$0.00286 + 2 \times 10^{-4} \ln(S) + 4.3 \times 10^{-5} \ln(c_{H2})$	-
		ξ_3	7.6×10^{-5}	-
		ξ_4	-1.93×10^{-4}	-
Membrane resistivity parameter			12.5	-
Fuel cell capacitance		C_{dl}	0.035 × 232	F
Cathode	Pressure	P_{O2}	3	atm
	Flow constant	k_a	0.065	mol/s/atm
	Volume	V_a	0.01	m^3
Anode	Pressure	P_{H2}	3	atm
	Flow constant	k_c	0.065	mol/s/atm
	Volume	V_c	0.005	m^3
Hydrogen enthalpy of combustion			285.5×10^3	kJ/mol
Thermal resistance			0.115	C/W
Total energy (for 6 h)			302.522	kW

Table 2. Battery's parameters [36,37].

Parameters	Value	Unit
Capacity	6.5	Ah
Rated voltage	1.2	V
Constant voltage	1.2848	V
Internal resistance	0.0046	Ω
Number of batteries	360	-
Exponential zone amplitude	0.144	V
Exponential zone time constant inverse	2.3077	$(Ah)^{-1}$
Polarization resistance constant	0.01875	Ω

Table 3. Bank of UCs' parameters [36,37].

Parameters	Value	Unit
Number of UC	80	
Rated voltage	2.7	V
Absolute maximum voltage	2.85	V
Absolute maximum current	1900	A
Rated capacitance	3000	F
Capacitance in the main cell	2100	F
	623	F
Capacitance in the slow cell	172	F
Resistance in the main cell	0.036×10^{-3}	Ω
Resistance in the slow cell	1.92	Ω

Moreover, to further evaluate the effectiveness of the proposed topology in effectively sharing power with each source, we consider the following algorithms in the comparative simulation:

- A1: Rule-based EMS for sharing power to each unit and MSM for splitting power to each PEMFC stack.

- A2: Optimization-based EMS with the daisy chain method for distributing commanded power for each PEMFC stack. In this setup, the PEMFC-1 (110-kW) first supplies power to the load until it reaches its maximum value; then, the PEMFC-2 (60-kW) enters the system.
- A3: Same as A2, but the PEMFC-2 (60-kW) supplies power to the load first until it reaches its maximum value; then, the PEMFC-1 (110-kW) enters the system.
- A4: Optimization-based EMS with equal distribution for sharing power to each PEMFC stack. In this setup, the reference power for each PEMFC stack is half that of the optimal power, i.e., $P^{ref}_{FC,i} = 0.5P^{opt}_{dFC}$.
- A5: Optimization-based EMS with MSM for sharing power to each PEMFC stack without fine-corrected optimal power. In this manner, the reference power of the dual PEMFC and battery are the same as their optimal power.
- A6 (proposed): Optimization-based EMS with MSM for sharing power to each PEMFC stack with fine-corrected optimal power.

Generally, all examined control strategies could fulfill the load demand, as shown in Figure 14. Herein, it should be noted that the optimal power of the dPEMFC and battery and the calculated power of the UCs under A2, A3, A4, and A5 are similar to each other since they are obtained regarding the optimization-based EMS. Only the power-sharing under A1 and A6 is different due to the different algorithms. Thus, to facilitate the observation, only four power efforts were plotted: demand load and power performance under A1, A2, and A6.

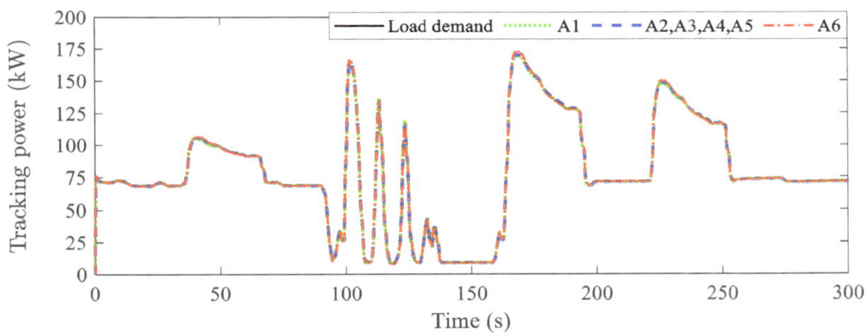

Figure 14. Power tracking qualification under different algorithms.

Figure 15 shows the total power efforts of the dPEMFC (top) and individual PEMFC power (bottom). Figure 16 displays the total efficiency of the dPEMFC system (top) and each primary supply's efficiency (bottom). Subsequently, the reference battery and UCs power were obtained concerning the frequency decoupling technique, as performed in Figure 17. As mentioned above, since the optimal power of the dPEMFC and battery and the calculated power of the UCs under A2, A3, A4, and A5 are the same as each other, only three power efforts of A1, A2, and A6 were plotted in the top sub-Figures 15 and 17.

As seen in Figure 15, as the heuristically designed rules, the power reference of the dPEMFC varied depending on the load demand, as shown. Although the rules were designed such that both PEMFC stacks were constrained to operate within the highest efficiency regions, they could not be manipulated at the optimal working point. Accordingly, the highest total efficiency could not be exhibited, as displayed in Figure 16, in which the efficiency of each PEMFC varied following their power behaviors. The average efficiency was about 0.488 for the dPEMFC, and 0.4833 and 0.498 when distributed for each PEMFC-1 and PEMFC-2, respectively. Moreover, by not operating at the optimal point, the battery and UCs were charged and discharged arbitrarily, which accordingly resulted in an overcharge, as shown in Figure 17.

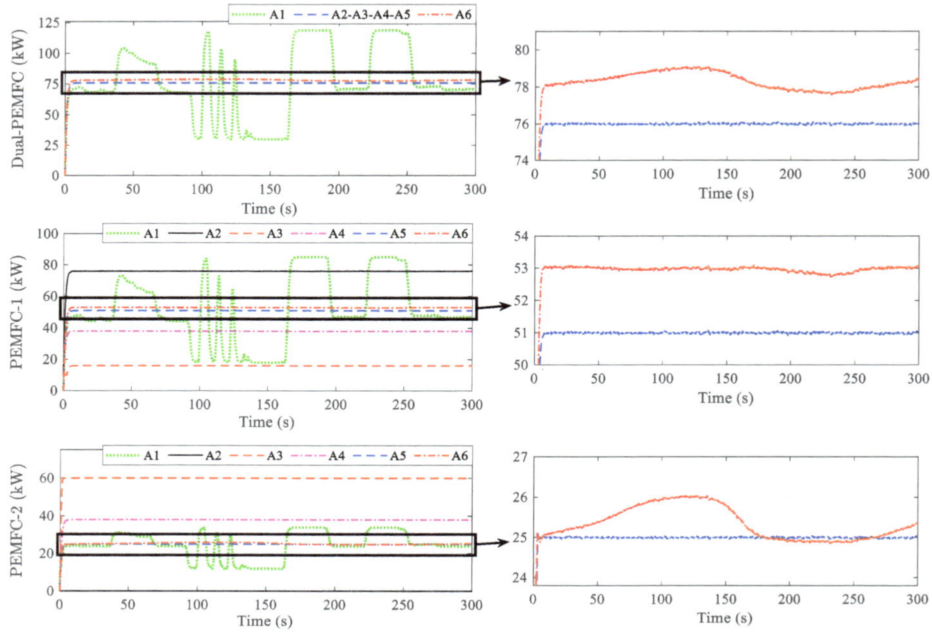

Figure 15. The power performance of the dPEMFC and individual PEMFC systems under different algorithms.

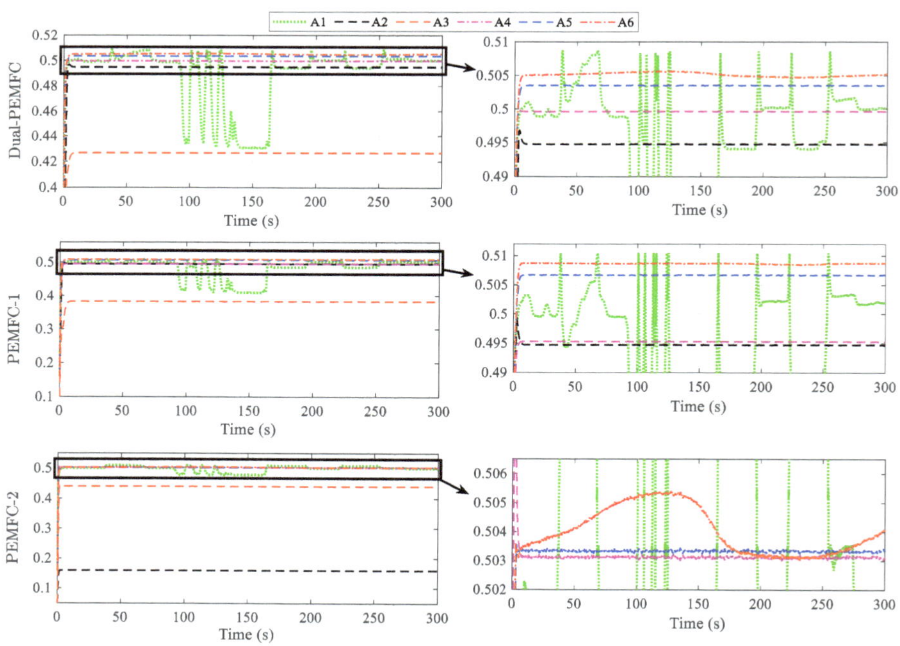

Figure 16. Efficiency of PEMFCs under difference algorithms: (**Top**) overall efficiency of dPEMFC, (**Middle**) efficiency of PEMFC-1, (**Bottom**) efficiency of PEMFC-2.

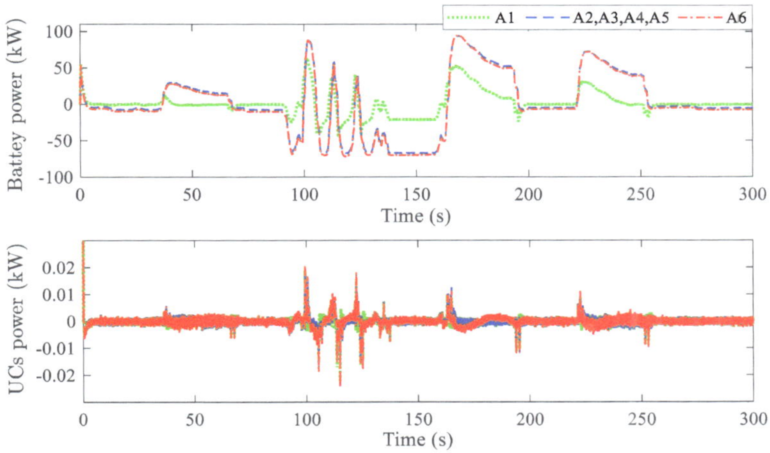

Figure 17. Reference power of the battery and UCs under different algorithms.

On the contrary, other algorithms (A2 to A6) could manipulate the dPEMFC system such that the adjacently highest efficiency was exhibited, as disclosed in Figure 15 thanks to the use of the optimization to seek out the optimal working point. Despite achieving the optimized reference power, in theory, the actual behaviors among them are different.

Despite being obtained through the optimization-based technique, the daisy chain approaches, A2 and A3, exhibited the worst performance. Since the dPEMFC optimized power was around 78 kW, if prioritizing using the 110 kW PEMFC (A2), it could sufficiently support the dPEMFC required to load and thus the 60 kW PEMFC was maintained with the minimum power to run the auxiliary systems (pump, fan, temperature system, and so on); thus, returning the lowest efficiency of about 0.16 and the overall efficiency was about 0.4945 as a result. Likewise, under A3, the 60 kW ran first to supply power to the system. In our design, its acceptable maximum power supply of 50 kW was selected for safety. Therefore, the 110 kW was entered to handle the remaining 28 kW. Consequently, the efficiency was just about 0.3835 for the 110 kW PEMFC and 0.4405 for the 60 kW PEMFC because those operating points are out of the high-efficiency regions with the overall efficiency being only 0.4268. Meanwhile, A4 could improve the overall efficiency a little (0.4996). With the halved power required of 39 kW for each PEMFC, as depicted in Figure 10, the 60 kW PEMFC unit operated in the high-efficiency region at which the efficiency of 0.5032 was exhibited. Meanwhile, the 110 kW operated in the low-efficiency region at which the efficiency of this source was approximately 0.495.

A5 and A6, owing to using the MSE, returned the best performance in which both PEMFC systems could operate in their highest efficiency regions compared to A2, A3, and A4. With the MSE, the optimized power was appropriately allocated to each device based on its characteristic and efficiency map by which the reference power for the PEMFC-1 was determined at about 50 kW and for the PEMFC-2 at around 28 kW.

However, the superiority of the proposed methodology, A6, is governed by not only the overall efficiency of the dPEMFC and hydrogen consumption but also the SOC regulation of the ESDs, as shown in Figures 17 and 18. As observed, the proposed methodology could well constrain the SOC of the battery and UCs compared to other optimization-based methods (A2 to A5). These accomplishments came from the fine-corrected power development to force strictly regulate the SOCs by adjusting the reference power of the dPEMFC and battery. The deviation of the SOC_{UC} from its reference SOC_{UC}^{ref} was corrected by manipulating the battery reference power; however, this may also cause the deviation of the SOC_{Bat} from its reference SOC_{Bat}^{ref} to increase. Therefore, the battery reference power was corrected by regulating the dPEMFC power as expressed in (50) and (51). The SOCs of the ESDs were critically controlled such that the final SOCs were the same as the initial ones at the end of the driving cycle (about 0.7692 for the battery's SOC and 0.6 for the UCs) As a result, with well-regulated ESD behaviors, only the proposed method A6 could satisfy the load demand regarding the real driving cycle [34]. In brief, the superiority of the proposed control algorithm compared to others is summarized in Table 4.

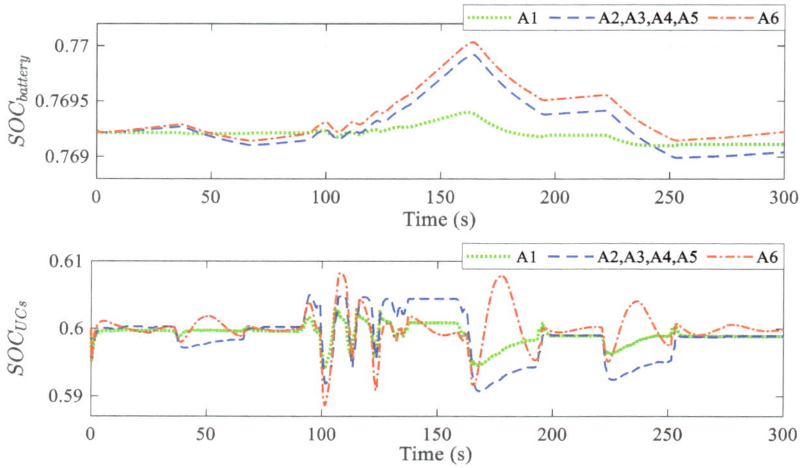

Figure 18. SOCs of the battery and UCs under different algorithms.

Table 4. Control performance summary.

EMSs	Average Power (kW)			Average Efficiency		
	dPEMFC	PEMFC-1	PEMFC-2	dPEMFC	PEMFC-1	PEMFC-2
A1	76.36	51.72	24.63	0.488	0.4833	0.5004
A2	76	76	Off	0.4947	0.4947	0.16
A3	76	16	60	0.4269	0.3835	0.4404
A4	76	38	38	0.4996	0.4953	0.5031
A5	76	51	25	0.5035	0.5067	0.5033
A6	78.6	52.93	25.5	0.5056	0.0587	0.05045

5. Conclusions

This paper proposed a novel topology of using the hybrid power sources of dual PEMFCs interconnected with battery and UCs to supply the HEV powertrain. Of the setup, the optimization-based hierarchical EMS was established to appropriately allocate optimal power for the dual PEMFC systems and battery such that each PEMFC unit could operate in the high-efficiency region. By using two PEMFC sources, the MSE was then initiated to split the optimal power for each unit properly. Moreover, as the risk of balancing the ESDs' behaviors, a fine-corrected layer was introduced to strictly regulate the SOCs of the battery and UCs varying around desirable values. In this design, the UCs' SOC was controlled by the battery power whereas the dPEMFC power manipulated the battery's SOC. As a result, not only the optimal power was exhibited and allocated to each supply but the ESDs' SOCs were also well constrained as demonstrated through comparative simulations. Moreover, with the optimal power obtained, the PEMFC systems operated in the highest efficiency regions, which consumed the lowest fuel; thus, prolonging the lifetime of all devices. However, some related problems, such as system degradation, remaining useful life, ESDs behavior constraints, state-of-health, and so on, have not yet been taken into consideration. Therefore, these existing regards motivate us to keep going for further developments of control strategies to enhance the overall system performance.

Author Contributions: Conceptualization, T.D.D. and H.V.A.T.; methodology, T.C.D. and H.V.A.T.; software, T.C.D. and H.V.A.T.; validation, T.C.D. and H.V.A.T.; formal analysis, T.D.D. and H.V.A.T.; investigation, T.C.D. and H.V.A.T.; resources, T.D.D. and H.V.A.T.; data curation, T.C.D. and H.V.A.T.; writing—original draft preparation, T.C.D. and H.V.A.T.; writing—review and editing, T.D.D.; visualization, H.V.A.T.; supervision, T.D.D.; project administration, T.D.D. and T.C.D.; funding acquisition, T.D.D. and T.C.D. All authors have read and agreed to the published version of the manuscript.

Funding: This research is funded by University of Economics Ho Chi Minh City, Vietnam (UEH).

Institutional Review Board Statement: Not applicable.

Informed Consent Statement: Not applicable.

Data Availability Statement: Data are contained within the article.

Conflicts of Interest: The authors declare no conflicts of interest. The funders had no role in the design of the study; in the collection, analyses, or interpretation of data; in the writing of the manuscript; or in the decision to publish the results.

References

1. Vessel Traffic Services. Available online: https://www.imo.org/en/OurWork/Safety/Pages/VesselTrafficServices.aspx (accessed on 24 June 2024).
2. How the Energy Crisis Started, How Global Energy Markets Are Impacting Our Daily Life, and What Governments Are Doing about It. Available online: https://www.iea.org/topics/global-energy-crisis (accessed on 24 June 2024).
3. Torreglosa, J.P.; González-Rivera, E.; García-Triviño, P.; Vera, D. Performance Analysis of a Hybrid Electric Ship by Real-Time Verification. *Energies* **2022**, *15*, 2116. [CrossRef]
4. Breucke, S.D.; Peeters, E.; Driesen, J. Possible applications of Plug-in Hybrid Electric Ships. In Proceedings of the 2009 IEEE Electric Ship Technologies Symposium, Baltimore, MD, USA, 20–22 April 2009; pp. 310–317. [CrossRef]
5. Trinh, H.A.; Truong, H.V.A.; Do, T.C.; Nguyen, M.H.; Phan, V.D.; Ahn, K.K. Optimization-based energy management strategies for hybrid construction machinery: A review. *Energy Rep.* **2022**, *8*, 6035–6057. [CrossRef]
6. Kanerva, S.; Klaus, V.; Krattiger, H. Fuel Cells in Commercial Marine Applications. In Proceedings of the ASNE TSS 2021 Virtual Conference, online, 26–28 January 2021; Available online: https://search.abb.com/library/Download.aspx?DocumentID=9AKK108466A2704&LanguageCode=en&DocumentPartId=&Action=Launch (accessed on 8 July 2024).
7. Dall'Armi, C.; Pivetta, D.; Taccani, R. Hybrid PEM Fuel Cell Power Plants Fuelled by Hydrogen for Improving Sustainability in Shipping: State of the Art and Review on Active Projects. *Energies* **2023**, *16*, 2022. [CrossRef]
8. Hu, W.; Shang, Q.; Bian, X.; Zhu, R. Energy management strategy of hybrid energy storage system based on fuzzy control for ships. *Int. J. Low-Carbon Technol.* **2021**, *17*, 169–175. [CrossRef]
9. Vafamand, N.; Boudjadar, J.; Khooban, M.H. Model predictive energy management in hybrid ferry grids. In Proceedings of the 6th International Conference on Power and Energy Systems Engineering (CPESE 2019), Okinawa, Japan, 20–23 September 2019.
10. Niu, L.; Xiao, L. Optimization of topology and energy management in fuel cell cruise ship hybrid power systems. *Intel. Mar. Technol. Syst.* **2024**, *2*, 12. [CrossRef]
11. Fan, A.; Li, Y.P.; Fang, S.D.; Li, Y.R.; Qiu, H. Energy management strategies and comprehensive evaluation of parallel hybrid ship based on improved fuzzy logic control. *IEEE Trans Transp. Electrif.* **2023**. *early access*. [CrossRef]
12. Kim, S.; Kim, J. Optimal Energy Control of Battery Hybrid System for Marine Vessels by Applying Neural Network Based on Equivalent Consumption Minimization Strategy. *J. Mar. Sci. Eng.* **2021**, *9*, 1228. [CrossRef]
13. Ge, Y.Q.; Zhang, J.D.; Zhou, K.X.; Zhu, J.T.; Wang, Y.K. Research on energy management for ship hybrid power system based on adaptive equivalent consumption minimization strategy. *J. Mar. Sci. Eng.* **2023**, *11*, 1271. [CrossRef]
14. Ghorashi Khalil Abadi, S.A.; Habibi, S.I.; Khalili, T.; Bidram, A. A model predictive control strategy for performance improvement of hybrid energy storage systems in DC microgrids. *IEEE Access* **2022**, *10*, 25400–25421. [CrossRef]
15. Jung, W.; Chang, D. Deep Reinforcement Learning-Based Energy Management for Liquid Hydrogen-Fueled Hybrid Electric Ship Propulsion System. *J. Mar. Sci. Eng.* **2023**, *11*, 2007. [CrossRef]
16. Peng, X.; Chen, H.; Guan, C. Energy Management Optimization of Fuel Cell Hybrid Ship Based on Particle Swarm Optimization Algorithm. *Energies* **2023**, *16*, 1373. [CrossRef]
17. Liu, H.Y.; Fan, A.L.; Li, Y.P.; Bucknall, R.; Chen, L. Hierarchical distributed MPC method for hybrid energy management: A case study of ship with variable operating conditions. *Renew. Sust. Energy Rev.* **2024**, *189*, 113894. [CrossRef]
18. Rafiei, M.; Boudjadar, J.; Khooban, M.H. Energy Management of a Zero-Emission Ferry Boat with a Fuel-Cell-Based Hybrid Energy System: Feasibility Assessment. *IEEE Trans. Ind. Electron.* **2021**, *68*, 1739–1748. [CrossRef]

19. Elkafas, A.G.; Rivarolo, M.; Gadducci, E.; Magistri, L.; Massardo, A.F. Fuel Cell Systems for Maritime: A Review of Research Development, Commercial Products, Applications, and Perspectives. *Processes* **2023**, *11*, 97. [CrossRef]
20. At What Power Rate Does a Ship Usually Operate? Available online: https://www.quora.com/At-what-power-rate-does-a-ship-usually-operate?top_ans=348256660 (accessed on 22 July 2024).
21. PEM Fuel Cell Carbon Stack-C260-S (120 kW). Available online: https://hyfindr.com/en/shop/products/category/components/fuel-cell-stacks/pem-stacks (accessed on 24 June 2024).
22. Zhang, C.; Zeng, T.; Wu, Q.; Deng, C.; Chan, S.H.; Liu, Z. Improved efficiency maximization strategy for vehicular dual-stack fuel cell system considering load state of sub-stacks through predictive soft-loading. *Renew. Energy* **2021**, *179*, 929–944. [CrossRef]
23. Zhou, S.; Fan, L.; Zhang, G.; Gao, J.; Lu, Y.; Zhao, P.; Wen, C.; Shi, L.; Hu, Z. A review on proton exchange membrane multi-stack fuel cell systems: Architecture, performance, and power management. *Appl. Energy* **2022**, *310*, 118555. [CrossRef]
24. Fernandez, A.M.; Kandidayeni, M.; Boulon, L.; Chaoui, H. An Adaptive State Machine Based Energy Management Strategy for a Multi-Stack Fuel Cell Hybrid Electric Vehicle. *IEEE Trans. Veh. Technol.* **2020**, *69*, 220–234. [CrossRef]
25. Wang, T.; Li, Q.; Wang, X.; Chen, W.; Breaz, E.; Gao, F. A Power Allocation Method for Multistack PEMFC System Considering Fuel Cell Performance Consistency. *IEEE Trans. Ind. Appl.* **2020**, *56*, 5340–5351. [CrossRef]
26. Peng, F.; Zhao, Y.; Li, X.; Liu, Z.; Chen, W.; Liu, Y.; Zhou, D. Development of master-slave energy management strategy based on fuzzy logic hysteresis state machine and differential power processing compensation for a PEMFC-LIB-SC hybrid tramway. *Appl Energy* **2017**, *206*, 346–363. [CrossRef]
27. Yan, Y.; Li, Q.; Chen, W.; Huang, W.; Liu, J. Hierarchical Management Control Based on Equivalent Fitting Circle and Equivalent Energy Consumption Method for Multiple Fuel Cells Hybrid Power System. *IEEE Trans. Ind. Electron.* **2020**, *67*, 2786–2797 [CrossRef]
28. Qiu, Y.; Zeng, T.; Zhang, C.; Wang, G.; Wang, Y.; Hu, Z.; Yan, M.; Wei, Z. Progress and challenges in multi-stack fuel cell system for high power applications: Architecture and energy management. *Green Energy Intell. Transp.* **2023**, *2*, 100068. [CrossRef]
29. Truong, H.V.A.; Trinh, H.A.; Do, T.C.; Nguyen, M.H.; Phan, V.D.; Ahn, K.K. An Enhanced Extremum Seeking-Based Energy Management Strategy with Equivalent State for Hybridized-Electric Tramway-Powered by Fuel Cell–Battery–Supercapacitors. *Mathematics* **2024**, *12*, 1849. [CrossRef]
30. Yan, Y.; Li, Q.; Chen, W.; Huang, W.; Liu, J.; Liu, J. Online Control and Power Coordination Method for Multistack Fuel Cells System Based on Optimal Power Allocation. *IEEE Trans. Ind. Electron.* **2021**, *68*, 8158–8168. [CrossRef]
31. Wang, T.; Li, Q.; Yin, L.; Chen, W.; Breaz, E.; Gao, F. Hierarchical Power Allocation Method Based on Online Extremum Seeking Algorithm for Dual-PEMFC/Battery Hybrid Locomotive. *IEEE Trans. Veh. Tech.* **2021**, *70*, 5679–5692. [CrossRef]
32. Do, T.-C.; Trinh, H.-A.; Ahn, K.-K. Hierarchical Control Strategy with Battery Dynamic Consideration for a Dual Fuel Cell/Battery Tramway. *Mathematics* **2023**, *11*, 2269. [CrossRef]
33. Xie, P.; Asgharian, H.; Guerrero, J.M.; Vasquez, J.C.; Araya, S.S.; Liso, V. A two-layer energy management system for a hybrid electrical passenger ship with multi-PEM fuel cell stack. *Int. J. Hydrogen Energy* **2024**, *50*, 1005–1019. [CrossRef]
34. Bassam, A.M.; Phillips, A.B.; Turnock, S.R.; Wilson, P.A. An improved energy management strategy for a hybrid fuel cell/battery passenger vessel. *Int. J. Hydrogen Energy* **2016**, *41*, 22453–22464. [CrossRef]
35. Odeim, F.; Roes, J.; Heinzel, A. Power Management Optimization of an Experimental Fuel Cell/Battery/Supercapacitor Hybrid System. *Energies* **2015**, *8*, 6302–6327. [CrossRef]
36. Do, T.C.; Truong, H.V.A.; Dao, H.V.; Ho, C.M.; To, X.D.; Dang, T.D.; Ahn, K.K. Energy Management Strategy of a PEM Fuel Cell Excavator with a Supercapacitor/Battery Hybrid Power Source. *Energies* **2019**, *12*, 4362. [CrossRef]
37. Truong, H.V.A.; Dao, H.V.; Do, T.C.; Ho, C.M.; To, X.D.; Dang, T.D.; Ahn, K.K. Mapping Fuzzy Energy Management Strategy for PEM Fuel Cell–Battery–Supercapacitor Hybrid Excavator. *Energies* **2020**, *13*, 3387. [CrossRef]
38. Sankar, K.; Aguan, K.; Jana, A.K. A proton exchange membrane fuel cell with an airflow cooling system: Dynamics, validation and nonlinear control. *Energy Convers. Manag.* **2019**, *183*, 230–240. [CrossRef]
39. Bizon, N.; Thounthong, P. Energy efficiency and fuel economy of a fuel cell/renewable energy sources hybrid power system with the load-following control of the fueling regulators. *Mathematics* **2020**, *8*, 151. [CrossRef]
40. Trinh, H.-A.; Phan, V.-D.; Truong, H.-V.-A.; Ahn, K.K. Energy Management Strategy for PEM Fuel Cell Hybrid Power System Considering DC Bus Voltage Regulation. *Electronics* **2022**, *11*, 2722. [CrossRef]
41. Martyushev, N.V.; Malozyomov, B.V.; Sorokova, S.N.; Efremenkov, E.A.; Qi, M. Mathematical Modeling of the State of the Battery of Cargo Electric Vehicles. *Mathematics* **2023**, *11*, 536. [CrossRef]
42. Wu, J.; Wang, X.; Li, L.; Du, Y. Hierarchical control strategy with battery aging consideration for hybrid electric vehicle regenerative braking control. *Energy* **2018**, *145*, 301–312. [CrossRef]
43. Uzunoglu, M.; Alam, M.S. Dynamic modeling, design and simulation of a PEM fuel cell/ultra-capacitor hybrid system for vehicular applications. *Energy Convers. Manag.* **2007**, *48*, 1544–1553. [CrossRef]
44. Zhang, L.; Hu, X.; Wang, Z.; Sun, F.; Dorrell, D.G. A review of supercapacitor modeling, estimation, and applications: A control/management perspective. *Renew. Sustain. Energy Rev.* **2018**, *81*, 1868–1878. [CrossRef]
45. Zubieta, L.; Bonert, R. Characterization of double-layer capacitors for power electronics applications. *IEEE Trans. Ind. Appl.* **2000**, *36*, 199–205. [CrossRef]
46. Lahyani, A.; Venet, P.; Guermazi, A.; Troudi, A. Battery/Supercapacitors Combination in Uninterruptible Power Supply (UPS). *IEEE Trans. Power Electron.* **2013**, *28*, 1509–1522. [CrossRef]

47. Trinh, H.-A.; Nguyen, D.G.; Phan, V.-D.; Duong, T.-Q.; Truong, H.-V.-A.; Choi, S.-J.; Ahn, K.K. Robust Adaptive Control Strategy for a Bidirectional DC-DC Converter Based on Extremum Seeking and Sliding Mode Control. *Sensors* **2023**, *23*, 457. [CrossRef]
48. Li, H.; Ravey, A.; N'Diaye, A.; Djerdir, A. A novel equivalent consumption minimization strategy for hybrid electric vehicle powered by fuel cell, battery and supercapacitor. *J. Power Sources* **2018**, *395*, 262–270. [CrossRef]
49. Torreglosa, J.P.; Jurado, F.; García, P.; Fernández, L.M. Hybrid fuel cell and battery tramway control based on an equivalent consumption minimization strategy. *Control Eng. Pract.* **2011**, *19*, 1182–1194. [CrossRef]

Journal of
*Marine Science
and Engineering*

MDPI

Article

Advanced Energy Management System for Generator–Battery Hybrid Power System in Ships: A Novel Approach with Optimal Control Algorithms

Eunbae Choi [1] and Heemoon Kim [2,*]

[1] Department of Marine System Engineering, Korea Maritime & Ocean University, Busan 49112, Republic of Korea; eunbae.choi@kmou.ac.kr

[2] Eco Friendly Propulsion System Technology Team, Korea Marine Equipment Research Institute, Yangsan-si 50592, Republic of Korea

* Correspondence: hmkim@komeri.re.kr

Abstract: Advancements in the reduction of carbon dioxide emissions from ships are driving the development of more efficient onboard power systems. The proposed non-equivalent parallel running operation system is explored in this study, which improves the efficiency of the main power generation source compared with traditional equal load-sharing methods used in power management systems. However, the asymmetric method reduces the efficiency of the auxiliary power sources. To address this issue, we propose a control method that integrates a battery system with an efficiency-based algorithm to optimize the overall system performance. The proposed approach involves establishing operation command values based on the characteristics of the power generation source and adjusting these commands according to the battery's state of charge (SOC). MATLAB/Simulink simulations confirmed the effectiveness of this method across various operating modes and revealed no operational issues. When applied to a ship's operating profile over 222 h, the method reduced fuel consumption by approximately 2.98 tons (5.57%) compared with conventional systems. Over 38 annual voyages, this reduction equates to savings of 115.96 tons of fuel or approximately 96.47 million Korean won. This study demonstrates that integrating an optimal efficiency algorithm into the energy management system significantly enhances both the propulsion and overall energy efficiency of ships.

Keywords: optimal efficiency algorithm; power management system; energy management system; carbon dioxide emissions; battery management system

1. Introduction

Various research projects on enhancing the energy efficiency of ship systems and reducing greenhouse gas emissions are being conducted [1–3]. In addition to advancements in propulsion systems, a significant amount of research has focused on improving the power system efficiency, including the use of onboard hybrid power sources [4–6]. Integrating photovoltaic generation, fuel cells, or renewable energy into conventional power systems has been shown to reduce greenhouse gas emissions and air pollutants, while offering cost savings compared with the use of traditional gas turbines and internal combustion engines [7,8]. However, these systems have high initial installation costs and technological limitations [5,9]. Hybrid systems combining diesel generators with battery systems have been introduced to enhance the efficiency of conventional electric propulsion systems [10,11]. Supercapacitors and variable-speed engine generators have demonstrated improvements in energy efficiency and a reduction in pollutant emissions [12–14]. Additionally, integrating a solar hybrid power system into a built-in energy storage system and a conventional diesel generator has been shown to reduce fuel consumption and carbon dioxide emissions [15,16]. A hybrid system that includes molten carbonate fuel cells (MCFCs), batteries, and diesel generators has been found to be effective at reducing carbon dioxide emissions compared with conventional diesel generators [17]. An efficient power management system is essential for integrating multiple energy sources and storage systems into a microgrid with renewable energy and storage devices [18]. Consequently, several control techniques and power management strategies have been developed, including a new power distribution control strategy based on logical thresholds [19,20]. The application of optimal power management methods to the electric power system of an electric propulsion vessel has been shown to minimize the operating costs and meet greenhouse gas (GHG) emission limit criteria [21,22].

In hybrid electric propulsion systems that incorporate various power generation sources, the design of the controller is crucial for the efficient operation and control of these sources based on load demands [23,24]. In addition, integrating the controller with an energy management system along with the existing power management system is essential [25–27]. One study focused on a hybrid electric vessel powered by a dual proton exchange membrane fuel cell (PEMFC), a battery, and an ultracapacitor (UC). In this study, an equivalent consumption minimization strategy (ECMS) was employed to optimize the total power output of a PEMFC [9,27]. A map search engine was used to maximize the efficiency and the power distribution to the battery, and the UC was managed based on the state of charge (SOC) of each component. Simulations demonstrated a high efficiency for this configuration [28].

Simulations of a hybrid system combining liquefied natural gas (LNG) and batteries on a tugboat, integrated with an energy management system using rule-based control, demonstrated that CO_2 emissions and daily fuel costs could be reduced when compared to a system without such a control strategy [29].

An analysis of a hybrid power system containing a fuel cell was conducted using an adaptive equivalent consumption minimization strategy (A-ECMS) and state-based and fuzzy-logic-based EMS (Energy Management System). The results indicate that the A-ECMS strategy can maintain a system efficiency of above 60% under most operating conditions and significantly reduce the fluctuations in the output power of the fuel cell [30].

A distributed variable sag slope control strategy was implemented to enhance the SOC equalization of vessels equipped with a fuel cell (FC) as the energy source and batteries, and supercapacitors as the energy storage system (ESS). This approach improved the speed and accuracy of the SOC equalization, optimized the characteristics of different energy storage devices, and reduced the degradation of these devices [31].

In a battery hybrid power system, a rule-based control method utilizing an ECMS is used to train a neural network. Simulations confirmed the accuracy of this method, demonstrating that the stability of the system was maintained by effectively controlling the speed, voltage, and current of the propulsion motor under varying battery SOC levels and rapidly changing ship loads [32].

A simulation of the battery hybrid method was conducted to compare and analyze its performance with those of conventional diesel-powered ships in terms of carbon dioxide emissions. The results demonstrate that the battery hybrid method effectively reduces carbon dioxide emissions compared with the use of a controller that implements load-sharing optimal control. Additionally, a life cycle assessment (LCA) confirmed that the proposed system is environmentally friendly throughout the energy generation process [33].

An energy management framework based on model predictive control (MPC) was developed for a ship's hybrid power generation system with a battery system by incorporating an advanced shipboard energy management strategy (EMS). Simulations that had the aim of ensuring safe voyages and a long battery life while considering battery capacity and SOC values showed that the proposed framework could achieve a 3.5% reduction in energy consumption [34].

A hybrid optimization algorithm combining chaotic algorithms and gray wolf optimization (GWO) was used to design an energy management strategy with non-linear model predictive control (NMPC). The study found that NMPC based on the GWO algorithm could reduce fuel consumption by approximately 26% and carbon emissions by approximately 56%, compared with other algorithms [35].

The operation of a ship's DC microgrid model, consisting of equipment, a controller, and a communication network, was simulated and the results with and without a secondary control strategy were compared. The results demonstrate that the secondary control strategy effectively addressed the problem of unbalanced SOC in the energy-storage module. The SOC of the battery gradually converges under the secondary strategy, proving that it is both reasonable and efficient [36].

A study was conducted on the electrification of low-tonnage vessels operating on short cycles with high-power demands; a situation which presents significant technical challenges. The study involved the real test case of the Seine River ferry with an installed propulsion power of 330 kW. The supercapacitor and battery-based hybrid structures were compared with those of a conventional propulsion system. The results showed that the hybrid structures achieved reductions in the CO_2 emissions of 18% and 29.7%, respectively, compared with the conventional method, corresponding to reductions of approximately 382 and 626 t of CO_2 over 20 years of operation [37].

For the optimal energy management system of the generator non-equivalent parallel running operation system proposed in this study, a load-sharing algorithm was designed based on the generator's optimal efficiency operation. This approach aims to improve upon previous methods that only partially enhance the efficiency. Using MATLAB/Simulink, we modeled an actual operational ship by linking the energy storage system to an existing power system to maximize the energy efficiency. An algorithm was developed to operate at the optimal efficiency point based on fuel consumption characteristics relative to the power output of the diesel generator. The goal of the control algorithm designed to operate at this optimal point is to maximize the energy efficiency of the ship. The proposed control method is compared with existing methods to verify its effectiveness, with the goal of reducing fuel consumption and greenhouse gas (GHG) emissions.

2. Methodology

The ship's specifications and system requirements were selected to assess the effectiveness of the proposed generator non-equivalent parallel running operation system (NEQP). The optimal operating point of the engine was determined based on the design of the control rules and operation modes. These were subsequently applied to the load profile of an actual ship, and the proposed NEQP method was compared with a conventional control method using MATLAB(R2021b)/Simulink. The optimal efficiency criteria for the generator were established to design rules for minimizing fuel consumption. An optimal operating point efficiency of 85% was selected considering the SFOC (Specific Fuel Oil Consumption) of the target ship and the stability of the generator engine. For the battery, charging and discharging limits were set to ensure safety and a long battery life. In this configuration, the generator provides a constant power output at the optimal efficiency point to satisfy the load requirements of the ship, and charges the battery whenever excess power is available. If the power output of the generator is insufficient for handling the system load, the controller is designed to release the energy stored in the battery to support part of the load.

Step 1. Selection of the system specifications.

The generator–battery hybrid system was evaluated using a vehicle carrier as the target ship and a 1500 kW battery system. The characteristics of, and basic information on, the diesel generators of the target ship are summarized in Table 1, and these data were used for the modeling. The target ship was equipped with three identical 1330 kW diesel generators produced by HYUNDAI-HiMSEN (Manufacturer: Hyundai Heavy Industries, Ulsan, Korea).

Table 1. Specifications of diesel generator on the vessel.

Maker	HYUNDAI-HiMSEN
Type	7H21/32
Engine power	1415 [kW]
Generator power	1330 [kW]
Engine speed	900 [rpm]
Engine set	3 [sets]

The Specific Fuel Oil Consumption (SFOC) required to calculate the fuel consumption based on the diesel generator's load is summarized in Table 2. The data were derived from the manufacturer's factory test operation report provided when the target ship was built.

Table 2. Specific fuel oil consumption value of diesel generator.

Engine power [%]	10	25	50	75	100
SFOC [g/kWh]	322.3	240.8	203.3	196.7	192.2

As illustrated in Figure 1, operating the generator at low loads results in a higher fuel oil consumption than operating it at relatively high loads. This indicates that the fuel efficiency varies with the load, with a noticeable difference between low- and high-load operations. For loads above 50%, there was little variation in the efficiency. However, for loads below 50%, the efficiency decreased sharply, as indicated by the steep slope. A simulation model was constructed considering these characteristics.

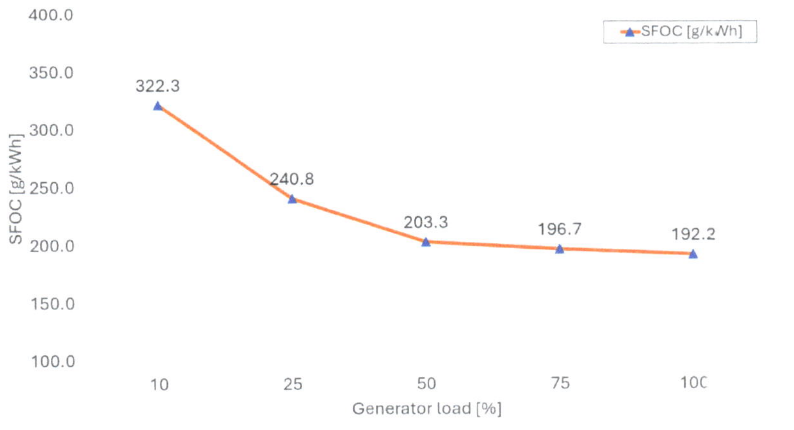

Figure 1. Specific fuel oil consumption value of diesel generator.

The optimal operating point of the engine, based on the lowest Specific Fuel Oil Consumption (SFOC) at 100% load, poses a risk of blackout due to the additional load when the generator runs at full capacity. To ensure safety, the Power Management System (PMS) activates a standby generator when the load exceeds 85%, reducing the strain on the main generator. Therefore, this paper designates 85% load as the point of highest efficiency with safety assured.

Step 2. Design of the energy management system.

To use the controller, the generator load-sharing criteria were set according to the system load size and battery SOC, which was designed by setting the rules for the energy-optimal control criteria for the optimal operation mode of the NEQP as an energy management system. As shown in Equation (1), the load is defined as the sum of the power output of the power generation source and the SOC of the battery. The command states of the generator output are divided into DG_{stop}, DG_{min}, DG_{opt}, and DG_{max}.

The optimal operating point (DG_{opt}) for efficient operation was selected, and the load of the system and the power balance of each generator and battery were equal to the sum of the power outputs of the three generators and the capacity of the battery with respect to the load of the ship.

$$W_{load} = W_{DG1} + W_{DG2} + W_{DG3} + W_{batt} \tag{1}$$

W_{load}: Required load [kW] on electrical power system
W_{DG1}: Output [kW] of No. 1 diesel generator
W_{DG2}: Output [kW] of No. 2 diesel generator
W_{DG3}: Output [kW] of No. 3 diesel generator
W_{batt}: Output [kW] of battery
DG_{stop}: 0 [%]
DG_{min}: 50 [%]
DG_{opt}: 85 [%]
DG_{max}: 100 [%]

The control rules were designed to ensure that each generator operates at its optimal efficiency point for each mode. The optimal energy control rules for minimizing the energy consumption while adhering to the control logic criteria based on the ship's load and battery SOC conditions are outlined in Table 3. As indicated in the table, the generators operate in the optimal mode in 10 of the 12 operating modes. The rules for each load zone were as follows: the battery is charged only when the SOC is below 30% and discharged only when the SOC is above 80%. When the SOC is between 30% and 80%, the battery is in a normal state, and charging/discharging rules are applied based on the system load. In this zone, all generators operate at their optimal efficiency. When the battery SOC is 80% or higher, charging is not allowed, and if the power load required by the ship is less than the generator's optimal output, the fuel consumption per unit horsepower will increase owing to the low-load operation.

Figure 2 shows the operating states of the power generator and battery based on the control rules.

Figure 3 presents a flowchart of the control rules based on the operational mode and control rules. Using the ship load and battery SOC as input values, the power output command of the generator was determined according to these control rules. Commands M1, M2, M3, Mmax, and Mvar vary based on the SOC of the battery, and the power output command of the generator is adjusted according to the current operating load.

Step 3. Selection of the load profile

To verify the reliability of the diesel generator–battery hybrid system model with the optimal efficiency algorithm-based control method, several scenarios were constructed using actual operational data from the target ship, and simulations were conducted. As shown in Figure 4, the sailing route of the target ship was as follows: after docking at the port of call for approximately 19.2 h, the ship departed and sailed for about 72 h. It then entered the next port and docked for 62.4 h for cargo loading and unloading, before departing again and sailing for 60 h. As depicted in Figure 5, the system load records from the actual ship, covering approximately 222 h (800,000 s) of operation, were extracted and converted into load profiles for constructing the simulation scenarios. The fuel consumption was then calculated using MATLAB/Simulink software (R2021b).

Table 3. Proposed NEQP optimal efficiency algorithm for the generator parallel operation system with battery.

SOC [%]	Load Condition \ Power	Mode	W_{DG1}	W_{DG2}	W_{DG3}	W_{batt}
0 ~ 30	$0 < W_{load} \leq DG_{opt}$	M1	DG_{opt}	DG_{stop}	DG_{stop}	$W_{load} - W_{DG1}$
	$DG_{opt} < W_{load} \leq DG_{opt} \times 2$	M2	DG_{opt}	DG_{opt}	DG_{stop}	$W_{load} - W_{DG1} - W_{DG2}$
	$DG_{opt} \times 2 < W_{load} \leq DG_{opt} \times 3$	M3	DG_{opt}	DG_{opt}	DG_{opt}	$W_{load} - W_{DG1} - W_{DG2} - W_{DG3}$
	$DG_{opt} \times 3 < W_{load} \leq DG_{max} \times 3$	Mmax	DG_{max}	DG_{max}	DG_{max}	$W_{load} - W_{DG1} - W_{DG2} - W_{DG3}$
30 ~ 80	$0 < W_{load} \leq DG_{min}$	M1	DG_{opt}	DG_{stop}	DG_{stop}	$W_{load} - W_{DG1}$
	$DG_{min} < W_{load} \leq DG_{opt} + DG_{min}$	M1	DG_{opt}	DG_{stop}	DG_{stop}	$W_{load} - W_{DG1}$
	$DG_{opt} + DG_{min} < W_{load} \leq (DG_{opt} \times 2) + DG_{min}$	M2	DG_{opt}	DG_{opt}	DG_{stop}	$W_{load} - W_{DG1} - W_{DG2}$
	$(DG_{opt} \times 2) + DG_{min} < W_{load} \leq DG_{max} \times 3$	M3	DG_{opt}	DG_{opt}	DG_{opt}	$W_{load} - W_{DG1} - W_{DG2} - W_{DG3}$
80 ~ 100	$0 < W_{load} \leq DG_{opt}$	Mvar	W_{load}	DG_{stop}	DG_{stop}	0
	$DG_{opt} < W_{load} \leq DG_{opt} \times 2$	M1	DG_{opt}	DG_{stop}	DG_{stop}	$W_{load} - W_{DG1}$
	$DG_{opt} \times 2 < W_{load} \leq DG_{opt} \times 3$	M2	DG_{opt}	DG_{opt}	DG_{stop}	$W_{load} - W_{DG1} - W_{DG2}$
	$DG_{opt} \times 3 < W_{load} \leq DG_{max} \times 3$	M3	DG_{opt}	DG_{opt}	DG_{opt}	$W_{load} - W_{DG1} - W_{DG2} - W_{DG3}$

W_{load}: Required load [kW] on electrical power system, W_{DG1}: output [kW] of No. 1 diesel generator, W_{DG2}: output [kW] of No. 2 diesel generator, W_{DG3}: output [kW] of No. 3 diesel generator, W_{batt}: output [kW] of battery, DG_1: No. 1 diesel generator, DG_2: No. 2 diesel generator, DG_3: No. 3 diesel generator, DG_{stop}: 0 [%], DG_{min}: 50 [%], DG_{opt}: 85 [%], DG_{max}: 100 [%], M1~3: Mode 1~3, Mvar: variable mode, Max: max mode.

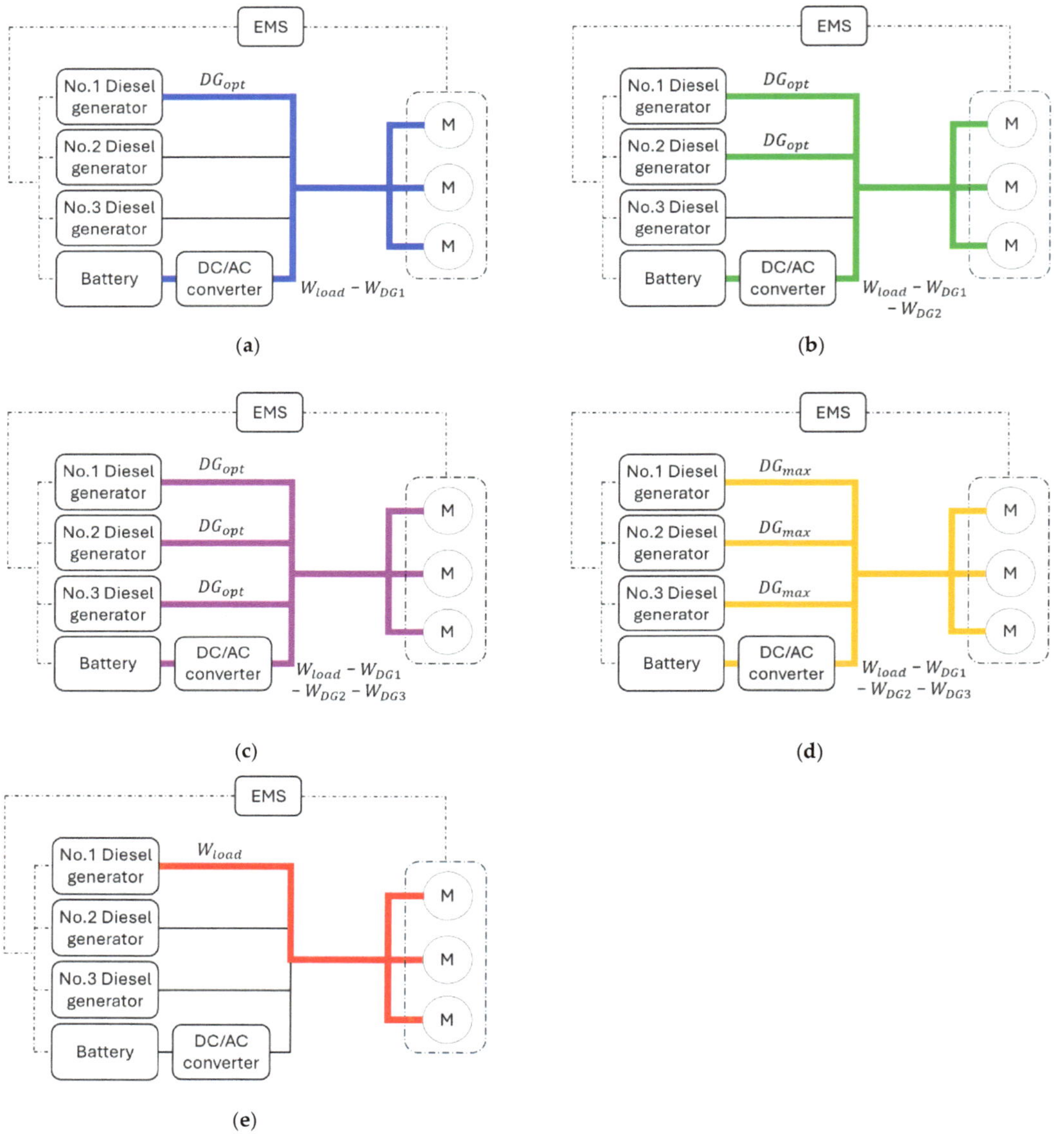

Figure 2. Operating status of proposed NEQP optimal efficiency algorithm. (**a**) M1, (**b**) M2, (**c**) M3, (**d**) Mmax, (**e**) Mvar.

The sections labeled "In Port" or "Sailing" include the loads required for sailing, operating equipment during cargo loading and unloading, and maintaining the ship's living quarters. These are the basic load sections at which a ship can operate normally. The three high-load sections labeled "Arrival" or "Departure" represent periods of maximum power consumption due to large loads such as bow thrusters when entering or leaving a port. Data were collected and stored in 10 s intervals during the voyage, with each 10 s interval assuming that the values retained the previous data.

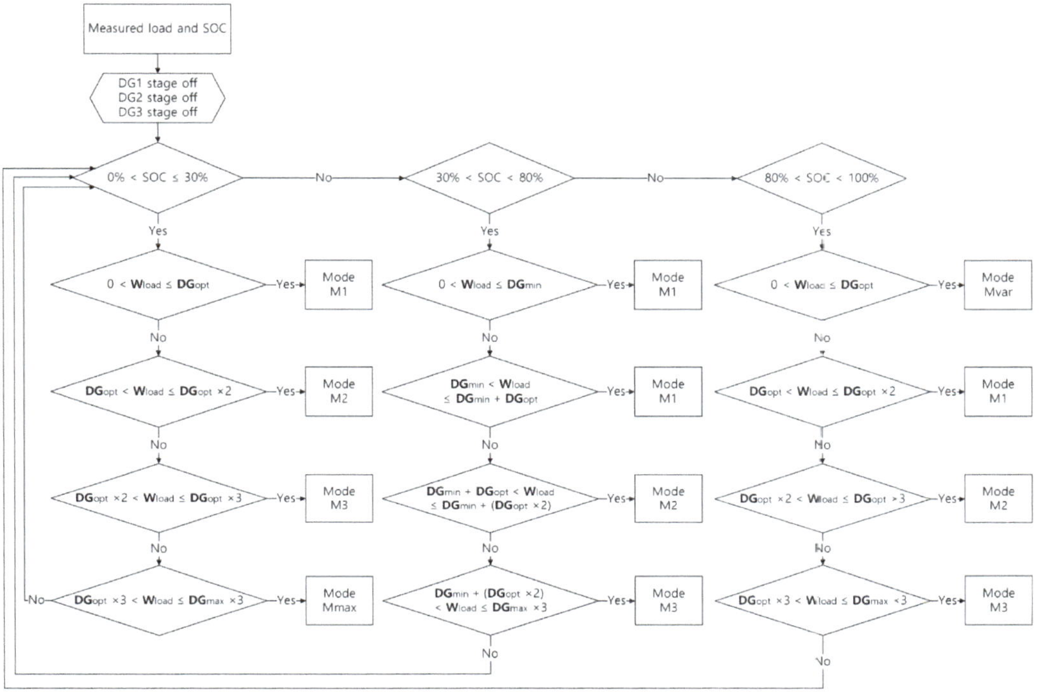

Figure 3. Block diagram of operation sequence for proposed NEQP optimal efficiency algorithm.

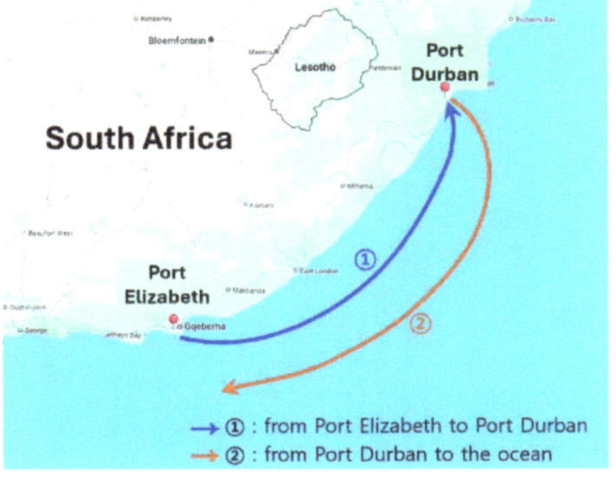

Figure 4. Voyage route of reference vessel.

Step 4. Simulation

The simulations were performed to verify the reliability of the diesel generator–battery hybrid system model with the optimal efficiency control method applied to the existing power system using MATLAB/Simulink. The controller inputs, which vary in real time, include the total load of the ship and the battery state of charge (SOC). The output was the power output command of the generator engine. As shown in Figure 6, the system was configured by integrating the battery system with the existing power system of the target ship, which consisted of three engine generators. Simulations were conducted using symmetric and asymmetric load-sharing methods, and the results were compared. The system load was managed using a MATLAB/Simulink block that facilitates real-time load variations.

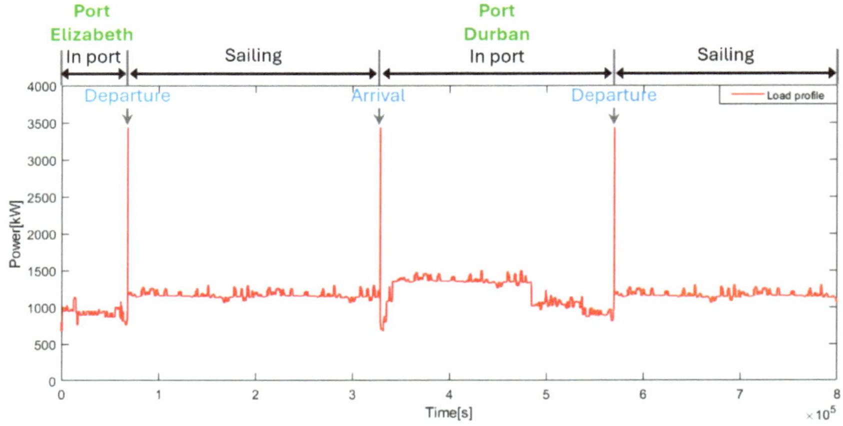

Figure 5. Operating load profile of reference vessel.

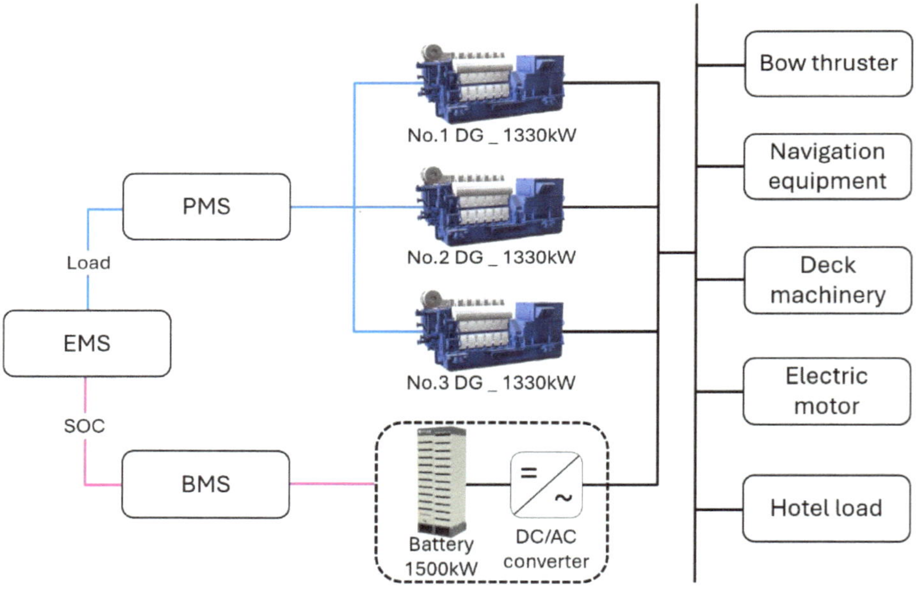

Figure 6. System configuration of generator–battery hybrid system.

As depicted in Figure 7, the generator operation characteristics of the target ship were applied to the model to extract data, such as the generator's power output, fuel consumption, voltage, and current relative to the load. The model included three generators with a battery capacity set greater than that of the generators. The initial SOC value [%] was set using a block. A scenario simulating the power load required for the actual operation of the target ship was input as the load. Additionally, a data collection system was modeled to extract and store the load data over time according to the scenario. A controller implementing both symmetric and asymmetric control methods was added to compare the proposed NEQP control method with existing methods.

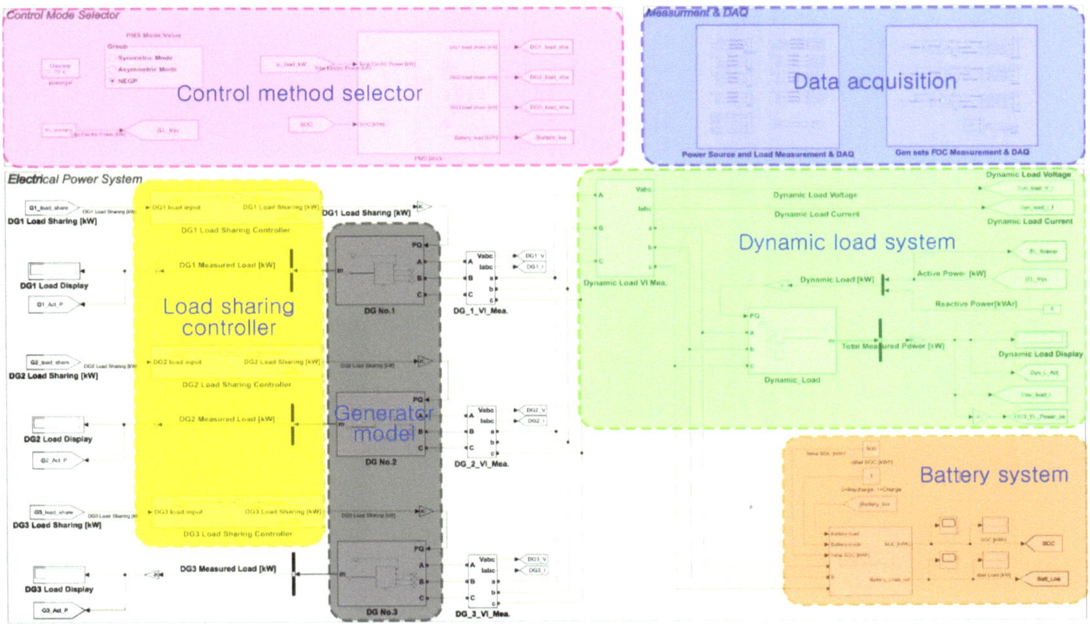

Figure 7. Simulation model of a generator–battery hybrid architecture.

3. Results

Based on the selected load profile, the simulation results for systems with symmetric load sharing, asymmetric load sharing, and the proposed NEQP controller were compared, as shown in Figure 8. In the symmetric load-sharing method, the system operated with medium-to-low loads in the 50–60% range, with power outputs equally distributed between the two generators, resulting in a lower efficiency. No. 3 DG was activated to handle peak loads during port arrivals and departures, demonstrating that the load was shared through parallel operation. Adding a sufficiently large battery to this system and connecting it to the grid can mitigate efficiency losses by using the stored energy for temporary peak and partial loads, thereby reducing the need for frequent parallel operations.

When operating with an asymmetric load-sharing controller, No. 1 DG functions at its optimal operating point of 85% (1130.5 kW), whereas the remaining load is managed by No. 2 and 3 in parallel operation. No. 2 DG operates at its lowest efficiency, handling a load of approximately 0–35% (0–400 kW) of its optimal capacity. No. 3 DG is activated to address the peak loads that arise during port arrivals and departures, with the load shared through a parallel operation. Consistently operating one generator at a high efficiency is advantageous for improving the overall efficiency and reducing fuel consumption. However, the remaining load is handled by another generator operating at a lower efficiency, and all changing system loads are handled in the low-load state. Thus, while a longer load scenario duration and a larger efficiency operating point based on load sharing can be beneficial, the system efficiency decreases as the operating point deviates from the optimal point compared with symmetric load operation. To enhance the efficiency of this system, a sufficiently large battery could be added and connected to the grid. In this scenario, the energy stored in the battery can be used for low and temporary peak loads handled by No. 2 DG, thereby compensating for the reduced efficiency.

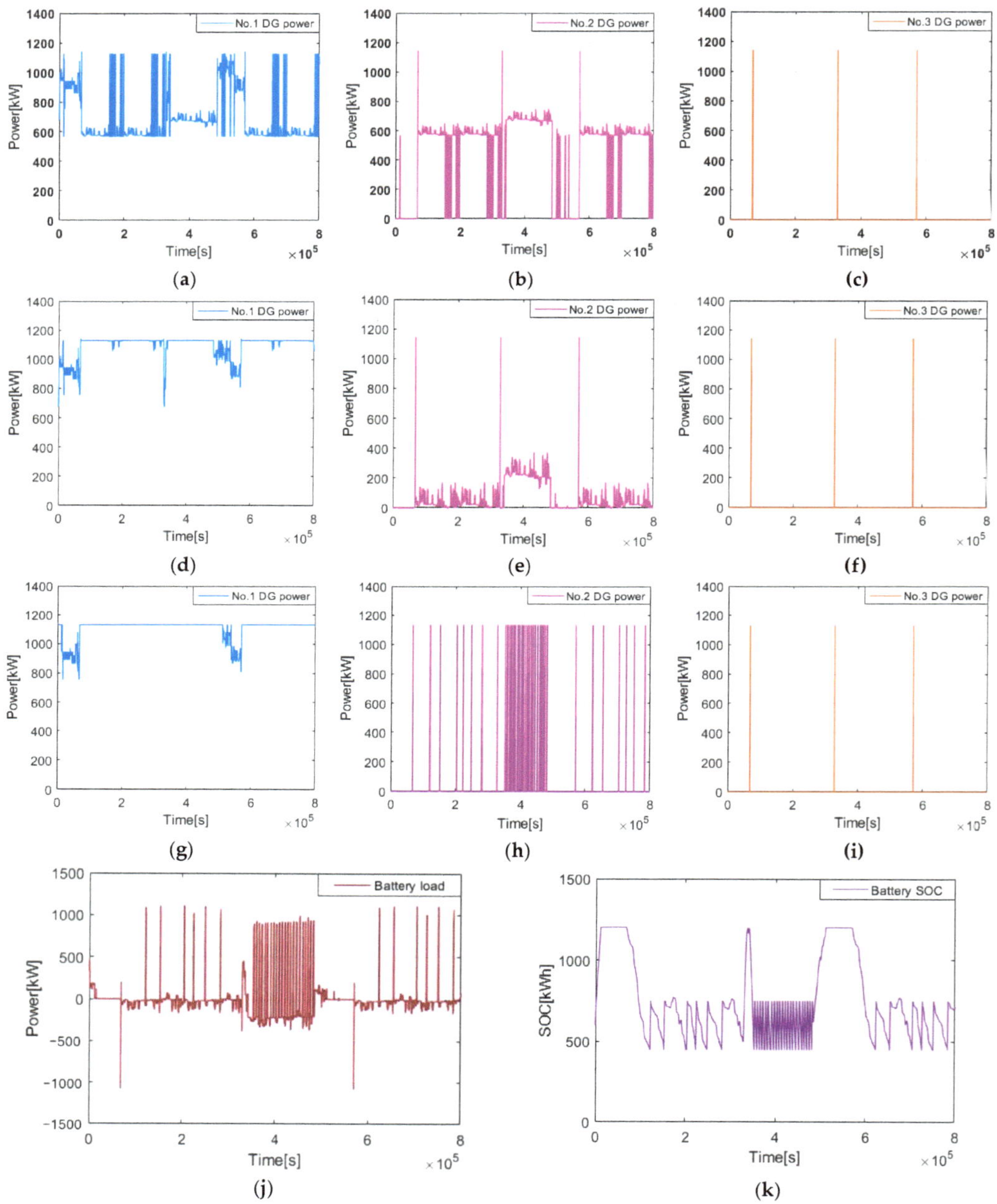

Figure 8. DG power, battery power, and SOC for operation modes. (**a**) No. 1 DG power in symmetric load sharing, (**b**) No. 2 DG power in symmetric load sharing, (**c**) No. 3 DG power in symmetric load sharing, (**d**) No. 1 DG power in asymmetric load sharing, (**e**) No. 2 DG power in asymmetric load sharing, (**f**) No. 3 DG power in asymmetric load sharing, (**g**) No. 1 DG power in proposed NEQP method, (**h**) No. 2 DG power in proposed NEQP method, (**i**) No. 3 DG power in proposed NEQP method, (**j**) Battery power in proposed NEQP method, (**k**) Battery SOC in proposed NEQP method.

The simulation of a generator–battery hybrid system, in which an additional battery is connected to the existing power system, demonstrates that the generators operate at their optimal points depending on the system load and battery SOC, whereas the battery assists with loads followed by charging and discharging. No. 1 DG operates at its optimal point of 85% (1130.5 kW), handling most of the power system load. The remaining load is managed through a parallel operation with Nos. 2 and 3, and some of the load is handled by the battery. All three generators operated at their optimal efficiency points. When No. 1 DG does not operate at its optimal point, the system, as shown in the battery SOC graph, is controlled to protect the battery when the SOC reaches 80%. In these cases, the generator operates at a variable load rather than at its optimal point, and the battery stops charging. During these periods, the system load is managed by a single generator, and the power output fluctuates according to the load changes. No. 3 DG is activated to handle peak loads during port arrivals and departures, with the load being shared through a parallel operation. The simulation results indicate that the generators operate at their optimal efficiency points in all zones, except those set for battery protection based on the battery SOC. The generators operate in parallel according to the load size, and the battery charges and discharges smoothly. When the generator operates at its optimal point, the charged battery discharges to manage the additional load above the optimal efficiency point. In the two zones in which the battery SOC reached the upper limit of 80% (1200 kWh), the battery stopped charging and discharged slightly but steadily, as indicated by the power input and output. When the SOC reached the lower limit of 30% (450 kWh), the battery immediately started charging again.

Table 4 provides details of the fuel oil consumption of each generator and the combined fuel oil consumption of all three generators throughout the simulation. At the end of the approximately 222-h simulation, the cumulative fuel consumption for the symmetric load-sharing controller was 53.49 t. The fuel consumption distribution was approximately 60% for generator 1, 40% for No. 2 DG, and approximately 0.1% for No. 3 DG, which handled the peak load. In total, 53.49 t of fuel oil were consumed.

Table 4. Total fuel oil consumption (unit: [ton]).

DG No. / Load Sharing Method	No. 1	No. 2	No. 3	Grand Total
Symmetric	32.17	21.25	0.07	53.49
Asymmetric	47.49	4.44	0.06	51.99
NEQP	47.89	2.55	0.06	50.5

For the asymmetric load-sharing controller, the total fuel consumption was 51.99 t. No. 1 DG operated at its optimal efficiency point and handled most of the system load, while No. 2 DG operated at a low efficiency and load and managed the remaining load. The cumulative fuel consumption of No. 1 DG was 47.49 t, which represented approximately 91.3% of the total cumulative fuel consumption. The cumulative fuel consumption of No. 2 DG was 4.44 t, accounting for approximately 8.5%, while that of No. 3 DG was approximately 0.1%. The total fuel oil consumption in the asymmetric load-sharing control mode was 51.99 t. The cumulative fuel consumption of No. 1 DG, which constitutes approximately 91.3% of the total fuel consumption, represents a significant reduction compared with that of the symmetric load-sharing control mode. This efficiency improvement results in fuel savings of approximately 1.5 t. By segmenting the system load into specific zones, one or more generators can operate at their optimal efficiency points, thereby reducing the fuel oil consumption and pollutant emissions. However, the remaining generators often operate at very low efficiencies in low-load zones, which can increase fuel consumption and pollutant emissions. These results may vary depending on the load profile and the operating time.

The final cumulative fuel consumption of the system with the proposed NEQP controller was 50.5 t. All generators operated at their optimal efficiency points and operated in conjunction with a battery to manage the power system load. No. 1 DG handled most of the power system load with a cumulative fuel consumption of 47.89 t, accounting for approximately 83% of the total fuel consumption. The cumulative fuel consumption of No. 2 DG was 2.55 t, or approximately 5%, while that of No. 3 DG was minimal at approximately 0.12%. Because all generators operated at their optimal efficiency points, the total fuel consumption was reduced compared with previous simulations. Despite No. 1 DG having the highest cumulative fuel consumption of 47.89 t, which is approximately 83% of the total, the overall fuel consumption was lower because Nos. 2 and 3 DG also operated efficiently, and the battery was connected to the grid to handle part of the load.

4. Discussion

4.1. Analysis of Results through Comparison

Figure 9 illustrates the fuel oil consumption of the generator–battery hybrid system with the proposed NEQP optimal efficiency control method compared with those of the conventional symmetric and asymmetric load-sharing control methods. The results demonstrate that the fuel consumption is the lowest with the proposed NEQP optimal efficiency algorithm.

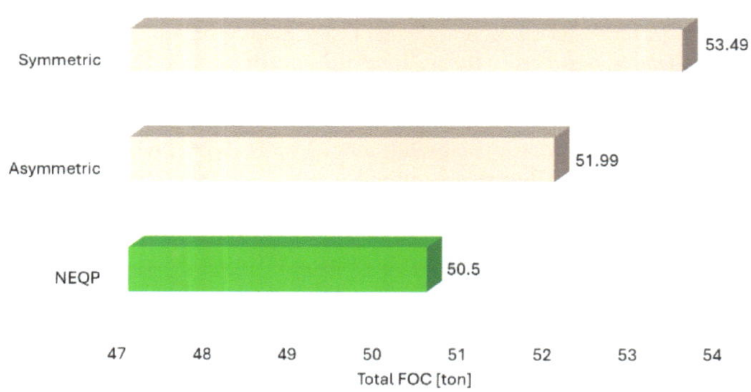

Figure 9. Total fuel oil consumption for the load sharing method.

Table 5 presents a comparison of the total fuel consumption for each control method based on the simulation results. The symmetric load-sharing method resulted in a total fuel consumption of 53.49 t, which is about 2.99 t higher than the 50.5 t consumed using the proposed NEQP control method. This indicates that the proposed method achieves a higher efficiency than conventional methods.

Table 5. Comparison of total fuel oil consumption for the load sharing method.

Load Sharing Method		**Symmetric**	**Asymmetric**
	Total FOC [ton]	53.49	51.99
NEQP	50.5	2.99 [ton] (5.59 [%])	1.49 [ton] (2.87 [%])

While adding a battery to the existing power system and applying NEQP is the most effective solution, the cost of adding the battery must be considered. In contrast, implementing symmetric and asymmetric load sharing methods in the existing power system may only require changes to the control logic, with no need for additional equipment installations.

4.2. Economic Benefits of Applying NEQP

Additionally, maintaining generators at their optimal operating points simplifies the prediction of maintenance cycles and is expected to yield significant economic benefits through reduced fuel and maintenance costs. Table 6 shows that if a ship using low-sulfur oil completes approximately 38 voyages per year, the NEQP optimal efficiency control method could save 115.96 t of fuel oil. This translates to an expected cost saving of approximately USD 69,807 or approximately KRW 96.47 million [38].

Table 6. Analysis of fuel savings and economic benefits per year depending on fuel type.

	Type of Fuel		
	High-Sulfur Fuel Oil (3.5 [%])	**Low-Sulfur Fuel Oil (0.5 [%])**	**LNG**
USD per ton [USD]	520	602	708
Savings of fuel per year [ton]	115.96	115.96	115.96
Savings of USD per year [USD]	60,299.2	69,807.9	82,099.7
Savings of KRW per year [10 K KRW]	8333.35	9647.45	11,346.18

This reduction translates to expected fuel savings of approximately 115.96 t for 38 voyages per year, equating to cost savings of approximately USD 69,807. The savings are anticipated to be even greater when using LNG, and the economic benefits are expected to increase if marine fuel oil prices rise in the future or if the supply price of carbon-free fuel exceeds that of LNG.

4.3. Limitations and Future Research

1. This study has limitations due to the use of simulated data based on the actual load profile of a ship's voyage. Therefore, it is essential to conduct further research by installing and validating the proposed control methods on software and equipment in actual operational ships.
2. Further research is needed to explore different system designs to determine if the C-rate of the battery system can handle peak loads during port arrival and departure, and if the ship's operational capability can be maintained without reserving auxiliary power for rapidly changing onboard loads. The use of dual-battery systems, super-capacitors, and advanced control methods should also be investigated to ensure the safety of the power systems under peak loads.
3. This paper acknowledges certain limitations, such as not considering the battery's price, weight, and safety in the event of a fire. Future research should address these issues through a comprehensive review of both economic and safety implications.
4. The study also focuses on long time-scale simulations, meaning it does not include components that capture transient states, such as Automatic Voltage Regulators (AVR). Future developments will need to address these transient states, presenting a new challenge for model enhancement.
5. The reduction in fuel consumption can lead to verified reductions in CO2 emissions, which presents another potential avenue for expanded research.

5. Conclusions

To improve the efficiency of the existing power system, three control methods were compared and analyzed. A model was developed using MATLAB/Simulink, and a simulation was conducted based on 222 h of real ship operation data.

1. When the symmetric load-sharing control method was applied, the number of generators operating was determined based on power loads of 85%, 170%, and 255%. It was observed that the load was evenly distributed among the generators operating in parallel. During the simulation, a total of 53.49 tons of fuel was consumed.
2. With the asymmetric load-sharing control method, the number of generators was also determined based on the size of the power load. However, in parallel operation, one generator operated at its optimal efficiency point, with the remaining load distributed among the other generators. This resulted in a total fuel consumption of 51.99 tons during the simulation.
3. When the NEQP control method proposed in this study was applied, the battery was charged or discharged depending on the defined load range and the battery's state of charge (SOC). This enabled the generators to maintain optimal efficiency for a longer period compared to the previously mentioned control methods. As a result, the total fuel consumption was reduced to 50.5 tons, the lowest among the control methods.
4. The simulation results demonstrated a reduction in generator fuel consumption of approximately 2.99 tons, or 5.59%, over 222 h of operation.

Author Contributions: Conceptualization, E.C.; Methodology, E.C.; Software, E.C.; Validation, H.K.; Formal analysis, E.C. and H.K.; Investigation, E.C. and H.K.; Resources, E.C.; Data curation, E.C. and H.K.; Writing—original draft, E.C.; Writing—review & editing, H.K.; Visualization, H.K.; Supervision, H.K.; Project administration, H.K.; Funding acquisition, E.C. All authors have read and agreed to the published version of the manuscript.

Funding: This research was supported by the Korea Institute of Marine Science and Technology Promotion and funded by the Ministry of Oceans and Fisheries (No. 20220603).

Institutional Review Board Statement: Not applicable.

Informed Consent Statement: Not applicable.

Data Availability Statement: Data presented in this article are available upon request from the corresponding authors.

Conflicts of Interest: The authors declare no conflicts of interest.

Abbreviations

DG: diesel generator, SOC: state of charge, SFOC: specific fuel oil consumption, NEQP: non-equivalent parallel running operation system.

References

1. Wang, K.; Wang, J.; Huang, L.; Yuan, Y.; Wu, G.; Xing, H.; Wang, Z.; Wang, Z.; Jiang, X. A comprehensive review on the predictio of ship energy consumption and pollution gas emissions. *Ocean Eng.* **2022**, *266*, 112826. [CrossRef]
2. Sharafian, A.; Blomerus, P.; Mérida, W. Natural gas as a ship fuel: Assessment of greenhouse gas and air pollutant reductio potential. *Energy Policy* **2019**, *131*, 332–346. [CrossRef]
3. Jimenez, V.J.; Kim, H.; Munim, Z.H. A review of ship energy efficiency research and directions towards emission reduction in th maritime industry. *J. Clean. Prod.* **2022**, *366*, 132888. [CrossRef]
4. Yuan, Y.; Wang, J.; Yan, X.; Shen, B.; Long, T. A review of multi-energy hybrid power system for ships. *Renew. Sustain. Energy Re* **2020**, *132*, 110081. [CrossRef]
5. Inal, O.B.; Charpentier, J.-F.; Deniz, C. Hybrid power and propulsion systems for ships: Current status and future challenges *Renew. Sustain. Energy Rev.* **2022**, *156*, 111965. [CrossRef]
6. Lan, H.; Wen, S.; Hong, Y.-Y.; David, C.Y.; Zhang, L. Optimal sizing of hybrid PV/diesel/battery in ship power system. *Appl Energy* **2015**, *158*, 26–34. [CrossRef]
7. Diab, F.; Lan, H.; Ali, S. Novel comparison study between the hybrid renewable energy systems on land and on ship. *Renew Sustain. Energy Rev.* **2016**, *63*, 452–463. [CrossRef]
8. Manickavasagam, K.; Thotakanama, N.K.; Puttaraj, V. Intelligent energy management system for renewable energy driven ship *IET Electr. Syst. Transp.* **2019**, *9*, 24–34. [CrossRef]

9. Ghenai, C.; Bettayeb, M.; Brdjanin, B.; Hamid, A.K. Hybrid solar PV/PEM fuel Cell/Diesel Generator power system for cruise ship: A case study in Stockholm, Sweden. *Case Stud. Therm. Eng.* **2019**, *14*, 100497. [CrossRef]
10. Geertsma, R.; Negenborn, R.; Visser, K.; Hopman, J. Design and control of hybrid power and propulsion systems for smart ships: A review of developments. *Appl. Energy* **2017**, *194*, 30–54. [CrossRef]
11. Zahedi, B.; Norum, L.E. Modeling and simulation of all-electric ships with low-voltage DC hybrid power systems. *IEEE Trans. Power Electron.* **2012**, *28*, 4525–4537. [CrossRef]
12. Lee, J.-H.; Lee, S.-H.; Sul, S.-K. Variable-speed engine generator with supercapacitor: Isolated power generation system and fuel efficiency. *IEEE Trans. Ind. Appl.* **2009**, *45*, 2130–2135.
13. Chen, H.; Zhang, Z.; Guan, C.; Gao, H. Optimization of sizing and frequency control in battery/supercapacitor hybrid energy storage system for fuel cell ship. *Energy* **2020**, *197*, 117285. [CrossRef]
14. Jenssen, M.; Suul, J.A. Model Predictive Control of a Variable Speed Diesel Generator Interfaced to an AC Ship Power System as a Virtual Synchronous Machine. In Proceedings of the 2021 IEEE International Future Energy Electronics Conference (IFEEC), Taipei, Taiwan, 16–19 November 2021; pp. 1–7.
15. Yuan, Y.; Wang, J.; Yan, X.; Li, Q.; Long, T. A design and experimental investigation of a large-scale solar energy/diesel generator powered hybrid ship. *Energy* **2018**, *165*, 965–978. [CrossRef]
16. Karatuğ, Ç.; Durmuşoğlu, Y. Design of a solar photovoltaic system for a Ro-Ro ship and estimation of performance analysis: A case study. *Sol. Energy* **2020**, *207*, 1259–1268. [CrossRef]
17. Inal, O.B.; Deniz, C. Assessment of fuel cell types for ships: Based on multi-criteria decision analysis. *J. Clean. Prod.* **2020**, *265*, 121734. [CrossRef]
18. Xu, L.; Wen, Y.; Luo, X.; Lu, Z.; Guan, X. A modified power management algorithm with energy efficiency and GHG emissions limitation for hybrid power ship system. *Appl. Energy* **2022**, *317*, 119114. [CrossRef]
19. Al-Falahi, M.D.; Tarasiuk, T.; Jayasinghe, S.G.; Jin, Z.; Enshaei, H.; Guerrero, J.M. AC ship microgrids: Control and power management optimization. *Energies* **2018**, *11*, 1458. [CrossRef]
20. Yu, W.; Zhou, P.; Wang, H. Evaluation on the energy efficiency and emissions reduction of a short-route hybrid sightseeing ship. *Ocean Eng.* **2018**, *162*, 34–42. [CrossRef]
21. Kanellos, F. Optimal power management with GHG emissions limitation in all-electric ship power systems comprising energy storage systems. *IEEE Trans. Power Syst.* **2013**, *29*, 330–339. [CrossRef]
22. Al-Falahi, M.D.; Nimma, K.S.; Jayasinghe, S.D.; Enshaei, H.; Guerrero, J.M. Power management optimization of hybrid power systems in electric ferries. *Energy Convers. Manag.* **2018**, *172*, 50–66. [CrossRef]
23. Baldi, F.; Ahlgren, F.; Melino, F.; Gabrielii, C.; Andersson, K. Optimal load allocation of complex ship power plants. *Energy Convers. Manag.* **2016**, *124*, 344–356. [CrossRef]
24. Su, C.-L.; Liao, C.-H. Ship electrical load analysis considering power generation efficiency. In Proceedings of the 2015 IEEE/IAS 51st Industrial & Commercial Power Systems Technical Conference (I&CPS), Calgary, AB, Canada, 5–8 May 2015; pp. 1–11.
25. Fan, A.; Li, Y.; Liu, H.; Yang, L.; Tian, Z.; Li, Y.; Vladimir, N. Development trend and hotspot analysis of ship energy management. *J. Clean. Prod.* **2023**, *389*, 135899. [CrossRef]
26. Guo, X.; Lang, X.; Yuan, Y.; Tong, L.; Shen, B.; Long, T.; Mao, W. Energy management system for hybrid ship: Status and perspectives. *Ocean Eng.* **2024**, *310*, 118638. [CrossRef]
27. Liu, H.; Fan, A.; Li, Y.; Vladimir, N. Testing methods for multi-energy ship energy management system: A systematic review. *Ocean Eng.* **2024**, *304*, 117889. [CrossRef]
28. Misyris, G.S.; Marinopoulos, A.; Doukas, D.I.; Tengnér, T.; Labridis, D.P. On battery state estimation algorithms for electric ship applications. *Electr. Power Syst. Res.* **2017**, *151*, 115–124. [CrossRef]
29. Roslan, S.B.; Tay, Z.Y.; Konovessis, D.; Ang, J.H.; Menon, N.V. Rule-based control studies of LNG–battery hybrid tugboat. *J. Mar. Sci. Eng.* **2023**, *11*, 1307. [CrossRef]
30. Ge, Y.; Zhang, J.; Zhou, K.; Zhu, J.; Wang, Y. Research on energy management for ship hybrid power system based on adaptive equivalent consumption minimization strategy. *J. Mar. Sci. Eng.* **2023**, *11*, 1271. [CrossRef]
31. Ma, Z.; Chen, H.; Han, J.; Chen, Y.; Kuang, J.; Charpentier, J.-F.; Aït-Ahmed, N.; Benbouzid, M. Optimal SOC control and rule-based energy management strategy for fuel-cell-based hybrid vessel including batteries and supercapacitors. *J. Mar. Sci. Eng.* **2023**, *11*, 398. [CrossRef]
32. Kim, S.; Kim, J. Optimal energy control of battery hybrid system for marine vessels by applying neural network based on equivalent consumption minimization strategy. *J. Mar. Sci. Eng.* **2021**, *9*, 1228. [CrossRef]
33. Kim, S.; Jeon, H.; Park, C.; Kim, J. Lifecycle environmental benefits with a hybrid electric propulsion system using a control algorithm for fishing boats in Korea. *J. Mar. Sci. Eng.* **2022**, *10*, 1202. [CrossRef]
34. Antonopoulos, S.; Visser, K.; Kalikatzarakis, M.; Reppa, V. MPC framework for the energy management of hybrid ships with an energy storage system. *J. Mar. Sci. Eng.* **2021**, *9*, 993. [CrossRef]
35. Chen, L.; Gao, D.; Xue, Q. Energy Management Strategy of Hybrid Ships Using Nonlinear Model Predictive Control via a Chaotic Grey Wolf Optimization Algorithm. *J. Mar. Sci. Eng.* **2023**, *11*, 1834. [CrossRef]
36. Zeng, H.; Zhao, Y.; Wang, X.; He, T.; Zhang, J. Modeling of Ship DC Power Grid and Research on Secondary Control Strategy. *J. Mar. Sci. Eng.* **2022**, *10*, 2037. [CrossRef]

37. Bennabi, N.; Menana, H.; Charpentier, J.-F.; Billard, J.-Y.; Nottelet, B. Design and comparative study of hybrid propulsions for a river ferry operating on short cycles with high power demands. *J. Mar. Sci. Eng.* **2021**, *9*, 631. [CrossRef]
38. Ship & Bunker. Bunker Prices, World Bunker Prices, Singapore. Available online: https://shipandbunker.com/prices/apac/sea/sg-sin-singapore (accessed on 15 July 2024).

Journal of
Marine Science and Engineering

MDPI

Article

A Study on Fishing Vessel Energy System Optimization Using Bond Graphs

Sang-Won Moon [1], Won-Sun Ruy [2,*] and Kwang-Phil Park [2]

1 Fishing Vessel Safety Policy Division & Fisheries Policy Division, Ministry of Oceans and Fisheries, Sejong 30110, Republic of Korea; swmoon7@korea.kr
2 Department of Autonomous Vehicle System Engineering, Chungnam National University, Deajeon 34134, Republic of Korea; kppark@cnu.ac.kr
* Correspondence: wsruy@cnu.ac.kr; Tel.: +82-42-821-6623; Fax: +82-42-823-5437

Abstract: Recently, environmental regulations have been strengthened due to climate change. This change comes in a way that limits emissions from ships in the shipbuilding industry. According to these changes, the trend of ship construction is changing installing pollutant emission reduction facilities such as scrubbers or applying alternative fuels such as low sulfur oil and LNG to satisfy rule requirements. However, these changes are focused on large ships. Small ships are limited in size. So, it is hard to install large facilities such as scrubbers and LNG propulsion systems, such as fishing boats that require operating space. In addition, in order to apply the pure electric propulsion method, there is a risk of marine distress during battery discharge. Therefore, the application of the electric–diesel hybrid propulsion method for small ships is being studied as a compromised solution. Since hybrid propulsion uses various energy sources, a method that can estimate effective efficiency is required for efficient operation. Therefore, in this study, a Bond graph is used to model the various energy sources of hybrid propulsion ships in an integrated manner. Furthermore, based on energy system modeling using the Bond graph, the study aims to propose a method for finding the optimal operational scenarios and reduction ratios for the entire voyage, considering the navigation feature of each different maritime region. In particular, the reduction gear is an important component at the junction of the power transmission of the hybrid propulsion ship. It is expected to be useful in the initial design stage as it can change the efficient operation performance with minimum design change.

Keywords: bond graph; hybrid propulsion; ship efficiency; energy system modeling; green ship

1. Introduction

Recently, environmental regulations, such as carbon neutrality, are on the rise due to climate change, and this trend is also applied to shipping industry [1]. As a result, regulations and rules governing ship emissions, ranging from NO_x and SO_x to CO_2, are being strengthened or newly established, primarily under the auspices of the Marine Environment Protection Committee (MEPC) of the International Maritime Organization (IMO). In response to these demands of the times, the construction trend of ships is changing with a propulsion method using eco-friendly fuels. Looking at the trend, it was initially shown that facilities such as scrubbers were added to existing ships to satisfy the rule and regulation on SO_x emission restrictions, etc., to install engines applied with low sulfur oil on new ships [2,3]. In addition, it has been expanded from using BOG (boil-off gas) of LNG carriers as propulsion fuel to pure LNG propulsion ships [4,5]. Recently, research on discovering new fuels such as LPG, DME, and ammonia and applying them to ships has been actively conducted [6–8].

However, this transition to eco-friendly ships is concentrated on medium and large-sized ships. The constraint in the transition of small ships to eco-friendly ships is due to the weak R&D base of small shipyards, but the main reason is the small size of the ship.

Compared to the existing heavy fuel oil (Bunker C oil), the size of the fuel tank is bound to increase because the energy density per volume is small. Therefore, there is a limit to the application of small ships with space restrictions [9]. In particular, the majority of the Republic of South Korea's 72,000 vessels, with most being fishing vessels (around 64,000), are subject to legal size restrictions aimed at preventing the overexploitation of marine resources [10,11]. Among them, coastal fishing boats (about 61,000 vessels) cannot be built in excess of 10 tons (tonnage) [12,13]. Additionally, fishing vessels, unlike cargo ships, require operational spaces for lowering and retrieving nets, and treating and sorting fish at sea. In other words, considering the ship owner's demand to have as wide a workspace as possible for fishing on the main deck, designs such as installing LNG tanks on deck are less field-soluble. In addition to the size restrictions and the necessity for operational space for fishing operations, small fishing vessels have distinctive features that contrast with large ships. In accordance with fishing methods, there are around 40 different types of vessels, each customized to the characteristics of the East, West, and South Sea areas surrounding the Republic of Korean Peninsula. As a result, the owner's needs for ship design also vary to consider target fish, which act as constraints during construction. Additionally, despite occupying a significant portion of Korea's domestic shipbuilding industry due to its strong domestic market, such as the construction of more than 2000 new ships per year, it is challenging to conduct research and development independently because most of them are built in small shipyards. It is challenging to acquire the same fundamental technology as an eco-friendly ship because these small shipyards are designed in a way that slightly modifies their existing designs.

In such circumstances, considering electric propulsion methods for small fishing vessels could be an option for transitioning to green ships [14]. However, considering that electric power consumption to operate facilities such as fishing gear takes one to two days per navigation, there is a risk of distress when the battery runs out, so there is a limit to the application. Therefore, as a national study in the Republic of Korea, research is being conducted to apply a hybrid propulsion method in which electric batteries are applied to existing diesel engines for small fishing vessels that are less than 10 tons [15].

As the hybrid propulsion method is applied, the propulsion system becomes more complex and creates another design problem. Previously, the design was carried out by estimating the main engine according to ship resistance on design speed, but when the hybrid propulsion method is applied, a new question arises as to what mode should be applied for each operation scenario to consider the entire navigation. For example, when moving to the port and the operating site, the target ship speed is different, and at this time, it is necessary to decide which of the hybrid propulsion methods to adopt. Hybrid propulsion in the automotive industry, now widely adopted, also distinguishes itself. Unlike cars, which typically find it difficult to determine fixed routes and target speeds, ships operate with defined routes and target speeds. To solve this problem, it can be used as a basis for determining the decision by measuring the efficiency for each mode of operation scenario. Currently, there are methods such as the Energy Efficiency Existing Ship Index (EEDI), which is the ratio for CO_2 emissions to a ship's transport capacity, to assess efficiency [16]. In addition, the efficiency of the ship can be reviewed as the ratio of effective power to indicated power, but it is difficult to calculate the hybrid power conversion process incorporating various energy sources. Therefore, it is necessary to consider a methodology that can comprehensively model a hybrid propulsion method with various energy sources and effectively analyze energy flow. In a previous study, changes in voltage and power were analyzed using a Bond graph in a ship propulsion system where a generator engine and a battery are connected by a single motor [17,18]. Previous research is actually a study on ships using only electricity as an energy source, which is different from the subject of this study on ships propelled by diesel engines and motors connected to electric batteries. In summary, there are not many studies that apply the Bond graph to ship propulsion systems, and similar prior studies are difficult to apply to hybrid propulsion ships with diesel engines and electric batteries, and there is no approach to efficiency estimation.

However, the Bond graph method used in previous studies can model targets in various areas such as machinery, electricity, and hydraulic pressure, effectively modeling the energy system of ships propelled by motors through electric batteries and diesel engines [19]. In addition, Bond graphs have the advantage of efficiently analyzing complex subjects by connecting each element to power, allowing for the calculation of energy flow, while also enabling the understanding of physical connection structures similar to the block diagram. In other words, since the Bond graph enables computational modeling and physical modeling at the same time, it can be used to analyze objects and, conversely, can be used as a means of optimization.

Therefore, in this study, a Bond graph was used to model the energy system for the propulsion system of a 9.77 tons hybrid propulsion (diesel engine + electric motor) coastal gill net fishing vessel. We try to judge the usefulness with a methodology that can analyze energy flow. Furthermore, the aim of the research extends beyond previous studies to establish a foundation for identifying the optimal operation mode of each hybrid propulsion system, and to select the optimal equipment specifications for this purpose. For reference, information related to the 9.77-ton coastal gill net vessel and relevant fishing regulations can be verified by searching on the National Institute of Fisheries Science website [20]. The general main specifications are shown in Table 1. Additionally, additional information on fishing vessels can be checked through the "Illustrations of Korean Fishing Vessels (2018)" published by the National Institute of Fisheries Science [21]. Previous research has focused on the performance evaluation of ships with a single energy source. In contrast, this study presents optimal operational power and equipment specifications for ships applying various energy sources. This is a key difference. In summary for this study, ships follow set navigation patterns, and the entire voyage can be divided into multiple navigation scenarios. When a hybrid propulsion system is applied, appropriate navigation modes and optimal power must be determined for each scenario. Additionally, suitable equipment capable of achieving these performances must be selected. This is illustrated in Figure 1.

Table 1. Normal principal dimension of 9.77 ton coastal gill net fishing vessel.

Tonnage	Length (m)	Breadth (m)	Depth (m)	Knots
9.77	11.5~17.46	2.52~4.52	0.79~1.66	10~12

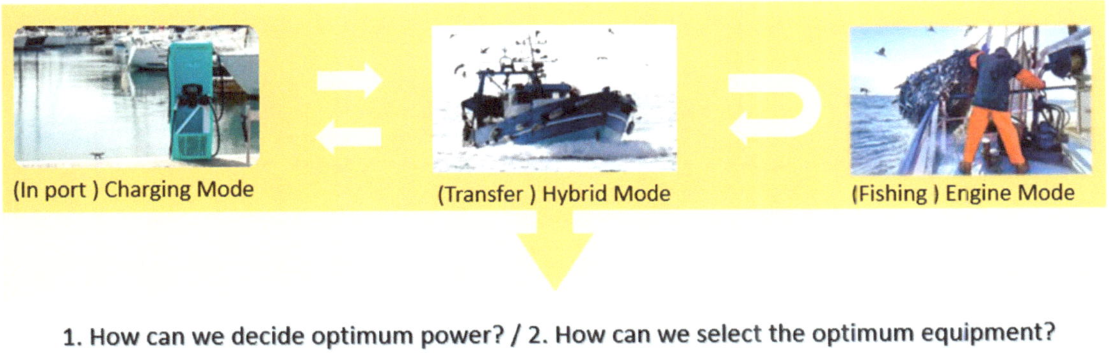

Figure 1. Concept of problem.

2. Fundamental of Bond Graph

2.1. Overview of Bond Graph

The Bond graph is a method proposed by Prof. Paynter in 1959 to structure each element of the analysis by connecting it with power [22]. It has started and developed by defining power exchange and energy flow between each element as a Bond graph methodology.

The core principle of the Bond graph can be said to be energy conversion based on the conservation law. Since energy is only transformed and preserved in shape, the Bond graph can be expressed in an integrated way for various domains such as machinery, electricity, heat, and fluid power. Therefore, it can be said that the Bond graph presents a method of effectively dealing with energy conversion based on a unified notation. In order to effectively understand the Bond graph based on these core principles, it is necessary to understand the variables, elements, causality between variables, and notation that make up the Bond graph [23,24].

2.2. Variable of Bond Graph

Variables in the Bond graph are categorized into power variables and energy variables. Power variables are again distinguished into effort (e) and flow (f), while energy variables are divided as momentum (p) and displacement (q). A power variable is a constituent variable of a bond that connects each element. All elements are connected by a bond, and the bond consists of effort (e) and flow (f). Power is defined as the product of effort (e) and flow (f) [power = effort (e) $\times$ flow (f)]. The reason why the variables are newly defined as effort (e) and flow (f) in the Bond graph is to unify and distinguish the variables that make up power for objects in various domains into effort (e) and flow (f). As some examples, variables for each domain are divided into effort (e) and flow (f) as shown in the following Table 2.

Table 2. Power variable and energy variables of various domains.

Description	Domain	Effort (e)	Flow (f)
Power variable	Mechanics translational	Force F [N]	Velocity v [m/s]
	Mechanics rotational	Angular moment M [N·m]	Angular velocity ω [rad/s]
	Electronics	Voltage u [V]	Current i [A]
	Hydraulic	Pressure P [N/m^2]	Volume flow Q [m^3/s]
Energy variable	Mechanics translational	Momentum P [N·s]	Displacement x [m]
	Mechanics rotational	Angular Momentum $p\omega$ [N·ms]	Angle θ [rad]
	Electronics	Linkage flux λ [V·s]	Charge q [A·s]
	Hydraulic	Pressure Momentum p_p [N/m^2 s]	Volume V [m^3]

The reason for distinguishing energy variables is momentum (p) and displacement (q) is for conversion between effort (e) and flow (f) in the Bond graph. After modeling with the Bond graph, effort (e) and flow (f) values in all connections (bonds) should be computed. At this time, the value regarding the specification of each element of effort (e) and flow (f) becomes known, while the energy storage element becomes unknown as a variable because it changes with time. Therefore, it is necessary to find the unknown term through the known term, and in this process, the conversion between effort (e) and flow (f) is required. Variables used in this transformation process are energy variables namely momentum (p) and displacement (q). Each definition is as below in Table 3, and a detailed transformation process will be described in causality between variables.

Table 3. Generalized momentum (p) and displacement (q).

Generalized Momentum (p)	Generalized Displacement (q)
$p(t) = p(t)_0 + \int_0^t e(t)dt)$	$q(t) = q(t)_0 + \int_0^t f(t)dt$

2.3. Element of Bond Graph

In the Bond graph, elements are divided into sources, storage, transform, converter, dissipator, and distribute. Examining each category is as follows. The source (S_e) is an element that supplies power to the entire system and includes, for example, engine torque and battery current. Storage (I) is an element that stores power and includes inertia, which stores rotational inertia, and capacity for electric power. Transform (TF) is an element that transforms effort to effort or flow to flow in a constant ratio, and a representative example is gear. The changes in Effort (e) and Flow (f) by Transform (TF) are shown in Table 4. Convert (GY) is an element that converts between effort to flow or flow to effort at a certain ratio. If motor is idealized, it can be said that this is the case. The change effort (e) and flow (f) by converter (GY) is defined as Table 4. Dissipator (R) is a resistive element that causes energy loss in the entire system. Typical examples are electrical resistance or friction loss. Distribute is an element that connects previous elements and distributes power. It is classified into 0 Junction and 1 Junction, and all elements connected to 0 Junction have the same effort, and all elements connected to 1 Junction have the same flow.

Table 4. Transform, convert and junction.

Transform		**Convert**	
$e_1 = TF \times e_2,\ f_2 = TF \times f_1$		$e_1 = GY \times f_2,\ e_2 = GY \times f_1$	
0 Junction		1 Junction	
$e_1 = e_2 = \cdots = e_n$ $f_1 = f_2 + \cdots + f_n$ $f_1 - f_2 - \cdots - f = 0$	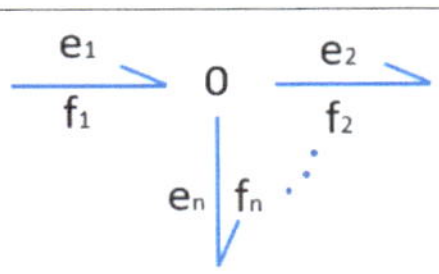	$f_1 = f_2 = \cdots = f_n$ $e_1 = e_2 + \cdots + e_n$ $e_1 - e_2 - \cdots - e_n = 0$	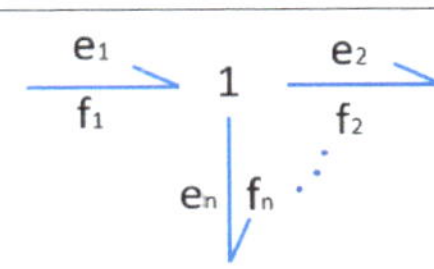

2.4. Causality of Bond Graph

The transformation on the Bond graph between variables follows causality. Causality is classified into integral and differential types, and the transformation between variables is schematically expressed in Figure 2. Note that when calculating effort and flow, the integral type or differential type should be applied according to the model characteristics, but it should be applied uniformly without mixing [25].

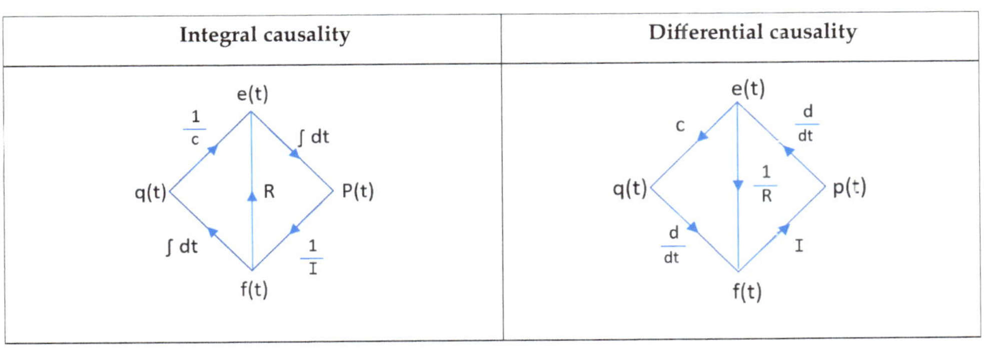

Figure 2. Integral causality vs. differential causality.

3. Analysis of the Research Vessel

3.1. Overview of the Research Vessel

The target vessels for this study are 9.77 tons of coastal gillnet fishing boats, which have the following main characteristics. Among the 72,000 vessels in the Republic of Korea, fishing boats account for 64,000, of which 61,000 coastal fishing boats' maximum size is limited to 10 tons (tonnage) in order to protect fishery resources in accordance with the "Fisheries Act", which is the domestic law [12,13]. In other words, the target vessels in this study are small in size, but they occupy an important position as they constitute most of the ships in the Republic of Korea [10]. In addition, gill netting is a fishing method that is caught by getting entangled in a net, and facilities such as a winch (fishing gear) are required to lift the net [21]. It is generally made of FRP (fiberglass-reinforced plastic), and it takes about 1 day for one voyage with 10 to 12 knots, and the number of crew is about two to three members [21]. Currently, the development of diesel and electric complex propulsion fishing vessels is in progress to convert to eco-friendly fishing boats (Ministry of Oceans and Fisheries of the Republic of Korea (hereinafter referred to as MOF), research project on the development of energy-efficient eco-friendly fishing vessels, '21~25'). This study modeled the energy system using a Bond graph based on the fishing vessel under development.

3.2. Operational Scenarios of the Research Vessel

Currently, the development of diesel and electric complex propulsion fishing vessels is in progress to convert to eco-friendly fishing boats (Ministry of Oceans and Fisheries of the Republic of Korea (hereinafter referred to as MOF), research project on the development of energy-efficient eco-friendly fishing vessels, '21~25'). This study modeled the energy system using a Bond graph. In order to model the energy system of the ship to be studied, the operation scenario was analyzed based on one voyage of the ship to identify the required power. The operation scenarios per voyage can be largely categorized into in-port operation, operation when moving to the operating site, and operation on site. In addition, since fishing vessels typically do not engage in ballasting, it is assumed that it is a reverse process at the time of departure in consideration of the consumption of fuel and fresh water and increased catches at the time of return. In other words, it is assumed that there are three operation scenarios (in-port navigation, navigation when moving to fishing grounds, and navigation during fishing operations) based on the fishing vessel under development. First, in order to model the propulsion system as a Bond graph and verify the modeling results, a load for each operation scenario is required. Therefore, the design load for each operation scenario of a ship under development (MOF, energy-efficient and eco-friendly fishing vessel development research project, '21~25') is used as a value for calculation verification, and the value is shown in Table 5. In addition, the load is shown together because it must be converted into a value obtained by dividing the torque by RPS in order to be applied as a resistance factor to the Bond graph modeling.

Table 5. Load cases according to operation scenario.

Operation Scenario	Torque [N·m] (A)	RPS (B)	Torque/RPS (A/B)
In port	5787.6	38.88	148.87
Transfer from port to fishing ground	10,403.4	53.83	193.27
Working (fishing)	8292.99	44.86	184.88

3.3. Research Vessel's Propulsion Modes

The target vessel for this study is a hybrid propulsion method that is propelled by a diesel engine and electric motor supplied with power from the battery. Basically, there may be engine propulsion, electric motor propulsion, and engine–electric motor complex propulsion methods. However, if the electric motor propulsion is used only as much as the battery is charged on shore, the use time of the electric motor propulsion is limited, so it is necessary to be able to charge the battery in consideration of the power required for each operation scenario when promoting the diesel engine. Therefore, there are four mode cases in the ship to be studied: hybrid mode, engine mode, motor mode, and charging mode which use both the engine and the motor.

3.4. Analysis of Research Vessel's Specifications

The ship to be studied was assumed to be based on the specifications of the diesel and electric complex propulsion fishing vessel under development (MOF, energy-efficient and eco-friendly fishing vessel development research project, '21~25'), and unconfirmed information was assumed by referring to the specifications of existing similar ships In addition, it was assumed that the power required for the fishing gear during operation was supplied by electricity produced by a generator engine and was independent of the battery for propulsion. The variables constituting the Bond graph modeling of the propulsion system of the ship to be studied are summarized in Table 6.

Table 6. Equipment specification for Bond graph modeling.

Description	Mechanical	Electrical
Source (*S*)	S_e(Engine Torque): 2461.6 [N·m]	S_e(Battery Voltage): 671.6 [V]
Storage (*I*)	J_F(Inertia of engine flywheel): 1.612 [kg·m^2] J_R(Inertia of motor rotor): 0.84 [kg·m^2]	*L*(Inductance on PTI *†): 0.000659 [*H*] *† Power Take In(PTI) of motor
Resister (*R*)	*T*(Load according to operation scenario): Refer to "Torque/PRS" on Table 5 α(Load of battery charing) |: 6.516 [N·m/(rad/s)] β(Resistance of stern tube) |: T | x | 0.005 *† [N·m/(rad/s)] *† Friction coefficient γ(Resistance of motor bearing) ||: 0.0038 | [N·m/(rad/s)]	R_E | (Resistance of motor coil on PTI) : 0.162 | [Ω] R_C | (Resistance of converter coil) 0.145 | [Ω]
Transformer (*TF*)	*TF*(Gear ratio): 1:3.5 ($N_2 > N_1$, $N_1 = m_1$, $N_2 = m_2$)	
Gyrator (*GY*)	*KC*(Convertor): 3.86 *† *† | 671.6 battery voltage(effort) change to 174 convertor ampere(flow) ⇒ 671.6 / 174 = 3.86	*KM*(Motor): 3.23 *† *† | 172.4 motor ampere(flow) changed to 557 motor torque(effort) ⇒ 557 / 172.4 = 3.23

4. Modeling Using Bond Graph

The Bond graph modeling was conducted in the flow as shown in Figure 3, and the detailed process of Bond graph modeling for the target ship is described sequentially as follows.

The propulsion system of the research vessel is described as a 3D image, the Scheme and Bond graph modeling results are described sequentially as follows. Figure 4 is the expression of the propulsion system of the research vessel as a 3D image, and it is revealed that it is only an illustration for effective explanation and not actual information used in this study.

Since it is difficult to model directly with a Bond graph in the state of Figure 4, each component from the engine and battery to propeller is converted into an element for Bond graph modeling, and the scheme is created in Figure 5 [26].

1. Analyzing the characteristics of the object
2. Decompose each piece of equipment into elements of the Bond graph and create a schematic
3. Based on the schematic, connect all the elements to complete the Bond graph
4. Set the storage element as a state variable and assume it as a given input
5. Calculate the effort and flow of each bond (Calculation step 1)
6. Organize equations for the state variables (Calculation step 2)
7. Solve the coupled nonhomogeneous differential equations for the state variable (Calculation step 3)
8. Substitute the calculated state variables into effort and flow and then compute the power for all bonds

Figure 3. Bond graph modeling flow chart.

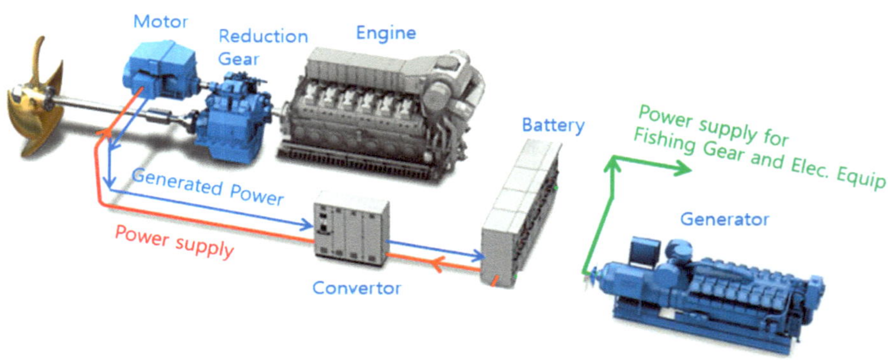

Figure 4. Three-dimensional modeling of energy system.

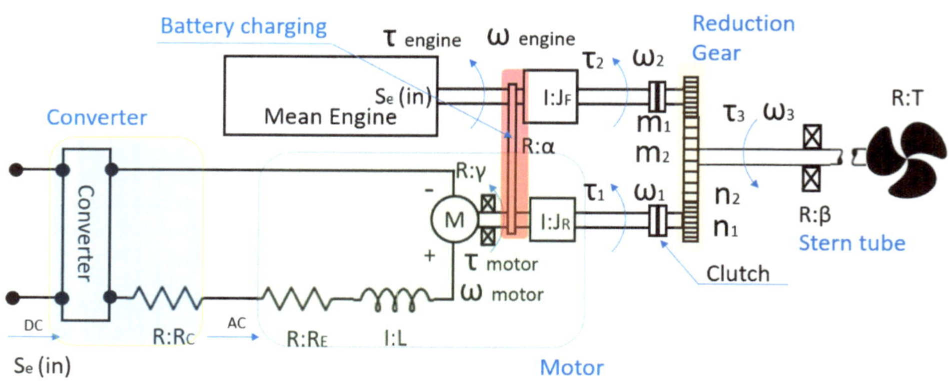

Figure 5. Scheme of energy system for Bond graph modeling.

Based on the above scheme, if the propulsion system of the target vessel for this study is idealized with a Bond graph, the following results are obtained [27].

5. Energy Flow Calculation

As depicted in Figure 6, the propulsion system of the research vessel was modeled using a Bond graph. Next, the effort and flow of each bond in the model represented in Figure 6 will be calculated to understand the energy variations. In this process, it is crucial to set the energy variable of the storage element as a state variable, assuming the unknowns as known variables. Subsequently, organizing equations for the state variables leads to a system of ordinary differential equations. By assigning initial values and solving, the effort and flow for all bonds can be obtained. Calculations for the most complex hybrid mode of the research vessel are presented below. The subscript numbers in the following calculation represent the respective bond numbers. Sequential calculations from bond 1 to 19 are as follows.

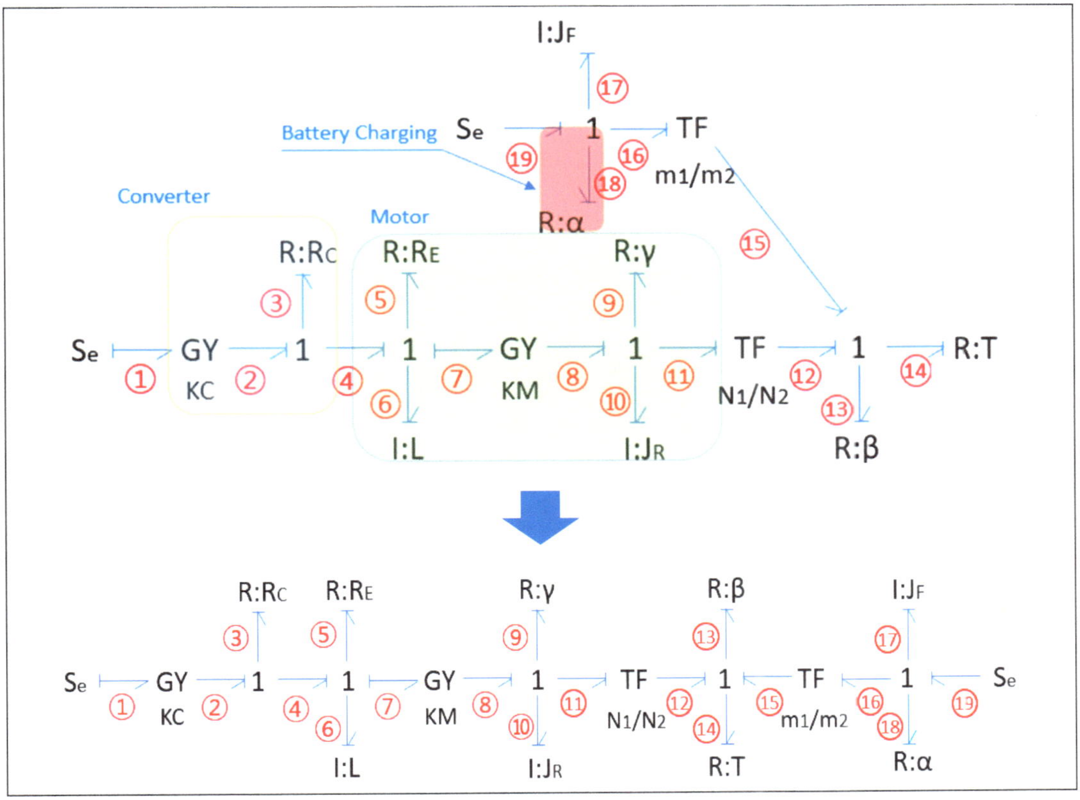

Figure 6. Bond graph modeling of energy system.

As shown in Figure 6, the propulsion system of the ship to be studied was modeled as a Bond graph. Next, the effort and flow of each bond according to the modeling of Figure 6 are calculated so that energy changes can be identified. The most important thing in this process is to assume that the unknown term is known by setting the energy variable of the storage element as a state variable. After that, if the equation for the state variable is summarized, it results in a simultaneous ordinary differential equation. At this time, if the initial value is substituted to obtain the solution, the effort and flow for all bonds can be obtained. The calculation for the most complex hybrid mode among the operation modes of the research vessel is as follows. In the calculation process below, the subscript number represent the respective bond numbers. Sequential calculations from bond 1 to 19 are as follows.

<Calculation for Bond graph modeling—Step1.>

$$e_1 = Se_{(Battery)}$$
$$f_1 : e_2 = KC \times f_1,\ f_1 = \frac{e_2}{KC} = \frac{1}{KC}\left(R_C \times \frac{p_6}{L} + R_E \times \frac{p_6}{L} + \dot{p_6} + KM \times \frac{p_{10}}{J_R}\right) \tag{1}$$

$$e_2 = e_3 + e_4 = R_C \times \frac{p_6}{L} + R_E \times \frac{p_6}{L} + \dot{p_6} + KM \times \frac{p_{10}}{J_R}$$
$$f_2 = \frac{p_6}{L}\ (\therefore f_2 = f_3 = f_4 = f_5 = f_6 = f_7),\ f_2 = \frac{e_1}{KC} = \frac{Se_{(Battery)}}{KC}\ (\therefore e_1 = KC \times f_2) \tag{2}$$

$$e_3 = R \times f_3 = R_C \times \frac{p_6}{L}$$
$$f_3 = \frac{p_6}{L} = \frac{Se_{(Battery)}}{KC} \tag{3}$$

$$e_4 = e_5 + e_6 + e_7 = R_C \times \frac{p_6}{L} + \dot{p_6} + KM \times \frac{p_{10}}{J_R}$$
$$f_4 = \frac{p_6}{L} = \frac{Se_{(Battery)}}{KC} \tag{4}$$

$$e_5 = R \times f_5 = R_E \times \frac{p_6}{L}$$
$$f_5 = \frac{p_6}{L} = \frac{Se_{(Battery)}}{KC} \tag{5}$$

$$e_6 = \dot{p_6}$$
$$f_6 = \frac{p_6}{I} = \frac{p_6}{L} = \frac{Se_{(Battery)}}{KC} \tag{6}$$

$$e_7 = KM \times f_8 = KM \times \frac{p_{10}}{J_R}$$
$$f_7 = \frac{p_6}{L} = \frac{Se_{(Battery)}}{KC} \tag{7}$$

$$e_8 = KM \times f_7 = KM \times \frac{p_6}{L}$$
$$f_8 = \frac{p_{10}}{J_R}\ (\therefore f_8 = f_9 = f_{10} = f_{11}) \tag{8}$$

$$e_9 = R \times f_9 = \gamma \times \frac{p_{10}}{J_R}$$
$$f_9 = \frac{p_{10}}{J_R} \tag{9}$$

$$e_{10} = \dot{p_{10}}$$
$$f_{10} = \frac{p_{10}}{I} = \frac{p_{10}}{J_R} \tag{10}$$

$$e_{11} : e_8 = e_9 + e_{10} + e_{11},\ e_{11} = e_8 - e_9 - e_{10} = KM \times \frac{p_6}{L} - \gamma \times \frac{p_{10}}{J_R} - \dot{p_{10}}$$
$$f_{11} = \frac{p_{10}}{J_R} \tag{11}$$

$$e_{12} = \frac{N_2}{N_1} \times e_{11} = \frac{N_2}{N_1}\left(KM \times \frac{p_6}{L} - \gamma \times \frac{p_{10}}{J_R} - \dot{p_{10}}\right)$$
$$f_{12} = \frac{N_1}{N_2} \times f_{11} = \frac{N_1}{N_2}\frac{p_{10}}{J_R}\ (\therefore f_{12} = f_{13} = f_{14} = f_{15}) \tag{12}$$

※ *Reduction Gear ratio*

$N_2 > N_1\ (N_1 = m_1, N_2 = m_2)$

$N_2 \times e_{11} = N_1 \times e_{12} \rightarrow e_{12} = \frac{N_2}{N_1} \times e_{11}$ / $N_1 \times f_{11} = N_2 \times f_{12} \rightarrow f_{12} = \frac{N_1}{N_2} \times f_{11}$

$$e_{13} = R \times f_{13} = \beta \frac{N_1}{N_2}\frac{p_{10}}{J_R}$$
$$f_{13} = \frac{N_1}{N_2}\frac{p_{10}}{J_R} = \frac{m_1}{m_2}\frac{p_{17}}{J_F} \tag{13}$$

$$e_{14} = R \times f_{14} = T\frac{N_1}{N_2}\frac{p_{10}}{J_R}$$
$$f_{14} = \frac{N_1}{N_2}\frac{p_{10}}{J_R} = \frac{m_1}{m_2}\frac{p_{17}}{J_F}\left(\rightarrow \frac{p_{10}}{J_R} = \frac{p_{17}}{J_F}\right) \tag{14}$$

$$e_{19} = Se_{(Engine)}$$
$$f_{19} = \frac{p_{17}}{J_F}\ (\therefore f_{16} = f_{17} = f_{18} = f_{19}) \tag{15}$$

$$\begin{aligned} e_{18} &= R \times f_{18} = \alpha \frac{N_1}{N_2} \frac{p_{17}}{J_F} \\ f_{18} &= \frac{p_{17}}{J_F} \text{ (used to charging mode only)} \end{aligned} \tag{16}$$

$$\begin{aligned} e_{17} &= \dot{p_{17}} \\ f_{17} &= \frac{p_{17}}{I} = \frac{p_{17}}{J_F} \end{aligned} \tag{17}$$

$$\begin{aligned} e_{16} : e_{19} &= e_{17} + e_{16}, e_{16} = e_{19} - e_{17} = Se_{(Engine)} - \dot{p_{17}} \\ f_{16} &= \frac{p_{10}}{J_R} \end{aligned} \tag{18}$$

$$\begin{aligned} e_{15} &= \frac{m_2}{m_1} \times e_{16} = \frac{m_2}{m_1}\left(Se_{(Engine)} - \dot{p_{17}}\right) \\ f_{15} &= \frac{m_1}{m_2} f_{16} = \frac{m_1}{m_2}\frac{p_{17}}{J_F} \\ e_{12} + e_{15} &= e_{13} + e_{14} \\ \frac{N_2}{N_1}\left(KM \times \frac{p_6}{L} - \gamma \times \frac{p_{10}}{J_R} - \dot{p_{10}}\right) &+ \frac{m_2}{m_1}\left(Se_{(Engine)} - \dot{p_{17}}\right) = \beta\frac{N_1}{N_2}\frac{p_{10}}{J_R} + T\frac{N_1}{N_2}\frac{p_{10}}{J_R} \end{aligned} \tag{19}$$

Based on the effort and flow calculations of each bond [Equations (1)~(19)], the equations for the state variables are summarized [Equation (22)], ultimately leading to the derivation of a matrix form system of coupled differential-algebraic equations [Equation (23)]. The relationship equation between the state variables p_{10} and p_{17} is derived using Equation (19). After that, rearranging Equation (20) with respect to p_{10} and similarly rearranging with respect to p_{17} results in the following.

<Calculation for Bond graph modeling—Step2.>

$$\begin{aligned} &e_{12} + e_{15} = e_{13} + e_{14} \\ &\frac{N_2}{N_1}\left(KM \times \frac{p_6}{L} - \gamma \times \frac{p_{10}}{J_R} - \dot{p_{10}}\right) + \frac{N_2}{N_1}\left(Se_{(Engine)} - \dot{p_{17}}\right) \\ &= \beta \frac{N_1}{N_2}\frac{p_{10}}{J_R} + T\frac{N_1}{N_2}\frac{p_{10}}{J_R} \quad (\therefore N_1 = m_1, N_2 = m_2) \end{aligned} \tag{20}$$

$$\begin{aligned} \dot{p_{10}} &= \left(\frac{1}{1+J_F/J_R}\right)\frac{KM}{KC}Se_{(Battery)} \\ &-\left(\frac{1}{1+J_F/J_R}\right)\frac{1}{J_R}\left[\gamma + (\beta+T)\left(\frac{N_1}{N_2}\right)^2\right]p_{10} + \left(\frac{1}{1+J_F/J_R}\right)Se_{(Engine)} \end{aligned} \tag{21}$$

$$\begin{aligned} \dot{p_{17}} &= \left(\frac{1}{J_R/J_F+1}\right)\frac{KM}{KC}Se_{(Battery)} \\ &-\left(\frac{1}{J_R/J_F+1}\right)\frac{1}{J_R}\left[\gamma + (\beta+T)\left(\frac{N_1}{N_2}\right)^2\right]p_{10} + \left(\frac{1}{J_R/J_F+1}\right)Se_{(Engine)} \end{aligned} \tag{22}$$

Equations (21) and (22) can be expressed as a coupled ordinary differential equation in matrix form, as follows:

<Calculation for Bond graph modeling—Step3.>

$$\begin{bmatrix} \dot{p_{10}} \\ \dot{p_{17}} \end{bmatrix} = \begin{bmatrix} \frac{-1}{(1+J_F/J_R)J_R}\left[\gamma + (\beta+T)\left(\frac{N_1}{N_2}\right)^2\right] & 0 \\ \frac{-1}{(J_R/J_F+1)J_R}\left[\gamma + (\beta+T)\left(\frac{N_1}{N_2}\right)^2\right] & 0 \end{bmatrix} \begin{bmatrix} p_{10} \\ p_{17} \end{bmatrix} + \begin{bmatrix} \frac{1}{(1+J_F/J_R)}\frac{KM}{KC} & \frac{1}{(1+J_F/J_R)} \\ \frac{1}{(J_R/J_F+1)}\frac{KM}{KC} & \frac{1}{(J_R/J_F+1)} \end{bmatrix} \begin{bmatrix} Se_{(Battery)} \\ Se_{(Engine)} \end{bmatrix} \tag{23}$$

Since the initial condition of the engine and motor is in a stationary state, if the coupled ordinary differential equations are solved by setting the initial value to 0 for time, the solution becomes the value of a state variable. In this way, since the effort and flow values of all bonds were obtained, the power of each bond can be computed. In summary, it means that the energy change for the propulsion system of the research vessel can be determined. The modeling results of hybrid mode, engine mode, motor mode, and charging mode are calculated for each mode in the same way as the previous calculation. However, since the load for each operation scenario is different in each mode, there are four modes (hybrid mode, engine mode, motor mode, and charging mode), and three operation scenarios (in-harbor navigation, transit to fishing ground navigation, and fishing operation navigation) exist. So, there are a total of 12 cases. The results of modeling in the Bond graph for hybrid mode, engine mode, motor mode, and charging mode are shown in Figures 7–10, respectively.

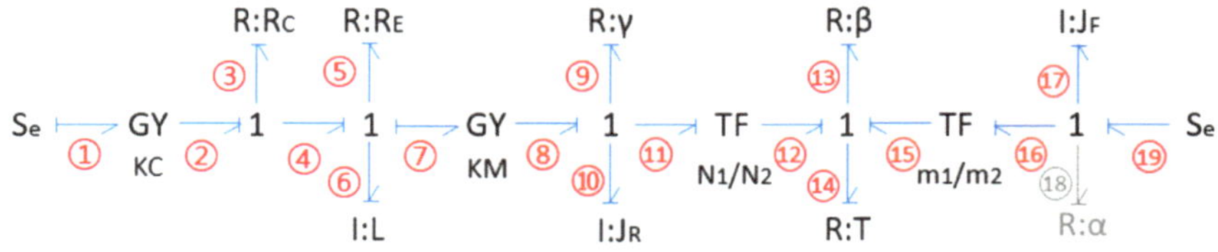

Figure 7. Bond graph modeling of hybrid mode.

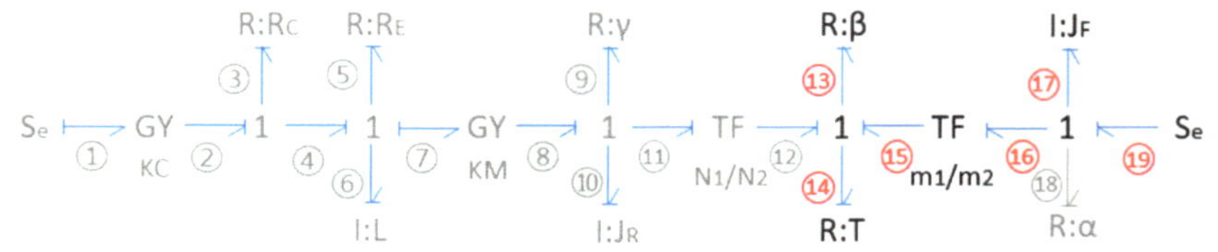

Figure 8. Bond graph modeling of engine mode.

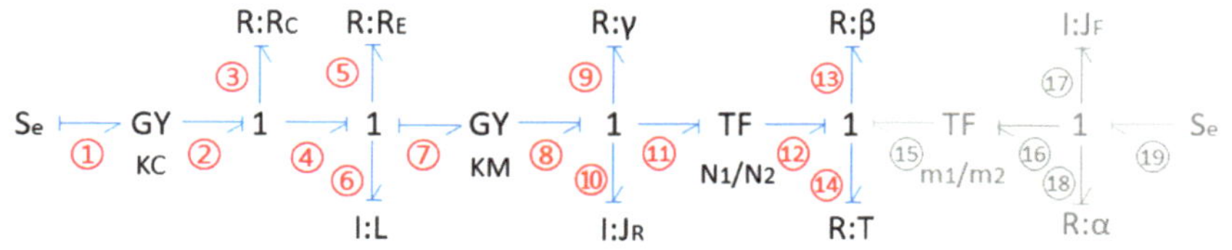

Figure 9. Bond graph modeling of motor mode.

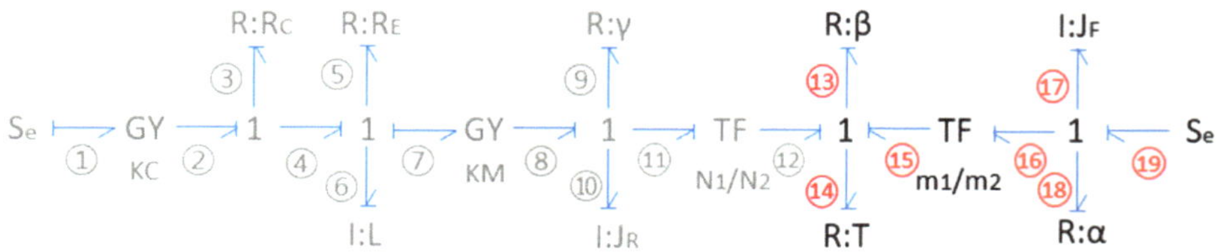

Figure 10. Bond graph modeling of charging mode.

6. Validation of Calculations

Next, the previous Bond graph modeling calculation result is verified. The verification method is determined by using the commercial software "20-sim 5.0" to model the same and comparing the value of the state variable [28]. However, it is essential to emphasize the difference between this study and commercial software before calculation verification. The commercial program only provides an environment for modeling by combining the concept of the Section 2. Bond graph and the elements introduced in the main contents but does not create the scheme of Figure 5 or provide modeling such as in Figure 6. In addition, the commercial program can be used universally, but it does not provide detailed analysis results for the propulsion system of the hybrid propulsion ship. Therefore, in this study, the performance estimation and energy flow can be calculated using the Bond graph modeling results for use in the initial design. The meaning of this study compared to the commercial program will be explained again in the results analysis. The following, Figure 11, shows the results of modeling hybrid modes using "20-sim 5.0" among the four modes.

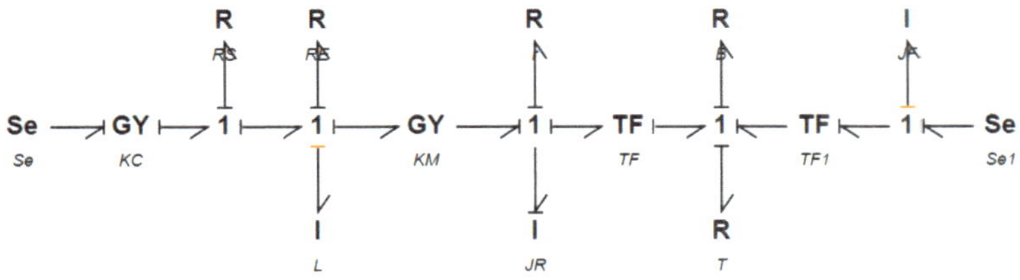

Figure 11. Bond graph modeling of hybrid mode by 20-sim 5.0.

In the case of using hybrid mode when moving to the fishing ground, the results of the study using the state variable value calculated through "20-sim 5.0" and the simultaneous ordinary differential equation derived through the energy flow calculation introduced in Section 5 (using the SciPy package(version 1.11.4) implemented by the Runge–Kutta 4th order method [29,30]) are compared as Figure 12.

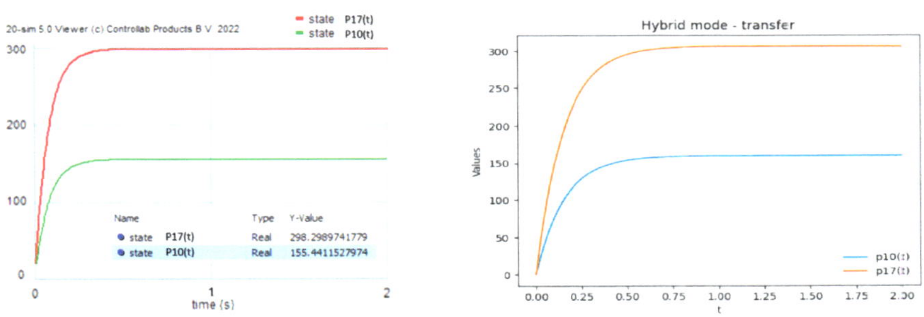

Figure 12. Result of 20-sim 5.0 (**left**) and study for Bond graph modeling (**right**).

The state variable corresponds to the variable bond 17(p_{17}) and the variable bond 10(p_{10}), and the physical meaning represents the energy stored in the rotational inertia of the engine flywheel and rotational inertia of the motor rotor, respectively. As a result of "20-sim 5.0", bond 17(p_{17}) is ⌈298.3⌋ and bond 10(p_{10}) is ⌈155.4⌋, and the results of this study calculation using Bond graph modeling were found to be bond 17(p_{17}) as ⌈306.4⌋ and bond 10(p_{10}) as ⌈160.3⌋, respectively. For hybrid mode, engine mode, motor mode, and charging mode, the load conditions of each operation scenario (in-port operation, transit to fishing grounds, and fishing operation) are substituted and the results of this study are compared as follows. Table 7 shows that the average error for all 12 cases is around 2.8%. The reason for these differences is difficult to precisely ascertain due to the unknown solver of the commercial program. However, it is estimated to be attributed to differences in the solver used for solving ordinary differential equations.

Table 7. Comparison of "20-sim 5.0" and study result.

Mode	Operation	Energy Variable No.	20-sim	Study Result	Error (%)
Hybrid	in port	p_{10}	201.8	208.2	3.2%
	transfer	p_{10}	155.4	160.3	3.2%
	working	p_{10}	162.5	167.6	3.1%
	in port	p_{17}	387.3	398.2	2.8%
	transfer	p_{17}	298.3	306.4	2.7%
	working	p_{17}	311.8	320.3	2.7%

Table 7. *Cont.*

Mode	Operation	Energy Variable No.	20-sim	Study Result	Error (%)
Engine	in port	p_{17}	315.4	324.9	3.0%
	transfer	p_{17}	242.9	250.3	3.0%
	working	p_{17}	253.9	261.6	3.0%
Motor	in port	p_{10}	37.5	38.6	2.9%
	transfer	p_{10}	28.9	29.8	3.1%
	working	p_{10}	30.2	31.1	3.0%
Charging	in port	p_{17}	207.8	211.9	2.0%
	transfer	p_{17}	173.6	177.4	2.2%
	working	p_{17}	179.2	183.0	2.1%
Average					2.8%

7. Formalization of the Optimization Problem

In this study, the Bond graph modeling of a hybrid propulsion system is used to propose that it can be used in the initial design. To this end, it is necessary to set the objective function and related design variables with the aim of improving the performance of the ship. First of all, the objective function aimed to evaluate the overall voyages by integrating three distinct operational scenarios as the voyages scenarios of the research vessel are divided into "in-port navigation", "transit between operation sites", and "operation at site". To achieve this, the value that can determine the performance in each operational scenario and the proportion (weighing factor) of the corresponding operational scenario in the total voyages were multiplied. In addition, in order to evaluate the overall voyages, the objective function was set to be maximized by summing up the values of each operational scenario. At this time, the value that can determine the performance in each operational scenario was set as the ratio of the required energy for propulsion to the energy input from the bond on the engine (or motor, or a combination of both) side in each operational scenario. This is expressed as an equation in the object function in Table 8.

Table 8. Formulation of optimization.

<Object Function>

$$\begin{aligned} F(x) &= max\left[a * f_{port}(x_1\ ,\ x_4\ ) + b * f_{transfer}(x_2\ ,\ x_4\ ) + c * f_{working}(x_3\ ,\ x_4)\right. \\ &= max\left[a * f_{port}(T_{port}\ ,\ TF) + b * f_{transfer}\left(T_{transfer}\ ,\ TF\right) + c * f_{working}\left(T_{working}\ ,\ TF\right)\right. \end{aligned}$$

x_1 $\left(T_{port}\right)$: Power Consumption during In-port Navigation
x_2 $\left(T_{transfer}\right)$: Power Consumption during Transit between Operation Sites
x_3 $\left(T_{working}\right)$: Power Consumption during Operation at Site
x_4 (TF): Reduction ratio
a: Weighted by Intra-port Navigation Time (In-port Navigation Time/Total Voyage Time)
b: Weighted by Transit to Operation Site Time (Transit to Operation Site Time/Total Voyage Time)
c: Weighted by Operation at Site Time (Operation at Site Time/Total Voyage Time)
f_{port}: Ratio of energy demanded during In-port Navigation to the energy input in the bond of the engine
$f_{transfer}$: Ratio of energy demanded during Transit to Operation Site to the energy input in the bond of the "engine + motor"
$f_{working}$: Ratio of energy demanded during Operation at Site to the energy input in the bond of the engine

<Constrains>

1. $(x_1 + Load_{\text{charging}}) <$ Engine Power$/rps^2$ .... physical conditions
2. $x_2 <$ (Engine Power + Motor Power)$/rps^2$.......................... physical conditions
3. $x_3 <$ Engine Power$/rps^2$... physical conditions
4. $1 < x_4 < 5.91$... system characteristics
5. Engine Power $< [a * x_1 + b * x_2 + c * x_3]$ .... system characteristics

The optimization variables for the above objective function were set to four types: "Reduction ratio", "Power consumption during in-port navigation", "Power consumption during transit to operation site", and "Power consumption during operation at site". When the engine and motor variables are set as optimization variables, there are many input variables for engine and motor specifications in the process of calculating the Bond graph, so if the variables of a specific part are optimized, it means that the engine and motor must be manufactured separately rather than off-the-shelf products. This is due to the lack of practicality, so in this study, a reduction gear was set as an optimization target. This is because the reduction gear is not only applied independently when calculating the Bond graph but also can effectively change the performance because it is at the contact point of all power transmission.

In addition, the design values of power consumption for each operational scenario are fixedly applied to the research target vessels under development (conducted by the MOF, energy-efficient and environment-friendly fishing vessel development research project, '21~25'). However, small fishing vessels have various operating patterns depending on the type of fish the vessels target and the maritime area. Therefore, in order to apply it practically to various ships, the optimized power consumption should be obtained by considering the operating time of each operational scenario. Therefore, the power consumption for each of the three operations—"power consumption during in-port navigation", "power consumption during transit to operation site", and "power consumption during operation at site"—were set as design variables for the optimization problem.

In summary, the core of this research formulation is to find the optimal power consumption for each operational scenario that can achieve maximum performance and the optimized reduction ratio when the operational scenario (in-port, transit to operation site, and operation at site) is determined according to type of fish the vessel target and maritime area.

Next, the constraints were set so that they did not violate physical conditions and system characteristics. First of all, the physical conditions are as follows. 1. During in-port navigation, the engine output is expected to exceed the sum of power consumption during in-port navigation and the charging load since the battery is charged during this state. 2. Since hybrid mode is used when moving to the operation site, the sum of engine and motor output will be greater than the power consumption during transit to the operation site. 3. Since only the engine is used during operation on the fishing ground, the engine output will be greater than the power consumption during operation in the working site.

Constraints according to the characteristics of the system are as follows. 4. In vessels, since the torque is increased and rotation speed is reduced through the reduction gear (to improve thrust and reduce cavitation), the reduction ratio will be greater than 1. Simultaneously, the reduction ratio should not exceed the maximum value for the reduction gear used in small vessels (up to a maximum of 5.91 for the tonnage of the specific vessel). 5. Since the entire voyage is evaluated, the representative value of the voyage is set to the sum of the product of the power consumption for each operational scenario and the ratio of usage time for each operational scenario (operational time for each scenario/total voyage time). This value is set to be greater than the engine output (if it is less than the engine output, it means the vessel can operate solely with the engine, rendering the hybrid mode unnecessary, indicating a mismatch with the system characteristics, and hence, setting a constraint). The above constraints are summarized as shown in Table 8 above.

8. Derivation of the Optimal Solution

The differential evolution method was employed for the optimization problem [31]. In order to find the optimal solution, it is necessary to be able to consider both physical constraints and constraints during Bond graph calculations. Moreover, in order to derive the global optimum, the differential evolution method was used because it had to be possible to find the optimal value by changing the initial value. Next is the evaluation criterion for convergence. The analysis result was judged to converge when the two criteria were satisfied. The first was judged to show the same result while increasing the population size in consideration of repetitive reproducibility. Second, the convergence was evaluated by monitoring how the optimal value responded to changes in inputs. As the purpose of this study is to use it for the initial design stage in consideration of the characteristics of different types of fish that vessels target and maritime areas, the input value is the usage time for each regional operational scenario. Therefore, the usage time for each regional operational scenario was investigated based on the interview with fishermen, and the results are shown in Table 9. In other words, it was applied as a criterion for determining whether the constraints to be activated change when the time value of the regional operational scenario is changed as shown in Table 9. This is because the objective function or constraint is inappropriate if only the same constraint operates according to the change in the input value. To summarize the contents so far, the problem can be expressed as shown in Figure 13.

Table 9. Operation scenario (A: operation time in port, B: operation time during transfer, C: operation time at working on site).

Description	Total Time (T) [h]	In Port		Transfer		Working	
		Time (A) [h]	Non-Dimensional (A/T)	Time (B) [h]	Non-Dimensional (B/T)	Time (C) [h]	Non-Dimensional (C/T)
West Sea	9.00	0.25	0.03	3.25	0.36	5.50	0.61
South Sea	13.75	1.25	0.09	3.50	0.25	9.00	0.65
East Sea	16.4	1.5	0.1	3.9	0.24	11	0.67

Accordingly, as a result of optimization, the "Reduction ratio", "Power consumption during in-port navigation", "Power consumption during transit to operation site", and "Power consumption during operation at site" for each operation scenario change in each region are shown in Table 10.

Table 10. Result of optimization for load and reduction ratio.

Description	Optimum Load [N·m/(rad/s)]			Reduction Ratio
	In Port	Transfer from Port to Fishing Ground	Working (Fishing)	
West Sea	194.6	196.4	204.0	3.72
South Sea	209.9	196.4	224.2	4.09
East Sea	206.6	196.4	219.7	4.01

Table 10 corresponds to the effort value of bond 14 on the Bond graph and is converted into a power unit by multiplying the flow value of bond 14 as shown in Table 11. Bond 14 flow can be calculated according to Equation (14).

Table 11. Result of optimization for load and reduction ratio(convert to KW).

Description	Optimum Load [kW]			Reduction Ratio
	In Port	Transfer from Port to Fishing Ground	Working (Fishing)	
West Sea	125.9	106.9	167.4	3.72
South Sea	123.6	117.4	184.0	4.09
East Sea	124.1	115.1	180.3	4.01

9. Analysis of Results

In order to analyze the trend for the optimization result, it is necessary to visualize the result. However, since there are four optimization design variables, they cannot be expressed at once. Therefore, it was visualized by dividing the contour graph into six sections so that the objective function value for the two optimization design variables (x,y) became z, and the results are shown in Figures 13–15. The results include the values of the objective function corresponding to changes in two optimization design variables, along with the optimal values of each design variable and the regions of the constraints. In each figure, "TF" is the reduction ratio, and "T" denotes the power [N·m/(rad/s)] consumed as the "T value" in Figure 6. "T_{c_pt}" and "T_{e_wk}" are the same " T value" in Figure 6 when calculating the Bond graph, but " T ", "T_{c_pt}", and "T_{e_wk}" mean the power consumed during transit, power consumption during in-port navigation, and power consumption during operation in site, respectively.

Let us first look at the general trend and analyze the differences according to the regional operational scenarios. The general trend is as follows.

The graphs ① to ③ of Figures 13–15 show the relationship between the reduction ratio (TF) and power consumption during in-port navigation (T_{c_pt}), transit power consumption (T), and operation in site power consumption (T_{e_wk}). It can be seen that the objective function tends to be maximized when the reduction ratio (TF) is maximized while minimizing each power consumption. This trend is consistent with the general trend of increasing the propulsion efficiency by increasing the reduction ratio while minimizing the power consumption for operation. In addition, when looking through the area of the constraints displayed in white, the transit power consumption (T) is not sensitive to the constraints, but the power consumption during in-port navigation (T_{c_pt}) and operation in site (T_{e_wk}) are affected by the constraints. In particular, it can be seen that the optimal point for power consumption during in-port navigation (T_{c_pt}) is located near the constraint region and is most affected by constraints. It can be said that the efficiency of operation is affected in the entire voyage according to the hybrid propulsion, which also serves the energy charged in charging mode within the port for movement to the fishing ground, concurrently employing the motor.

The graphs ④ to ⑥ of Figures 13–15 are the results of analyzing the effects on power consumption during intra-port navigation (T_{c_pt}), transit power consumption (T), and operation power consumption (T_{e_wk}). When evaluated by the slope of the contour lines, it can be seen that the slope is sharp in the relationship between operation power consumption (T_{e_wk}) and transit power consumption (T), and the impact of thetransit power consumption (T) is large. Second, the power consumption during in-port navigation (T_{c_pt}) and transit power consumption (T) have similar effects when looking at the slope of the graph. Third, looking at the relationship between power consumption during in-port navigation (T_{c_pt}) and operation in site power consumption (T_{e_wk}), it can be seen that the slope of the contour lines is gentle, and the influence of power consumption during in-port navigation (T_{c_pt}) has a significant impact.

In addition, the difference according to the operational scenario of each maritime region is analyzed as follows. The contour lines for each sea area have similar results. However, in the case of the West Sea, the operating distance and operating time are short depending on the sea boundary with China. Therefore, the preparation time for departure from the port is short, and looking at the fifth graph in Figure 13 (West Sea), it can be seen that the slope of the contour lines of the power consumption during in-port navigation (T_{c_pt}) and transit power consumption (T) is gentle compared to other maritime regions. In other words, in this hybrid propulsion method, how much battery is charged in the port is an important part of efficient operation for the entire voyage.

In addition, it was observed that the optimal reduction ratio was derived differently depending on the operation scenario of each maritime region. It is judged that this can contribute to the performance improvement of the ship by reviewing the operation time during the initial design stage and determining the optimal reduction ratio.

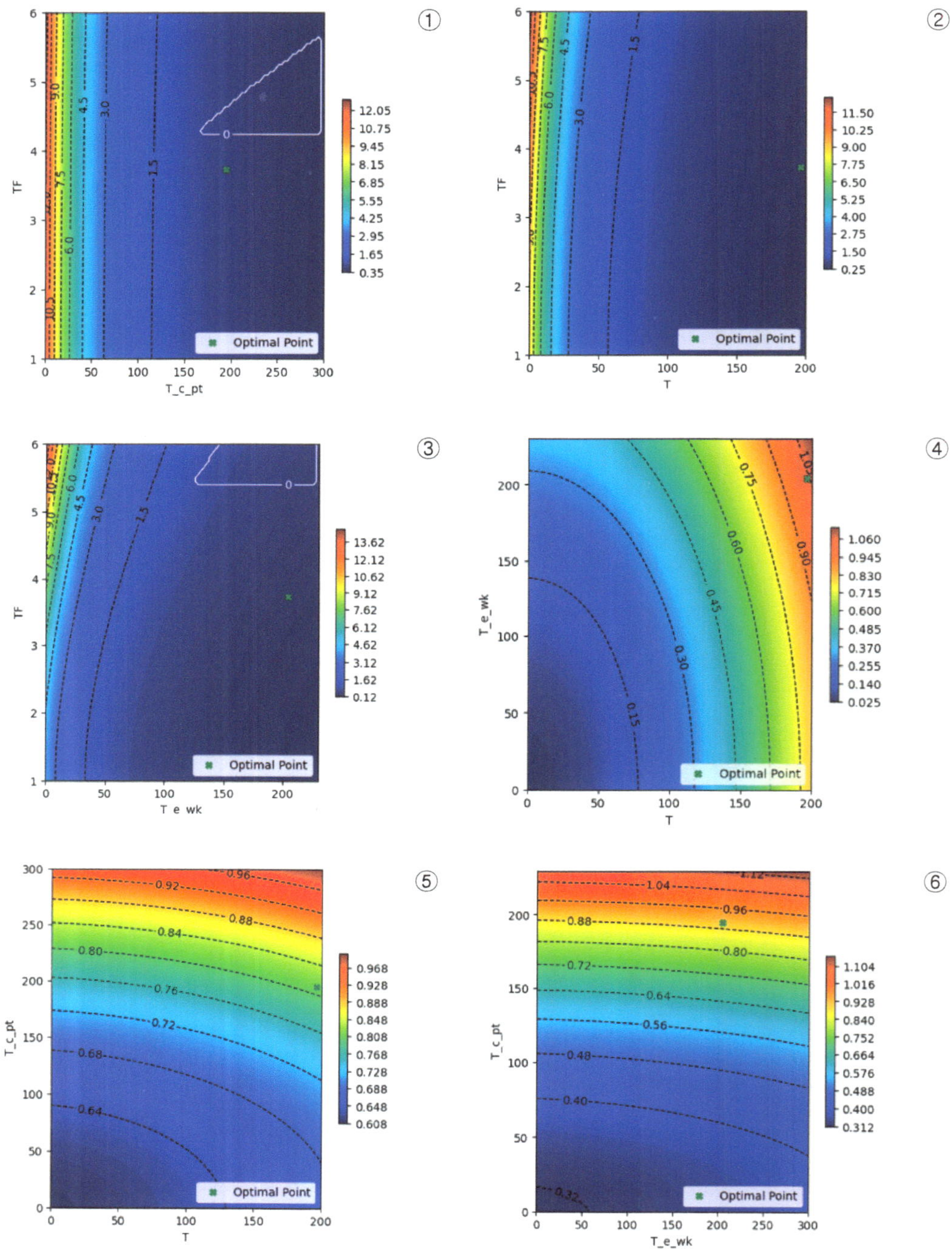

Figure 13. Contour of optimization for West Sea.

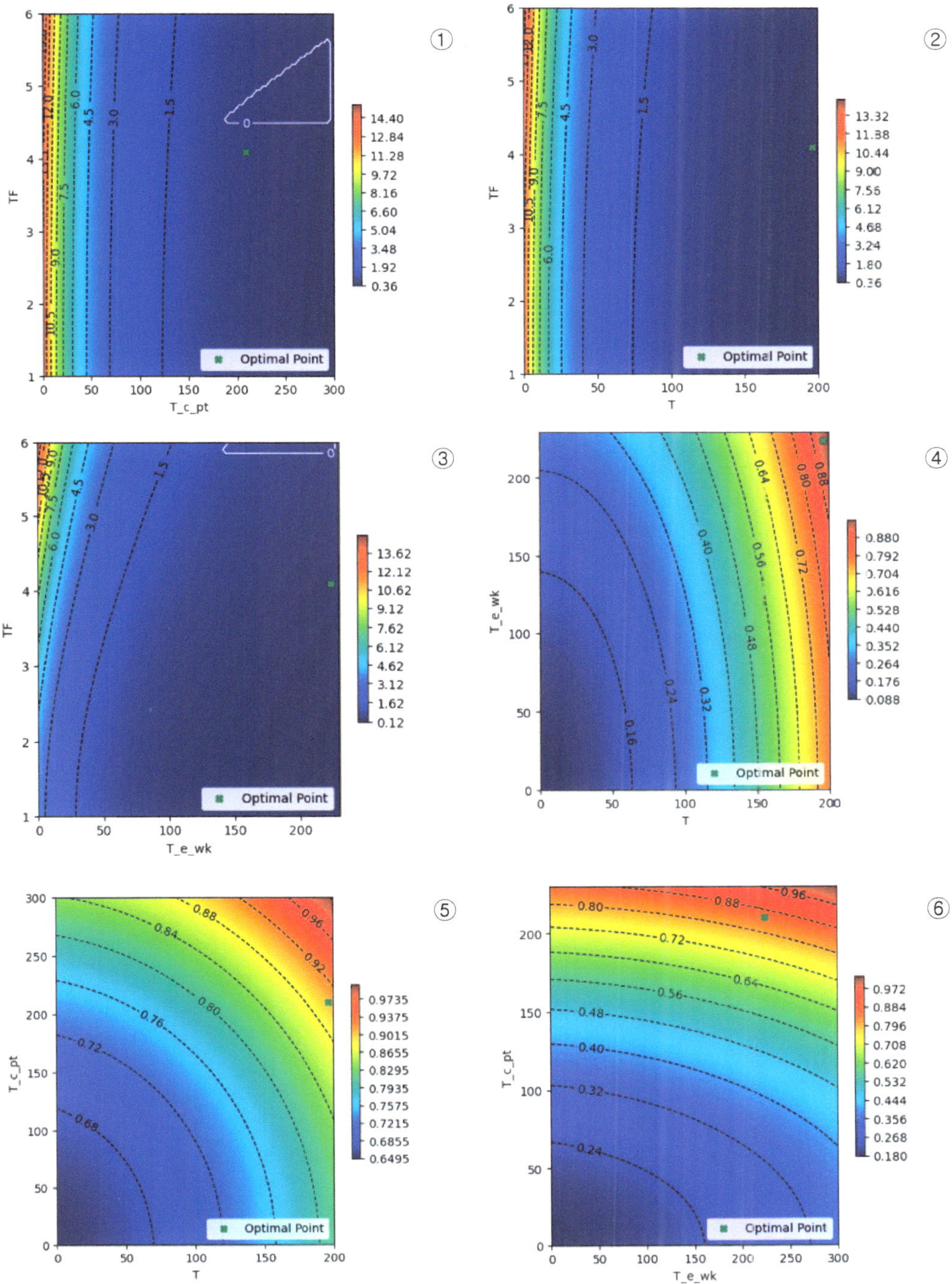

Figure 14. Contour of optimization for South Sea.

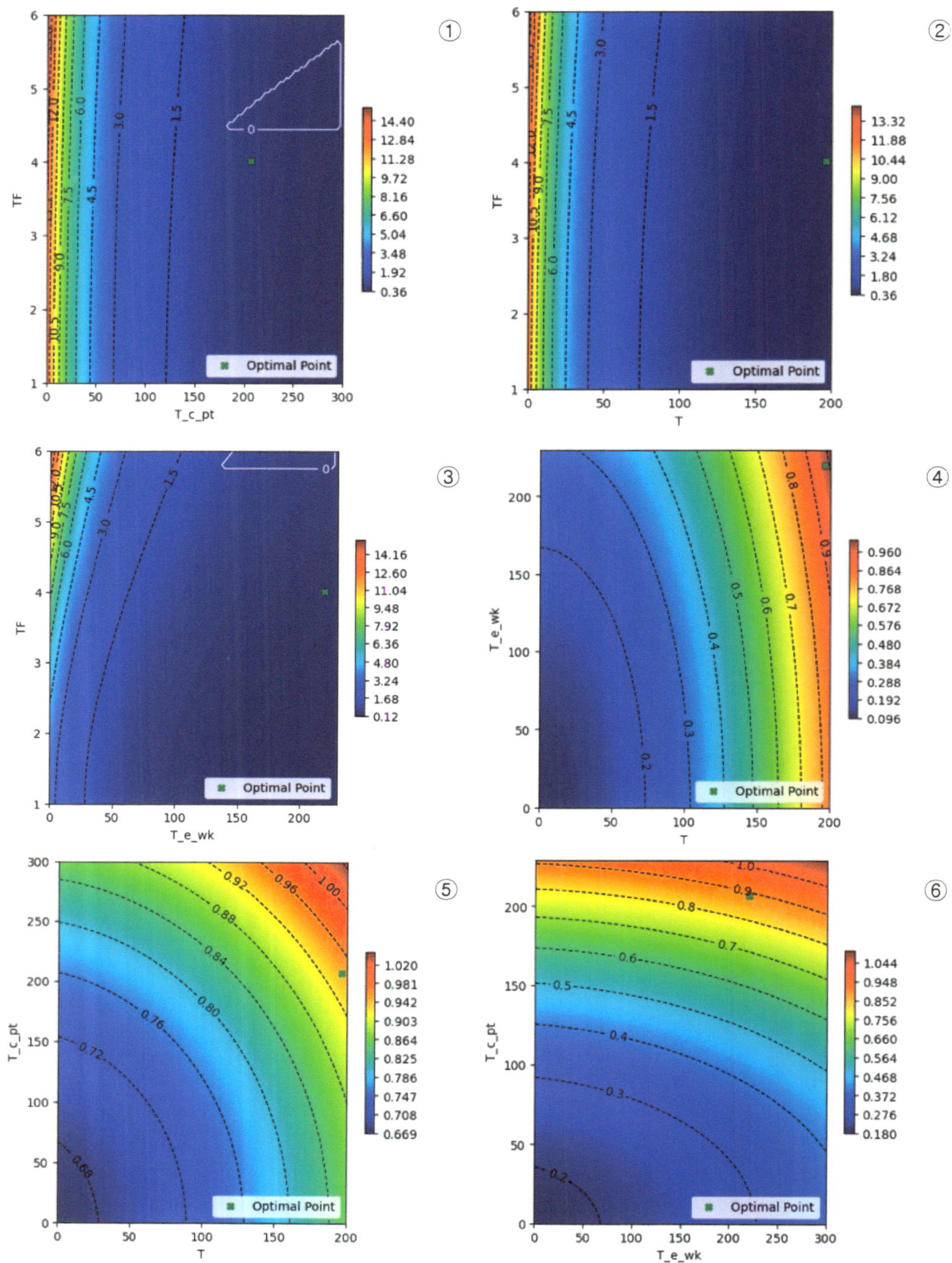

Figure 15. Contour of optimization for East Sea.

10. Conclusions

The supply of hybrid propulsion (diesel and electric complex propulsion) ships is expected to increase in accordance with the demand for eco-friendly ships. This is because small ships have a limitation of space to install an LNG propulsion system, etc., and the risk of accidents at sea is increased when the battery is exhausted to apply only battery propulsion.

However, the design of most small ships, including fishing vessel, is generally completed in a way that slightly changes the design by reflecting the requirements of the order based on the design information of reference ship. Also, it is very rare to have an R&D base. Therefore, if hybrid propulsion ships such as the research target ship (MOF, energy-saving and eco-friendly fishing boat development research project, '21~25') are distributed, it is expected that there will be difficulties in changing the design because there is no reference ship and R&D base.

Therefore, in the results of this study, when the ratio of the usage time for each operation scenario per vessel is input, the optimized reduction ratios and operation point (power consumption) can be derived. Since the reduction ratio is a key design variable of the reduction gear and is at the contact point (Node) when all power is transmitted, an efficient performance change can be expected while minimizing design changes compared to modifying other factors in the propulsion system.

In addition, considering the design conditions of small ships with limited scale and R&D capabilities, it is expected to be practical because the performance of the ship can be changed in consideration of the operating characteristics of each maritime regions and catching methods for target fish with minimal design change in the reference ship. In addition, this study is expected to be useful in contributing to the widespread adoption of hybrid propulsion ships because it can recommend efficient operational points for the entire voyage to users by presenting the required power for each operation scenario.

In summary, the transition to eco-friendly ships, such as small vessels like fishing boats, is expected to involve the application of hybrid electric propulsion systems. The studies related to hybrid electric propulsion vessels are currently underway to develop models optimized for specific tonnages and vessel types. However, utilizing the findings of this study could maximize the utility of eco-friendly vessels by enabling more efficient navigation, considering the characteristics of various maritime regions and vessel types. Nevertheless, due to the lack of comprehensive data on equipment specifications for various model such as engines and motors, this study was limited to optimizing specifications for reduction gear only. It is anticipated that future studies will be able to expand optimization targets to include engines, motors, and other equipment if more information becomes available. This would enable subsequent studies to select various equipment.

Author Contributions: The first author S.-W.M. led the planning, model development, and interpretation of the research. The corresponding W.-S.R. supervised the optimization algorithm and the project based on the model development. All authors discussed the results and exchanged opinions regarding the manuscript. The K.-P.P. proposed the conceptual methodology of the Bond graph. All authors have read and agreed to the published version of the manuscript.

Funding: This research received no external funding.

Institutional Review Board Statement: Not applicable.

Informed Consent Statement: Not applicable.

Data Availability Statement: The data in this study are available from the corresponding authors on request.

Acknowledgments: This work was supported by research fund of Chungnam National University.

Conflicts of Interest: The authors declare no competing financial interests.

Abbreviations

e_i	Effort (power variable)
f_i	Flow (power variable)
GY	Convert element (convert effort to flow, flow to effort)
I, C	Storage element
p_i	Momentum (energy variable)
q_i	Displacement (energy variable)
R	Dissipator element
s_e, s_f	Source element
TF	Transform element (transform effort to effort, flow to flow)
0 *Junction*	Distribution element (all effort is distributed as same)
1 *Junction*	Distribution element (all flow is distributed as same)

References

1. Ahn, J. A Study on Improvement for Greenship Certification Scheme to Achieve Net-Zero. *J. Soc. Nav. Archit. Korea* **2022**, *5* 372–384. [CrossRef]
2. Park, J.; Lee, K.-S.; Lee, K.; Kim, J.; Choi, J.; Lee, M. *A Study on the Performance Evaluation of An Electrospray Scrubber for Simultaneou Removal of PM/NOx/Sox*; Summer conference paper collection; The Society of Air-Conditioning and Refrigerating Engineers Korea: Seoul, Republic of Korea, 2022; pp. 356–357.
3. Yeo, I.; Park, C.G. *Study on Performance Evaluation of SOx-NOx Wet Scrubber Wastewater Treatment System in Ship*; Winter conferenc paper collection; The KSFM Journal of Fluid Machinery: Jeju Island, Republic of Korea, 2022.
4. Choia, M.-S. A Study on Prospects for the Introduction of LNG Fueled Ships. *J. Int. Trade Commer.* **2023**, *19*, 329–333.
5. Kim, J.-S.; Kim, D.-Y.; Kim, Y.-T. *A Fundamental Study on Boil-Off Gas Re-Liquefaction Systems for LNG-Fueled Ship*; Winter Conferenc Paper Collection; The KSFM Journal of Fluid Machinery: Jeju Island, Republic of Korea, 2022.
6. Lee, W.-J.; Kim, J.-S.; Yeo, S.-J.; Noh, J.-H.; Kim, T.-G.; Lee, J.-W.; Lee, J.-U.; Jeon, H.-M. Demand analysis and environment assessment for LPG fuelled ships. In Proceedings of the Spring Academic Conference, Online, 24–25 June 2021.
7. Lee, C.; Oh, J.; You, B.; Kim, D. A Numerical Study on Reduction of Nitrogen Oxide and Black Carbon According to Combustio Parameters of Marine Engine with DME fuels. In Proceedings of the Spring Academic Conference, Online, 25–26 June 2020.
8. Jee, J.-H. A Study on Risk Assessment using What-If Method for Ammonia Fueled Ship. *J. Fishries Mar. Sci. Educ.* **2023**, *3* 783–795. [CrossRef]
9. Hwang, J.; Jeon, H.; Hur, J.; Yoon, K.; Kim, J. A Study on eco-friendly ship propulsion system applicable to fishing vessels. Proceedings of the Fall Academic Conference, Pusan, Republic of Korea, 24–25 November 2022.
10. Number of Fishing Vessel by Tonnage and Type of Construction Material. Available online: https://kosis.kr/statHtml/statHtm do?orgId=146&tblId=DT_MLTM_5002762&vw_cd=MT_ZTITLE&list_id=K2_1&seqNo=&lang_mode=ko&language=kor&ob var_id=&itm_id=&conn_path=MT_ZTITLE/ (accessed on 2 September 2023).
11. Fisheries ACT, [Enforcement Date 12. Jan, 2023.] [Act No.18755, 11. January 2022, Whole Amendment], Article 58 (Restrictions c Bottoms of Fishing Vessels). Available online: https://www.law.go.kr/%20engLsSc.do?menuId=1&subMenuId=21&tabMenuId 20=117&query=%EC%88%98%EC%25%2082%B0%EC%97%85%EB%B2%95# (accessed on 2 September 2023).
12. Enforcement Decree of Fisheries ACT. [Enforcement Date 12 January 2024.] [Act No.34119, 9. March, 2024., Partial Amen ment], Article 37. Available online: https://www.law.go.kr/lsSc.do?section=&menuId=1&subMenuId=15&tabMenuId= &eventGubun=060101&query=%EC%88%98%EC%82%B0%EC%97%85%EB%B2%95#J37:0 (accessed on 2 September 2023).
13. Enforcement Decree of Fisheries ACT. [Enforcement Date 12. Jan, 2024.] [Act No.34119, 9. March, 2024., Partial Amendmen Article 37, Appendex 6. Available online: https://www.law.go.kr/sBylInfoPLinkR.do?lsiSeq=258361&lsNm=%EC%88% 8%EC%82%B0%EC%97%85%EB%B2%95+%EC%8B%9C%ED%96%89%EB%A0%B9&bylNo=0006&bylBrNo=00&bylCls=BE bylEfYd=20240112&bylEfYdYn=Y/ (accessed on 2 September 2023).
14. Bae, C.-S.; Yang, W.-J. Analysis of the Importance of Eco-friendly Ship Dissemination Policy using the Analytic Hierarchy Proce *J. Korean Soc. Mar. Environ. Saf. Res. Pap.* **2022**, *28*, 117–124. [CrossRef]
15. Kang, C.E.; Baek, C.H.; Kim, S.H.; Lee, C.J. A Study of the Development Safety Criteria for Hybrid Electrical Propulsion Fishi Boats. *J. Korean Soc. Mar. Environ. Saf. Res. Pap.* **2023**, *29*, 207–214.
16. You, Y.-J.; Park, H.-R. Development of a framework to estimate the EEOI of a ship considering the hydrodynamic characteristi and engine mode. *J. Soc. Nav. Archit. Korea* **2018**, *55*, 457–465. [CrossRef]
17. Ghimire, P.; Reddy, N.P.; Zadeh, M.K.; Pedersen, E.; Thorstensen, J. Dynamic Modeling and Real-Time Simulation of a Sh Hybrid Power System Using a Mixed-Modeling Approach. In Proceedings of the 2020 IEEE Transportation Electrificati Conference & Expo (ITEC), Chicago, IL, USA, 23–26 June 2020. originally scheduled.
18. Ghimire, P.; Zadeh, M.; Pedersen, E.; Thorstensen, J. Dynamic Modeling, Simulation, and Testing of a Marine DC Hybrid Pow System. *IEEE Trans. Transp. Electrif.* **2021**, *7*, 905–919. [CrossRef]
19. Moon, S.-W.; Ruy, W.-S. A Study on the Modeling of Ship Energy System Using Bond Graph. *J. Soc. Nav. Archit. Korea* **2024**, 19–28. [CrossRef]

20. National Institute of Fisheries Science. Available online: https://www.nifs.go.kr/ (accessed on 2 September 2023).
21. Jung, S.-J.; Kim, I.-O.; Park, C.-D.; Kim, S.-H.; Jeong, G.-C. *Illustration of Korea Fishing Vessels*; National Fisheries Research and Development Institute: Quezon City, Philippines, 2018.
22. Bond Graph.com. About Bond Graph. Available online: https://www.scribd.com/document/48821925/Samantaray-2001-www-bondgraphs-com-about/ (accessed on 2 September 2023).
23. Merzouki, R.; Samantaray, A.K.; Pathak, P.M.; Bouamama, B.O. *Intelligent Mechatronic Systems Modeling, Control and Diagnosis*; Springer: Berlin/Heidelberg, Germany, 2013.
24. Borutzky, W. *Bond Graph Methodology Delopment and Analysis of Multidisciplinary Dynamic System Models*; Springer: Berlin/Heidelberg, Germany, 2010.
25. Kim, J.-S.; Park, J.-S. Basic concepts of Bond graph modeling techniques and it's applications. *J. Korean Soc. Mech. Eng.* **1993**, *33*, 22–32.
26. Pedersen, T.A. Bond Graph Modeling of MarinePower Systems. Ph.D. Thesis, Norwegian University of Science and Technology, Trondheim, Norway, 2009.
27. Ayala-Jaimes, G.; Gonzalez-Avalos, G. MOSFET Modelling for A Three-Level Inverter Circuit: A Hybrid Bond graph Approach. In Proceedings of the 48th Annual Conference of the IEEE Industrial Electronics Society, Brussels, Belgium, 17–20 October 2022.
28. 20-sim 5.0. Available online: https://www.20sim.com/ (accessed on 2 September 2023).
29. Runge-Kutta Method. Available online: https://ko.wikipedia.org/wiki/%EB%A3%BD%EA%B2%8C-%EC%BF%A0%ED%83%80_%EB%B0%A9%EB%B2%95/ (accessed on 2 September 2023).
30. SciPy Library. Available online: https://docs.scipy.org/doc/scipy/reference/generated/scipy.optimize.minimize.html/ (accessed on 2 September 2023).
31. Differential Evolution. Available online: https://docs.scipy.org/doc/scipy/reference/generated/scipy.optimize.differential_evolution.html/ (accessed on 2 September 2023).

MDPI AG
Grosspeteranlage 5
4052 Basel
Switzerland
Tel.: +41 61 683 77 34

Journal of Marine Science and Engineering Editorial Office
E-mail: jmse@mdpi.com
www.mdpi.com/journal/jmse

www.ingramcontent.com/pod-product-compliance
Lightning Source LLC
Chambersburg PA
CBHW041402210926
51850CB00013B/3157
* 9 7 8 3 7 2 5 8 4 1 4 9 3 *